LONDON MATHEMATICAL SOCIETY LECTURE NOTE SERIES

1. General cohomology theory and K-theory, P.HILTON
4. Algebraic topology: a student's guide, J.F.ADAMS
5. Commutative algebra, J.T.KNIGHT
8. Integration and harmonic analysis on compact groups, R.E.EDWARDS
9. Elliptic functions and elliptic curves, P.DU VAL
10. Numerical ranges II, F.F.BONSALL & J.DUNCAN
11. New developments in topology, G.SEGAL (ed.)
12. Symposium on complex analysis, Canterbury, 1973, J.CLUNIE & W.K.HAYMAN (eds.)
13. Combinatorics: Proceedings of the British combinatorial conference 1973, T.P.McDONOUGH & V.C.MAVRON (eds.)
14. Analytic theory of abelian varieties, H.P.F.SWINNERTON-DYER
15. An introduction to topological groups, P.J.HIGGINS
16. Topics in finite groups, T.M.GAGEN
17. Differentiable germs and catastrophes, Th.BROCKER & L.LANDER
18. A geometric approach to homology theory, S.BUONCRISTIANO, C.P.ROURKE & B.J.SANDERSON
20. Sheaf theory, B.R.TENNISON
21. Automatic continuity of linear operators, A.M.SINCLAIR
23. Parallelisms of complete designs, P.J.CAMERON
24. The topology of Stiefel manifolds, I.M.JAMES
25. Lie groups and compact groups, J.F.PRICE
26. Transformation groups: Proceedings of the conference in the University of Newcastle upon Tyne, August 1976, C.KOSNIOWSKI
27. Skew field constructions, P.M.COHN
28. Brownian motion, Hardy spaces and bounded mean oscillation, K.E.PETERSEN
29. Pontryagin duality and the structure of locally compact abelian groups, S.A.MORRIS
30. Interaction models, N.L.BIGGS
31. Continuous crossed products and type III von Neumann algebras, A.VAN DAELE
32. Uniform algebras and Jensen measures, T.W.GAMELIN
33. Permutation groups and combinatorial structures, N.L.BIGGS & A.T.WHITE
34. Representation theory of Lie groups, M.F.ATIYAH et al.
35. Trace ideals and their applications, B.SIMON
36. Homological group theory, C.T.C.WALL (ed.)
37. Partially ordered rings and semi-algebraic geometry, G.W.BRUMFIEL
38. Surveys in combinatorics, B.BOLLOBAS (ed.)
39. Affine sets and affine groups, D.G.NORTHCOTT
40. Introduction to H_p spaces, P.J.KOOSIS
41. Theory and applications of Hopf bifurcation, B.D.HASSARD, N.D.KAZARINOFF & Y-H.WAN
42. Topics in the theory of group presentations, D.L.JOHNSON
43. Graphs, codes and designs, P.J.CAMERON & J.H.VAN LINT
44. Z/2-homotopy theory, M.C.CRABB
45. Recursion theory: its generalisations and applications, F.R.DRAKE & S.S.WAINER (eds.)
46. p-adic analysis: a short course on recent work, N.KOBLITZ
47. Coding the Universe, A. BELLER, R. JENSEN & P. WELCH
48. Low-dimensional topology, R. BROWN & T.L. THICKSTUN (eds.)
49. Finite geometries and designs, P. CAMERON, J.W.P. HIRSCHFELD & D.R. HUGHES (eds.)
50. Commutator Calculus and groups of homotopy classes, H.J. BAUES
51. Synthetic differential geometry, A. KOCK
52. Combinatorics, H.N.V. TEMPERLEY (ed.)
53. Singularity theory, V.I. ARNOLD
54. Markov processes and related problems of analysis, E.B. DYNKIN
55. Ordered permutation groups, A.M.W. GLASS
56. Journees arithmetiques 1980, J.V. ARMITAGE (ed.)
57. Techniques of geometric topology, R.A. FENN
58. Singularities of differentiable functions, J. MARTINET
59. Applicable differential geometry, F.A.E. PIRANI and M. CRAMPIN
60. Integrable systems, S.P. NOVIKOV et al.

London Mathematical Society Lecture Note Series. 55

Ordered Permutation Groups

A.M.W. GLASS
Professor, Bowling Green State University,
Bowling Green, Ohio

CAMBRIDGE UNIVERSITY PRESS

CAMBRIDGE

LONDON NEW YORK NEW ROCHELLE

MELBOURNE SYDNEY

CAMBRIDGE UNIVERSITY PRESS
Cambridge, New York, Melbourne, Madrid, Cape Town, Singapore, São Paulo

Cambridge University Press
The Edinburgh Building, Cambridge CB2 8RU, UK

Published in the United States of America by Cambridge University Press, New York

www.cambridge.org
Information on this title: www.cambridge.org/9780521241908

First published 1981
Re-issued in this digitally printed version 2008

A catalogue record for this publication is available from the British Library

Library of Congress Catalogue Card Number: 81–16996

ISBN 978-0-521-24190-8 paperback

To my wife, Rona,
and my mother, Vicky;

and in memory of
my father, Jack (1912-1969)
and my aunt, Ethel (1902-1980).

TABLE OF CONTENTS

PREFACE

In the past thirty years, groups of order-preserving permutations of totally ordered sets have been extensively studied. The purpose of these notes is to provide a uniform, systematic account of this research and its applications. In the first half of this book (Parts I and II), I attempt a streamlined (and I trust intuitive) presentation of the main results in the structure theory, taking full advantage of recent research. In Chapters 3 and 5, the study of such permutation groups is reduced to an investigation of the basic building blocks of the subject, "primitive" order-preserving permutation groups. These are classified and examined in Chapters 2 and 4. The second half of the book is devoted to various applications of the structure theory; e.g., to the construction of infinite simple groups. Most of the chapters in it can be read quite independently of each other. I have chosen the topics to illustrate the use of the structure theory in a wide variety of settings, but readily admit that the selection is strongly influenced by my own quirks and prejudices. The total order on a set naturally lifts to a lattice order on its group of order-preserving permutations. Since every lattice-ordered group can be embedded (as a group and lattice simultaneously) in such a lattice-ordered group--the Cayley-Holland Theorem, see Appendix I--the theory can be used to study lattice-ordered groups. For example, I will show: every lattice-ordered group can be embedded in a simple divisible lattice-ordered group in which any two elements greater than the identity are conjugate; and my favourite application: there is a finitely presented lattice-ordered group with insoluble word problem (Whereas analogues of both these results for groups can be proved without recourse to permutation groups, I know of no proof for lattice-ordered groups which avoids using groups of order-preserving permutations of totally ordered sets).

My main aim throughout has been to impart the intuition and excitement of the subject. If any of this comes across, I will be most satisfied.

With the exception of Chapter 1, I have begun each chapter with an account of the goals and main theorems it contains (though often not stated in the generality in which they are proved). I hope that this will help the reader to get a good overall picture of the subject, as well as whet his or her appetite; it should also make it easier to locate results in the book. I have included several theorems which have not previously been published, so these notes hold something new even for the specialist. Since anyone who has read Parts I and II and a few of the applications is at the forefront of the subject, I have included a list of unsolved problems which I think are of especial interest. I am sure that the reader will find others to work on. Caution: in order to give as uncluttered and intuitive an account as possible, I have often put more stringent hypotheses on theorems than are really necessary. This eliminates some technicalities. The interested reader should consult the literature to find out the full story (an annotated bibliography is included to make this easier).

As a result of my earlier book of the same title (published in 1976 by Bowling Green State University and now out of print), David Tranah approached me to write a monograph on the subject for Cambridge University Press. These notes are the consequence. Unlike the 1976 version, they are not encyclopaedic but are designed instead for a much broader audience. I have confined my attention to those theorems in the structure theory which have been most fruitful for applications, and provided simpler proofs to some when more recent research has made this possible. Also, many of the applications given here are the result of work done in the last five years, and so did not appear in the 1976 version.

This book was obtained by photocopying the typed version of the manuscript. Consequently, everyone reading it will appreciate the excellence of the typist, Linda Shellenbarger, and the deep gratitude I owe her. I am also beholden to Bruce Lyle for his superb illustrations; Ashok Kumar Arora and Manfred Droste for reading the handwritten version and making suggestions for needed improvements; Todd Feil for reading the typed version; and Rona Glass for the index.

When it comes to thanking mathematicians, I do not know where to begin. I have had the great fortune to find, without exception, only kindness and generous assistance from teachers, lecturers, professors and colleagues on both sides of the Atlantic. Since a list of names would be far too long, let me just humbly thank all I have come in contact with. However, I would especially like to thank W. Charles Holland who, first as a thesis director and then as a colleague, has so willingly shared many insights and speculations with me. Stephen McCleary has also been a constant source of encouragement and contagious enthusiasm. In addition, I would like to thank my colleagues in the mathematics department here for their friendliness and interest in my work. Like everyone in the States, they have made my stay here most enjoyable. It gives me great pleasure to acknowledge my thanks to them in print.

Finally, I would like to thank David Tranah at Cambridge University Press for his encouragement, help, patience and good humour throughout the ordeal of writing these notes. It is quite accurate to say that without his urging, this book would never have been written. I trust it will not be held against him for too long!

Bowling Green, Ohio, U.S.A
Whitmonday, 1981

BACKGROUND TERMS AND NOTATION

1. *SET THEORY.*

Let A and B be sets. We will write $A \subseteq B$ if A is a subset of B, and $A \subsetneq B$ if A is a proper subset of B. We will use $A\backslash B$ for $\{a \in A: a \notin B\}$. If f is a function from A into B, let $Af = \{af: a \in A\}$, the *range* (or *image*) of f; and if $C \subseteq A$, let $f|C = \{(a,b) \in f: a \in C\}$, the *restriction* of f to C. We will use $\emptyset$ for the empty set, ω for the least infinite ordinal, ω_1 for the least uncountable infinite ordinal, etc. So $\omega = \{0,1,2,\ldots\}$. We will often identify ω with $\aleph_0$, ω_1 with $\aleph_1$, etc. We will write $|A|$ for the cardinality of A, and say A is *countable* if $|A| = \aleph_0$.

2. *PARTIALLY ORDERED SETS.*

A set Ω with a reflexive, antisymmetric, transitive relation $\leq$ defined on it is called a *partially ordered set*, or *p.o. set*. As usual, we will write $\alpha < \beta$ for $\alpha \leq \beta$ & $\alpha \neq \beta$. A p.o. set in which $\alpha \leq \beta$ only if $\alpha = \beta$ is said to be *trivially ordered*. In contrast, a p.o. set Ω is said to be *totally ordered* or *linearly ordered* (or a *chain*, for short) if for all $\alpha,\beta \in \Omega$, $\alpha \leq \beta$ or $\beta \leq \alpha$. A p.o. set Ω in which each $\alpha,\beta \in \Omega$ have a least upper bound (denoted by $\alpha \vee \beta$) and a greatest lower bound (denoted by $\alpha \wedge \beta$) is called a *lattice*. If Ω is a lattice and $\alpha \vee (\beta \wedge \gamma) = (\alpha \vee \beta) \wedge (\alpha \vee \gamma)$ for all $\alpha,\beta,\gamma \in \Omega$, then Ω is said to be a *distributive lattice*; equivalently, if $\alpha \wedge (\beta \vee \gamma) = (\alpha \wedge \beta) \vee (\alpha \wedge \gamma)$ for all $\alpha,\beta,\gamma \in \Omega$. If Ω is a p.o. set and $T \subseteq \Omega$, we say that T is *dense* in Ω if whenever $\alpha < \beta$ in Ω, there is $\tau \in T$ such that $\alpha < \tau < \beta$ (cf., dense in itself, page 83). We will call Ω

dense if it is dense in Ω. If $\alpha < \beta$ in a p.o. set Ω, we write $[\alpha,\beta] = \{\gamma \in \Omega: \alpha \leq \gamma \leq \beta\}$ for the closed interval, and $(\alpha,\beta) = \{\gamma \in \Omega: \alpha < \gamma < \beta\}$ for the open interval. If $\gamma \in [\alpha,\beta]$, we say that γ lies *between* α and β; if $\gamma \in (\alpha,\beta)$, that γ lies *strictly between* α and β. A subset T of a p.o. set Ω is said to be an *interval* of Ω (or *convex* in Ω) if whenever $\tau_1,\tau_2 \in T$ and $\gamma \in \Omega$ lies between τ_1 and τ_2, then $\gamma \in T$. If Ω is a p.o. set, Ω^* is the set Ω with the reverse ordering; i.e., $\alpha \leq \beta$ in Ω^* precisely when $\alpha \geq \beta$ in Ω. So Ω^* is also a p.o. set. If I is a chain and $\{\Omega_i: i \in I\}$ is a family of p.o. sets, then $\overleftarrow{\bigcup}\{\Omega_i: i \in I\}$ is the set $\bigcup\{\Omega_i: i \in I\}$ ordered by: $\alpha \leq \beta$ if $\alpha \in \Omega_i$, $\beta \in \Omega_j$ and $i < j$ or $(i = j \ \& \ \alpha \leq \beta$ in $\Omega_i)$; $\overrightarrow{\bigcup}\{\Omega_i: i \in I\}$ is the same set but ordered by: $\alpha \leq \beta$ if $\alpha \in \Omega_i$, $\beta \in \Omega_j$ and $i > j$ or $(i = j \ \& \ \alpha \leq \beta$ in $\Omega_i)$.

We will use $\mathbb{Z}$, $\mathbb{Q}$ and $\mathbb{R}$ for the chains of integers, rationals and reals respectively (under the natural order). $\mathbb{Z}^+$, $\mathbb{Q}^+$ and $\mathbb{R}^+$ will denote the sets of strictly positive integers, rationals and reals, respectively.

If Ω_1 and Ω_2 are lattices, then a *lattice isomorphism* is a one-to-one map of Ω_1 onto Ω_2 that preserves the lattice operations. In the special case that Ω_1 and Ω_2 are chains, lattice isomorphisms are called *ordermorphisms*, and the chains are said to be ordermorphic.

3. *GROUP THEORY.*

If G is a group and H is a subgroup of G, we will write $H \lhd G$ if H is normal. We will write $H \rtimes G$ for a group with normal subgroup H, having quotient by H isomorphic to G; i.e., an extension of H by G. If C is a subgroup of a group G, $R(C) = \{Cg: g \in G\}$, the set of *right cosets* of C in G. All isomorphisms are onto. $\mathbb{Z}_2$ will denote the two element group. We will use e for the identity element of any group.

4. *MODEL THEORY.*

Our language will be the first order language of groups, lattices with e, or ℓ-groups. So our formulae will be built up in the usual way from equality of group words, lattice words or ℓ-groups words by using $\neg$ (not), & or $\bigwedge$ (and), or or $\bigvee$ (or), $\forall$ (for all) and $\exists$ (there exists). Note ($\exists$ integer n) is not permitted; nor are infinite conjunctions, etc. A formula without free variables is called a *sentence*. We will write $G \models \theta$ if the sentence θ holds in the group (lattice, ℓ-group) G, and $G \equiv H$ if the groups (lattices with e, ℓ-groups) satisfy the same sentences of the language of groups (lattices with e, ℓ-groups).

5. *CONVENTIONS ADOPTED.*

Capital Greek letters are used for chains, small Greek letters for elements of them (exception: α, κ, λ, μ, ν, ξ, η, ζ are sometimes used in other contexts, e.g., as ordinals, members of index sets; ψ, ϕ, θ are usually reserved for mappings). C, G, H will denote groups, lattices with e, or ℓ-groups and c, f, g, h, k, x, y, z will denote elements of them. I and J are reserved for general index sets; i, j for elements of these index sets. Script letters are used for equivalence relations; $\underline{k}$ is reserved for the set of covering pairs of congruences (see Sections 1.6, 1.7 and Chapter 3), and K for an element of $\underline{k}$.

We will sometimes use // to denote the end of a proof; sometimes Q.E.D.; sometimes nothing.

EXPLANATION OF DIAGRAMS.

The picture

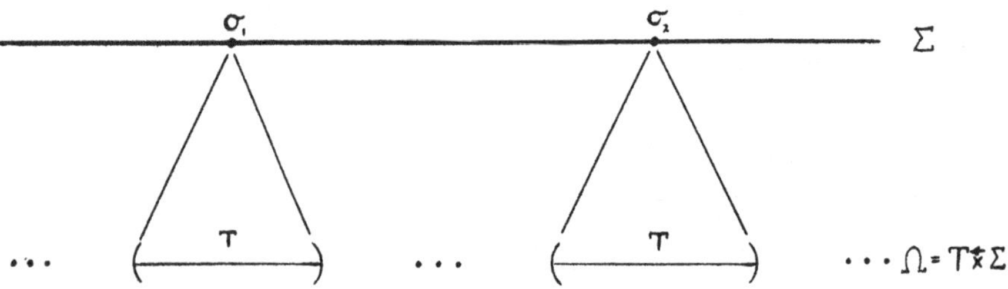

will be used to illustrate the chain Ω obtained from Σ by replacing each $\sigma \in \Sigma$ by a copy of the chain T; i.e., $\Omega = \underline{T \overleftarrow{\times} \Sigma} = \{(\tau,\sigma): \tau \in T, \sigma \in \Sigma\}$ ordered by: $(\tau_1,\sigma_1) < (\tau_2,\sigma_2)$ if $\sigma_1 < \sigma_2$ or $(\sigma_1 = \sigma_2 \ \& \ \tau_1 < \tau_2)$. Let Σ, T_1, T_2 be chains, with Σ_1, Σ_2 a partition of Σ.

The picture

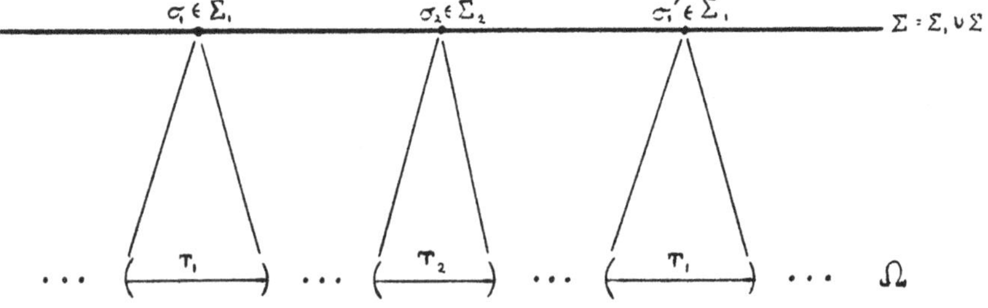

will be used to illustrate the chain Ω obtained by replacing each $\sigma_i \in \Sigma_i$ by a copy of T_i $(i = 1,2)$. That is,
$\Omega = \{(\tau_1,\sigma_1): \tau_1 \in T_1, \sigma_1 \in \Sigma_1\} \cup \{(\tau_2,\sigma_2): \tau_2 \in T_2, \sigma_2 \in \Sigma_2\}$
ordered by: $(\tau,\sigma) < (\tau',\sigma')$ if $\sigma < \sigma'$ (in Σ) or ($\sigma = \sigma'$ & $\tau < \tau'$).

For such layered diagrams, we will use the horizontal (rather than the diagonal) to depict the identity map; e.g.,

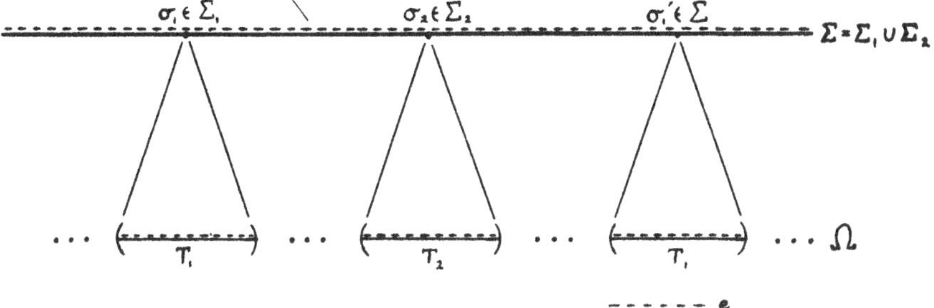

PART I

OPENING THE INNINGS

CHAPTER 1
INTRODUCTION

The theory of permutation groups (especially in the finite case) has been developed over a long period of time. One motivation for the study was geometric. Another was provided by Cayley's Theorem that every group is (isomorphic to) a subgroup of a permutation group. More recently, infinite permutation groups have been studied and used to provide examples (and counterexamples) in infinite group theory.

On the other hand, ordered permutation groups have been studied for a relatively short time. Again, one motivation to examine them was geometric. Another was provided in 1963 by W. Charles Holland's analogue of Cayley's Theorem: Every lattice-ordered group is (isomorphic to) a subgroup of the lattice-ordered group of all order-preserving permutations of a chain (linearly ordered set). A further reason was provided by model theory: If T is a first order theory having infinite models and $\langle \Omega, \leq \rangle$ is a chain, then there is a model $\mathfrak{A}$ of T containing Ω such that each order-preserving permutation of Ω extends to an automorphism of $\mathfrak{A}$ (Ehrenfeucht and Mostowski [56]). Hence any study of the group of automorphisms of models of such theories should begin with a study of groups of order-preserving permutations of chains.

In this chapter we will devote ourselves to basic definitions and results. Many of these are routine translations of theorems concerning permutation groups to the ordered permutation group setting. The presentation we give is self-contained and does not require prior knowledge of the theory of permutation groups. However, anyone wishing to become conversant with the roots of the subject could do worse than refer to H. Wielandt's works [64], [67] and [69]. In contrast, some of the theory we present here (especially in Sections 1.8, 1.10 and 1.11) is peculiar to the ordered case, though we have included it in this chapter only if it is rather straightforward.

1.1. $A(\Omega)$.

Let Ω be a chain. We define *$A(\Omega)$* to be the set of all order-preserving permutations of Ω; i.e., $g \in A(\Omega)$ if and only if g is a one-to-one function from Ω onto Ω that satisfies: if $\alpha, \beta \in \Omega$ and $\alpha < \beta$, then $\alpha g < \beta g$. $A(\Omega)$ is a group under composition: if $f,g \in A(\Omega)$ define fg by: $\alpha(fg) = (\alpha f)g$ $(\alpha \in \Omega)$. Further, we can define an order $\leq$ on $A(\Omega)$ via: $f \leq g$ if $\alpha f \leq \alpha g$ for all $\alpha \in \Omega$. This order on $A(\Omega)$ is called the *pointwise order*. It is immediate that if $f,g,h \in A(\Omega)$ and $f \leq g$, then $fh \leq gh$ and $hf \leq hg$. Moreover, any pair of elements $g_1, g_2 \in A(\Omega)$ has a supremum (least upper bound) and infimum (greatest lower bound) in $A(\Omega)$--which we denote by *$g_1 \vee g_2$* and *$g_1 \wedge g_2$* respectively--namely: $\alpha(g_1 \vee g_2) = \max\{\alpha g_1, \alpha g_2\}$ and $\alpha(g_1 \wedge g_2) = \min\{\alpha g_1, \alpha g_2\}$ $(\alpha \in \Omega)$. (Since Ω is a chain, $\alpha g_1 \leq \alpha g_2$ or $\alpha g_2 \leq \alpha g_1$.)

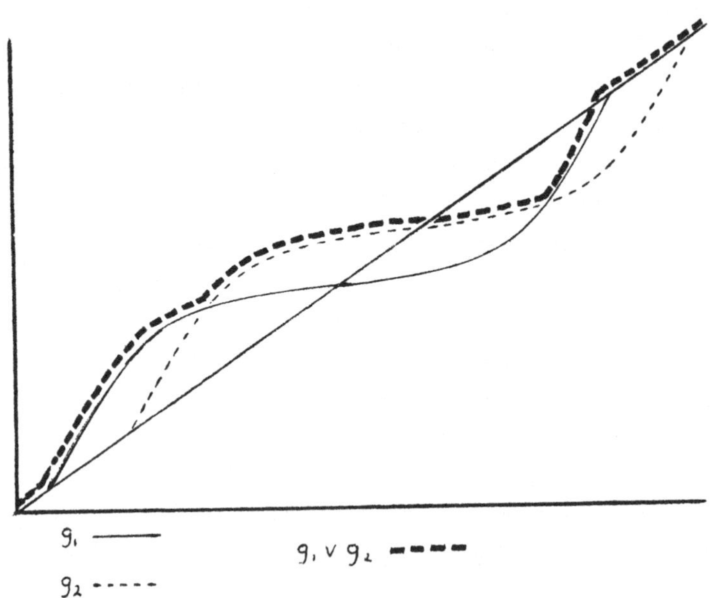

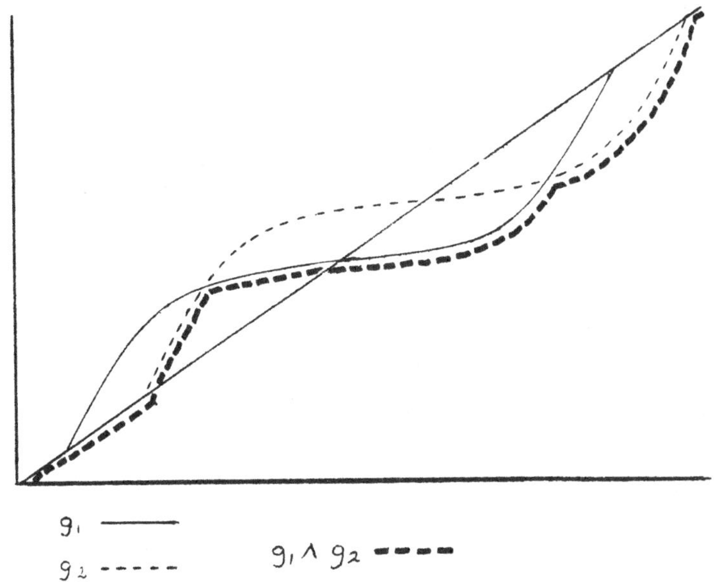

EXAMPLE 1.1.1. Let $\Omega = \mathbb{Z}$, the chain of integers (in the natural ordering). The order-preserving permutations of $\mathbb{Z}$ are just translations by integers. If $t_n \in A(\mathbb{Z})$ denotes translation by n, then the map $n \mapsto t_n$ is a group isomorphism of $\mathbb{Z}$ onto $A(\mathbb{Z})$ that preserves order; i.e., if $n \leq m$, then $t_n \leq t_m$ in the pointwise ordering ($\alpha t_n = \alpha + n \leq \alpha + m = \alpha t_m$ for all $\alpha \in \mathbb{Z}$). In this example, for each $\alpha, \beta \in \Omega$, there is a unique $g \in A(\Omega)$ such that $\alpha g = \beta$, and the pointwise order is a total order.

EXAMPLE 1.1.2. Let $\Omega = \mathbb{R}$, the chain of real numbers (in the natural ordering). Let $g_0 \colon \alpha \mapsto 2\alpha$ ($\alpha \in \mathbb{R}$). Then $g_0 \in A(\mathbb{R})$. However, $1g_0 = 2$ (so $e \not\geq g_0$) and $(-1)g_0 = -2$ (so $e \not\leq g_0$). Hence the pointwise order is not a total order; so there is no ordermorphism of $\mathbb{R}$ onto $A(\mathbb{R})$. Let $f_0 \colon \alpha \mapsto \alpha + 1$ ($\alpha \in \mathbb{R}$). Then $f_0 \in A(\mathbb{R})$ and $f_0 g_0 \neq g_0 f_0$. Thus $A(\mathbb{R})$ is not an Abelian group. If $\alpha_1, \beta_1, \alpha_2, \beta_2 \in \mathbb{R}$ with $\alpha_1 < \beta_1$ & $\alpha_2 < \beta_2$, there is $h_0 \in A(\mathbb{R})$ such that $\alpha_1 h_0 = \alpha_2$ & $\beta_1 h_0 = \beta_2$; e.g., h_0 is the straight line passing through (α_1, α_2) and (β_1, β_2).

EXAMPLE 1.1.3. Let $\mathbb{Z}_i$ $(i = 1,2)$ be disjoint copies of $\mathbb{Z}$ and let $\Omega = \mathbb{Z}_1 \overset{\rightarrow}{\cup} \mathbb{Z}_2$. Any two points of $\mathbb{Z}_2$ have only a finite

number of elements of Ω between them. Hence if $g \in A(\Omega)$, the images under g of any two points of $\mathbb{Z}_2$ still enjoy this property. Therefore $\mathbb{Z}_2 g = \mathbb{Z}_1$ or $\mathbb{Z}_2$. If $\alpha_i \in \mathbb{Z}_i$ $(i = 1,2)$, then $\alpha_1 < \alpha_2$; so $\alpha_1 g < \alpha_2 g$. It follows that $\mathbb{Z}_i g = \mathbb{Z}_i$ $(i = 1,2)$. Consequently, by Example 1.1.1, there is an order-preserving group isomorphism between $A(\Omega)$ and $\mathbb{Z} \oplus \mathbb{Z}$ if we order $\mathbb{Z} \oplus \mathbb{Z}$ by: $(m_1,n_1) \leq (m_2,n_2)$ if $m_1 \leq m_2$ & $n_1 \leq n_2$ (in $\mathbb{Z}$).

1.2. ACTIONS OF GROUPS ON CHAINS.

Let Ω be a chain and G a group. An *action* of G on Ω is a triple (G,Ω,θ) where $\theta: G \to A(\Omega)$ is a homomorphism. The *kernel* of θ, *ker(θ)*, is called the *lazy subgroup* (if $g \in \ker(\theta)$, $\alpha(g\theta) = \alpha$ for all $\alpha \in \Omega$). If θ is one-to-one (i.e., the lazy subgroup is $\{e\}$), we call the action *faithful* and say that (G,Ω,θ) is a *representation* of G. It is clear that if (G,Ω,θ) is an action, $(G/\ker(\theta),\Omega,\theta^*)$ is a faithful action where $(\ker(\theta)g)\theta^* = g\theta$ $(g \in G)$.

EXAMPLE 1.2.1. Consider $(\mathbb{R},\mathbb{R},\theta)$ where $\alpha(r\theta) = \alpha + r$ $(\alpha, r \in \mathbb{R})$. Then $(\mathbb{R},\mathbb{R},\theta)$ is a faithful action, and so a representation of $\mathbb{R}$. Observe that the group $\mathbb{R}\theta$ (the group operation is composition) is isomorphic to the additive group of reals. Hence $\mathbb{R}\theta \neq A(\mathbb{R})$. Indeed, the pointwise ordering on $\mathbb{R}\theta \subseteq A(\mathbb{R})$ is total (if $r,s \in \mathbb{R}$, $r\theta \leq s\theta$ or $s\theta \leq r\theta$ where $\leq$ is the pointwise ordering).

LEMMA 1.2.2. *If* (G,Ω,θ) *is an action of* G *on a chain* Ω, *then the map* $(\alpha,g) \mapsto \alpha(g\theta)$ *of* $\Omega \times G$ *onto* Ω *satisfies (i)* $(\alpha,e) \mapsto \alpha$, *(ii) the image of* (α,g) *is less than the image of* (β,g) *whenever* $\alpha < \beta$, *and (iii)* (α,fg) *has the same image as* (β,g) *where* β *is the image of* (α,f). *Conversely, any map from* $\Omega \times G$ *into* Ω *satisfying (i), (ii) and (iii) gives rise to a homomorphism of* G *into* $A(\Omega)$ *(and hence to an action of* G *on* Ω*).*

Proof: That any action gives rise to a map satisfying (i), (ii) and (iii) is utterly obvious. Conversely, define θ by: $\alpha(g\theta)$ is the image of (α,g) $(\alpha \in \Omega,\ g \in G)$. By (i), $(\alpha,g^{-1}g) \mapsto \alpha$ so by (iii) $\beta(g\theta) = \alpha$ where $\beta = \alpha(g^{-1}\theta)$. Hence, by (ii), $g\theta \in A(\Omega)$. Thus $\theta: G \to A(\Omega)$. By (iii), θ is a homomorphism.

We will often write (G,Ω) instead of (G,Ω,θ) when θ is clear from context. In particular, if $G \subseteq A(\Omega)$, θ will be assumed to be the identity (unless specifically stated to the contrary). In this case we will endow G with the pointwise ordering and say that (G,Ω) is an *ordered permutation group*. So an ordered permutation group (G,Ω) is a subgroup of $A(\Omega)$ together with its inherited action on Ω. If (G,Ω) is an ordered permutation group and for each $g_1,g_2 \in G$, the elements $g_1 \vee g_2$ and $g_1 \wedge g_2$ of $A(\Omega)$ actually belong to G, we will say that (G,Ω) is an *ℓ-permutation group*. So for any chain Ω, $(A(\Omega),\Omega)$ is an ℓ-permutation group.

EXAMPLE 1.2.3. Let $\Omega = \mathbb{R}$ and G be the group of order-preserving permutations of $\mathbb{R}$ that are finitely piecewise linear. That is, $g \in G$ if $g \in A(\mathbb{R})$ and there are $\alpha_0 < \dots < \alpha_n$ in $\mathbb{R}$ such that $g|(-\infty,\alpha_0]$, $g|[\alpha_i,\alpha_{i+1}]$ and $g|[\alpha_n,\infty]$ are linear $(0 \leq i < n)$. Then $(G,\mathbb{R})$ is an ℓ-permutation group. Note G is not Abelian ($f_0,g_0 \in G$ where f_0,g_0 are defined in Example 1.1.2).

If (G,Ω) is an action, we may think of Ω as an algebra with unary operations g $(g \in G)$, and binary operations $\vee$ and $\wedge$ given by $\vee(\alpha,\beta) = \max\{\alpha,\beta\}$ and $\wedge(\alpha,\beta) = \min\{\alpha,\beta\}$. A *subalgebra* (or *G-invariant set*) will then be a subset Δ of Ω such that $\Delta g = \Delta$ for all $g \in G$. Note that $\emptyset$ is a G-invariant set. If $\alpha \in \Omega$, the subalgebra generated by α is denoted by αG and is called the *G-orbit of* α. So $\alpha G = \{\alpha g\colon g \in G\}$.

THEOREM 1.2.4. *Let* (G,Ω) *be an action. Every G-invariant set is the union of the G-orbits it contains. Moreover, the set of G-orbits of points of* Ω *partitions* Ω. *Thus the collection of G-invariant sets forms a complete Boolean algebra under union and intersection.*

Proof: If $\beta \in \alpha G$, then $\beta = \alpha g$ for some $g \in G$. Hence $\beta G = \alpha g G = \alpha G$. So the G-orbits partition Ω. The rest of the theorem is now obvious ($\emptyset$ and Ω are the 0 and 1 of the complete Boolean algebra).

If $\Omega = \alpha G$ for some (and hence every) $\alpha \in \Omega$, then the action of G on Ω is said to be *transitive*. Examples 1.1.1, 1.1.2, 1.2.1 and 1.2.3 are transitive ℓ-permutation groups but Example 1.1.3 shows that $(A(\Omega),\Omega)$ need not be transitive. If $(A(\Omega),\Omega)$ is transitive, we will simply say that Ω is *homogeneous*. (Caution: This is not the model theory usage of the word; homogeneous for us is "1-homogeneous" of model theory.)

Throughout the book we will assume that

IF (G,Ω) IS TRANSITIVE, THEN Ω IS INFINITE.

This eliminates the trivial case that $G = \{e\}$ and $|\Omega| = 1$.

1.3. PARTIALLY ORDERED GROUPS.

A group G together with a partial order $\leq$ on the elements of G is called a *p.o.group* (*partially ordered group*) if whenever $f,g,h \in G$, $f \leq g$ implies $fh \leq gh$ and $hf \leq hg$. If, in addition, $\leq$ is a lattice (i.e., $g_1 \vee g_2$ (the supremum or least upper bound of g_1 and g_2) and $g_1 \wedge g_2$ (the infimum or greatest lower bound of g_1 and g_2) exist for all $g_1, g_2 \in G$), we say that G is a *lattice-ordered group*, or *ℓ-group* for short. Note that if G is an ℓ-group, $(f \vee g)h = fh \vee gh$, $h(f \vee g) = hf \vee hg$ and dually for $\wedge$; also $f \vee g = (f^{-1} \wedge g^{-1})^{-1}$ and $f \wedge g = (f^{-1} \vee g^{-1})^{-1}$. If G is a p.o. group in which $\leq$ is a total order, we say that G is a *totally ordered group*, or *o-group* for short. If G is a p.o. group and $e \leq g \in G$, we say that g is *positive*; if $e < g \in G$, we say that g is *strictly positive*.

As examples, $\mathbb{Z}$ and $\mathbb{R}$ with the usual orders are o-groups, and G is an ℓ-group under the pointwise ordering whenever (G,Ω) is an ℓ-permutation group. In particular, $A(\Omega)$ is an ℓ-group for every chain Ω. Another example of an ℓ-group is $\mathbb{Z} \oplus \mathbb{Z}$ ordered as in Example 1.1.3. If (G,Ω) is an ordered permutation group, then G is a p.o. group.

If G is an ℓ-group and H is a subgroup of G that is closed under the lattice operations ($\vee$ and $\wedge$), then H is said to be an *ℓ-subgroup* of G. Hence an ordered permutation group (H,Ω) is an ℓ-permutation group precisely when H is an ℓ-subgroup of $A(\Omega)$.

An *o-homomorphism* from a p.o. group G into a p.o. group H is a group homomorphism from G into H which preserves the partial ordering. If G and H are ℓ-groups and the o-homomorphism preserves the lattice operations, we will call the o-homomorphism an *ℓ-homomorphism*. If ψ is one-to-one and both ψ and ψ^{-1} are o-homomorphisms, we say that ψ is an *o-embedding*. If ψ is a one-to-one ℓ-homomorphism, we say that ψ is an *ℓ-embedding*. It is trivial to show that any ℓ-embedding is an o-embedding. An (ℓ-)o-embedding of G onto H is called an *(ℓ-)o-isomorphism* between G and H. We will frequently use the following fact (which is easy to verify): A homomorphism $\psi\colon G \to H$ is an ℓ-homomorphism if $(g \vee e)\psi = g\psi \vee e$ for all $g \in G$. In Example 1.1.3, $A(\Omega)$ and $\mathbb{Z} \oplus \mathbb{Z}$ are ℓ-isomorphic.

Let (G,Ω) and (H,T) be ordered (ℓ-)permutation groups. If $\phi\colon \Omega \to T$ is an order-preserving map and $\psi\colon G \to H$ is an o-(ℓ-)homomorphism, then (ψ,ϕ) is said to be an *o-(ℓ-)homomorphism* of (G,Ω) into (H,T) if $(\alpha\phi)(g\psi) = (\alpha g)\phi$ $(\alpha \in \Omega,\ g \in G)$. If, in addition, ϕ is one-to-one and ψ is an o-(ℓ-)embedding we obtain an *ordered (ℓ-)permutation group embedding* of (G,Ω) into (H,T), and an *o-(ℓ-)isomorphism* if ϕ and ψ are also onto. If $H = G$ and ψ is the identity, then ϕ is said to be an *o-(ℓ-)homomorphism* if (ψ,ϕ) is.

1.4. CONGRUENCES.

Let X be a partially ordered set and $Y \subseteq X$. Y is convex (in X) if $y_1, y_2 \in Y$ and $y_1 \leq x \leq y_2$ imply $x \in Y$ (see p. vii).

Let (G,Ω) be an action. A *congruence* of (G,Ω) is an equivalence relation $\mathcal{C}$ on Ω such that each equivalence class is convex and $\alpha g\,\mathcal{C}\,\beta g$ whenever $\alpha\,\mathcal{C}\,\beta$ $(\alpha,\beta \in \Omega,\ g \in G)$.

If $\mathcal{C}$ is an equivalence relation on Ω, we will write $\underline{\alpha\mathcal{C}}$ for $\{\beta \in \Omega\colon \alpha\,\mathcal{C}\,\beta\}$, the equivalence class of $\mathcal{C}$ which contains α, and $\underline{\Omega/\mathcal{C}}$ for $\{\alpha\mathcal{C}\colon \alpha \in \Omega\}$.

If (G,Ω) is an action with congruence $\mathcal{C}$, then we can linearly order $\Omega/\mathcal{C}$ by: $\alpha\mathcal{C} < \beta\mathcal{C}$ if $\sigma < \tau$ for all $\sigma \in \alpha\mathcal{C}$ and $\tau \in \beta\mathcal{C}$.

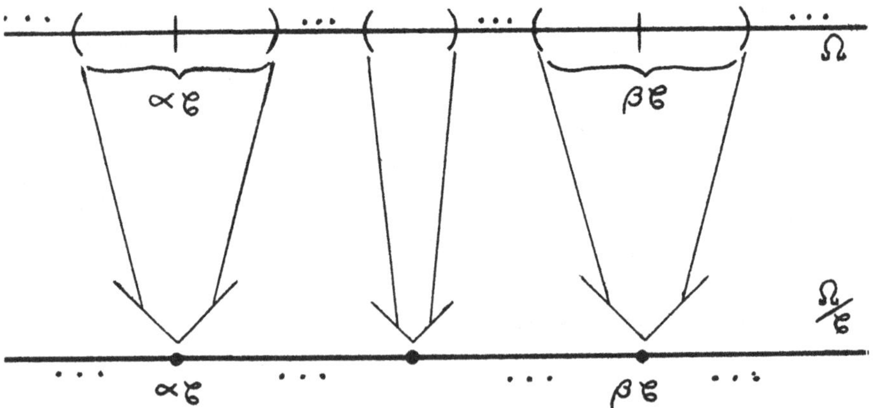

There is a natural action $(G,\Omega/\mathcal{C})$ given by: $(\alpha\mathcal{C})g = (\alpha g)\mathcal{C}$ $(\alpha \in \Omega,\ g \in G)$, and $\phi\colon \Omega \to \Omega/\mathcal{C}$ is an o-(ℓ-)homomorphism if (G,Ω) is an ordered (ℓ-)permutation group. Moreover, if L is the lazy subgroup of the action $(G,\Omega/\mathcal{C})$, then $(G/L,\Omega/\mathcal{C})$ is an ordered (ℓ-)permutation group and the natural map of (G,Ω) onto $(G/L,\Omega/\mathcal{C})$ is an ordered (ℓ-)permutation group homomorphism.

If (G,Ω) and (G,T) are ordered permutation groups and $\phi: \Omega \to T$ is an o-homomorphism, the *kernel* of ϕ is the equivalence relation $\mathcal{K}$ defined by: $\alpha \mathcal{K} \beta$ if $\alpha\phi = \beta\phi$. Clearly, $\mathcal{K}$ is a congruence of (G,Ω). So if $T = \Omega/\mathcal{C}$ and ϕ is the natural map of Ω onto $\Omega/\mathcal{C}$, the kernel of ϕ is $\mathcal{C}$.

The following two theorems are straightforward--either by direct proof or universal algebra.

THEOREM 1.4.1 *(Homomorphism Theorem). Let (G,Ω) be an ordered permutation group.*

(i) The kernel of an o-homomorphism is a congruence of (G,Ω).

(ii) Let $\phi: \Omega \to T$ be an o-homomorphism with kernel $\mathcal{K}$ and $\eta: \Omega \to \Omega/\mathcal{K}$ the natural map. Then there is an o-embedding $\phi': \Omega/\mathcal{K} \to T$ such that $\eta\phi' = \phi$.

(iii) If ϕ is as in (ii), then $\Omega\phi$ is a G-invariant subset of T; and if ϕ is onto, ϕ' is an ordermorphism.

THEOREM 1.4.2 *(Correspondence Theorem). Let (G,Ω) be an ordered permutation group and $\mathcal{C}$ a congruence of (G,Ω). The natural map $\Omega \to \Omega/\mathcal{C}$ induces a one-to-one correspondence between the congruences of (G,Ω) which contain $\mathcal{C}$ and the congruences of $(G,\Omega/\mathcal{C})$; it induces a one-to-one correspondence between the G-invariant subsets of Ω which are unions of $\mathcal{C}$ classes and the G-invariant subsets of $\Omega/\mathcal{C}$.*

THEOREM 1.4.3. *Let (G,Ω) be an ordered permutation group and N be a convex normal subgroup of G. Define $\mathcal{C}_N$ by: $\alpha \mathcal{C}_N \beta$ if $\alpha h_1 \leq \beta \leq \alpha h_2$ for some $h_1, h_2 \in N$. Then $\mathcal{C}_N$ is a congruence of (G,Ω) and $(G/N,\Omega/\mathcal{C}_N)$ is an action where $(\alpha\mathcal{C}_N)(gN) = (\alpha g)\mathcal{C}_N$ $(\alpha \in \Omega,\ g \in G)$. Conversely, if $\mathcal{C}$ is any congruence of (G,Ω), let $N = \{g \in G: \alpha g \mathcal{C} \alpha$ for all $\alpha \in \Omega\}$. Then*

N is a convex normal subgroup of G that is an ℓ-subgroup if (G,Ω) is an ℓ-permutation group. Moreover, in this case, $(G/N,\Omega/\mathcal{C})$ is an ℓ-permutation group where $Ng \geq Nf$ if $g \geq hf$ for some $h \in N$.

Proof: Suppose $\alpha\mathcal{C}_N\beta$ and let $g \in G$. Let $f \in N$ with $\alpha \leq \beta \leq \alpha f$ (without loss of generality). Then $\alpha g \leq \beta g \leq \alpha fg = \alpha g(g^{-1}fg)$. Since $N \lhd G$, $g^{-1}fg \in N$; so $\alpha g\mathcal{C}_N\beta g$. Hence $\mathcal{C}_N$ is a congruence of (G,Ω). If $Nh = Ng$, then $hN = gN$; so $g^{-1}h \in N$. But $\alpha h \leq \beta h \leq \alpha fh = (\alpha g)g^{-1}fg \cdot g^{-1}h$. Thus $(\beta\mathcal{C}_N)(Nh) = (\alpha\mathcal{C}_N)(Ng)$ if $\alpha\mathcal{C}_N\beta$ and $Nh = Ng$. It now follows easily that $(G/N,\Omega/\mathcal{C}_N)$ is an action (use Lemma 1.2.2).

Conversely, if $\mathcal{C}$ is a congruence of (G,Ω) and $N = \{g \in G: \alpha g\,\mathcal{C}\,\alpha \text{ for all } \alpha \in \Omega\}$, then clearly N is a subgroup of G. If $f \in G$ and $g \in N$, then as $(\alpha f^{-1})g\,\mathcal{C}(\alpha f^{-1})$ for all $\alpha \in \Omega$, $f^{-1}gf \in N$. Hence $N \lhd G$. Since N is the lazy subgroup of $(G,\Omega/\mathcal{C})$, $(G/N,\Omega/\mathcal{C})$ is faithful. The "ℓ-" part is now straightforward.

We next give some examples. The justifications of the claims in each are routine and are left to the reader.

EXAMPLE 1.4.4. Let G be a p.o. group and H a convex subgroup of G. Let $R(H)$ be the set of right cosets of H in G ordered by: $Hg_1 \leq Hg_2$ if $hg_1 \leq g_2$ for some $h \in H$. Then $R(H)$ is a partially ordered set. If $R(H)$ is a chain (under this ordering) then $(G,R(H),\theta)$ is an action where $(Hf)g\theta = Hfg$ $(f,g \in G)$. This action is faithful precisely when $\bigcap_{g \in G} g^{-1}Hg = \{e\}$. In the special case that $H = \{e\}$ and $R(H)$ is a chain (hence $G = R(H)$ is an o-group), the faithful transitive action (G,G,θ) is called the *right regular representation* of G. Examples 1.1.1 and 1.2.1 are right regular representations.

EXAMPLE 1.4.5. Let (G,Ω) be any ordered permutation group. Then (G,Ω) has two *trivial* congruences, denoted throughout the book by $\mathcal{S}$ and $\mathcal{U}$: $\alpha\mathcal{S}\beta$ if $\alpha = \beta$ (the *singleton* congruence) and $\alpha \mathcal{U} \beta$ if $\alpha, \beta \in \Omega$ (the *universal* congruence).

EXAMPLE 1.4.6. Let $\Omega = \mathbb{R} \setminus \mathbb{Z}$. Then Ω is homogeneous and $(A(\Omega), \Omega)$ has a non-trivial congruence $\mathcal{C}$ defined by: $\alpha \mathcal{C} \beta$ if $\alpha, \beta \in (n, n+1)$ for some integer n.

1.5. STABILISERS AND BLOCKS.

Let (G, Ω) be an action and $\Delta \subseteq \Omega$. The *stabiliser* (or *isotropy*) *subgroup* on Δ, $G_\Delta = \{g \in G : \delta g = \delta$ for all $\delta \in \Delta\}$. So G_Δ is the set of elements of G that fix Δ pointwise. If $\Delta = \{\alpha\}$, we will write G_α for $G_{\{\alpha\}}$. On the other hand, let $G_{(\Delta)} = \{g \in G : \Delta g = \Delta\}$; i.e., $G_{(\Delta)}$ is the set of all elements of G which map Δ to Δ (not necessarily fixing Δ pointwise). Note that G_Δ and $G_{(\Delta)}$ are subgroups of G and $G_\Delta \subseteq G_{(\Delta)}$. Moreover, if (G, Ω) is an ordered (ℓ-)permutation group, then G_Δ and $G_{(\Delta)}$ are convex (ℓ-)subgroups of G. In Examples 1 1.1 and 1.2.1, $G_\alpha = \{e\}$ for all $\alpha \in \Omega$, and $G_\Delta = G_{(\Delta)} = \{e\}$ if $\Delta \neq \emptyset, \Omega$. However, in Example 1.1.3, if $\alpha \in \Omega$, $G_\alpha \cong \mathbb{Z}$ and $\mathbb{Z} \cong G_{\mathbb{Z}_i} \subsetneq G_{(\mathbb{Z}_i)} = G$ $(i = 1,2)$. In Example 1.1.2, if $\Delta = \mathbb{Z}$, $G_\Delta \neq G_{(\Delta)}$ as is easily seen $(f_0 \in G_{(\Delta)} \setminus G_\Delta)$. In Example 1.2.3, if $\Delta = (0,1)$, $G_\Delta \neq G_{(\Delta)}$.

A central fact which we will use repeatedly (and should therefore be engraved on the reader's mind) is:

The Fundamental Triviality:
$$\begin{array}{ccc} \alpha & \xrightarrow{\;g\;} & \alpha g \\ \downarrow f & & \downarrow f \\ \alpha f & \xrightarrow[f^{-1}gf]{} & \alpha g f \end{array}$$
commutes

if (G, Ω) is an action. Hence $f^{-1} G_\alpha f = G_{\alpha f}$.

By the fundamental triviality, $\bigcap_{g \in G} g^{-1} G_\alpha g = G_{\alpha G}$. Hence if (G, Ω) is transitive, $\bigcap_{g \in G} g^{-1} G_\alpha g = G_{\alpha G} = G_\Omega = \{e\}$. So if (G, Ω) is a transitive ordered (ℓ-)permutation group and $R(G_\alpha)$ is a chain, $(G, R(G_\alpha))$ is also a transitive ordered (ℓ-)permutation group.

THEOREM 1.5.1. *Let (G,Ω) be a transitive ℓ-permutation group and $\alpha \in \Omega$. Then $R(G_\alpha)$ is a chain and (G,Ω) is ℓ-isomorphic to $(G,R(G_\alpha))$.*

Proof: Let $f,g \in G$. Without loss of generality, $\alpha f \leq \alpha g$. Hence $fg^{-1} \vee e \in G_\alpha$; i.e., $(f \vee g)g^{-1} \in G_\alpha$. Thus $G_\alpha g = G_\alpha(f \vee g)$, so $G_\alpha f \leq G_\alpha g$. Consequently, $R(G_\alpha)$ is a chain. Let $\beta \in \Omega$. Choose $g \in G$ so that $\alpha g = \beta$. Let $\phi\colon \Omega \to R(G_\alpha)$ be defined by $\beta\phi = G_\alpha g$. If $\alpha f = \beta = \alpha g$, then $fg^{-1} \in G_\alpha$ so $G_\alpha f = G_\alpha g$. Therefore ϕ is well-defined. It is clearly one-to-one and onto. If $\beta_1 \leq \beta_2$, let $g_i \in G$ with $\alpha g_i = \beta_i$ $(i = 1,2)$. Now $\alpha(g_1 \vee g_2) = \beta_2 = \alpha g_2$, so $\beta_1\phi = G_\alpha g_1 \leq G_\alpha(g_1 \vee g_2) = \beta_2\phi$. Hence ϕ preserves order. Finally, $(\beta h)\phi = (\alpha gh)\phi = G_\alpha gh = (G_\alpha g)h = (\beta\phi)h$ $(h \in G)$. Consequently, ϕ is an ℓ-isomorphism. //

Let (G,Ω) be an action and $\emptyset \neq \Delta \subseteq \Omega$. Δ is a *block* (of (G,Ω)) if Δ is convex and for each $g \in G$, either $\Delta g = \Delta$ or $\Delta g \cap \Delta = \emptyset$. Any non-empty convex G-invariant subset of Ω is of course a block.

THEOREM 1.5.2. *Let (G,Ω) be an action. The classes of congruences of (G,Ω) are blocks and conversely, every block is the class of a congruence of (G,Ω).*

Proof: Let $\mathcal{C}$ be a congruence of (G,Ω) and Δ be a $\mathcal{C}$-class, say $\Delta = \delta\mathcal{C}$. Then Δ is convex by definition and $\Delta g = \delta\mathcal{C}g = \delta g\mathcal{C}$, a $\mathcal{C}$-class. Since the $\mathcal{C}$-classes partition Ω, Δ is a block.

Conversely, let Δ be a block. If Δ is a class of a congruence of (G,Ω), then so is Δg for each $g \in G$. Hence we define $\mathcal{C}$ so that its non-singleton classes are Δg $(g \in G)$. Formally, $\alpha\mathcal{C}\beta$ if $\alpha = \beta$ or $\alpha,\beta \in \Delta g$ for some $g \in G$. Since Δ is a block, the $\mathcal{C}$ classes are convex and partition Ω (so $\mathcal{C}$ is an equivalence relation). If $\alpha\mathcal{C}\beta$ and $f \in G$, either $\alpha = \beta$ or $\alpha,\beta \in \Delta g$. In the first case, $\alpha f = \beta f$; in the second, $\alpha f,\beta f \in \Delta gf$. Hence $\mathcal{C}$ is a congruence of (G,Ω) having Δ as a class.

Note that in defining $\mathcal{C}$ in the converse part of the proof, we gave a minimal congruence. Instead, we could have used a maximal one: $\alpha\mathcal{K}\beta$ if $\alpha,\beta \in \Delta g$ for some $g \in G$, or $\alpha,\beta \notin \bigcup_{h\in G} \Delta h$ and α,β belong to the same interval Λ of Ω maximal with respect to $\Lambda \cap \bigcup\{\Delta h\colon h \in G\} = \emptyset$. Thus the $\mathcal{C}$ defined in Theorem 1.5.2 is not in general unique, but it is unique if (G,Ω) is transitive. This makes transitive permutation groups nicer to deal with, a fact borne out most strikingly in Chapter 3.

Let A and B be chains. Then $A \overleftarrow{\times} B$ is the set $\{(\alpha,\beta)\colon \alpha \in A,\ \beta \in B\}$ ordered by: $(\alpha_1,\beta_1) < (\alpha_2,\beta_2)$ if $\beta_1 < \beta_2$ or $(\beta_1 = \beta_2\ \&\ \alpha_1 < \alpha_2)$. Note $A \overleftarrow{\times} B$ is a chain.

EXAMPLE 1.5.3. Let $\{\alpha_1,\alpha_2\}$ be linearly ordered by $\alpha_1 < \alpha_2$ where $\alpha_1,\alpha_2 \notin \mathbb{Z}$. Let $T = \mathbb{Z} \overleftarrow{\cup} \{\alpha_1,\alpha_2\}$ and $\Omega = T \overleftarrow{\times} \mathbb{Z}$. Let $\Delta = \{(n,0)\colon n \in \mathbb{Z}\}$. Then Δ is a block of $(A(\Omega),\Omega)$ that is a class of the congruences $\mathcal{C}$ and $\mathcal{K}$ of $(A(\Omega),\Omega)$ given by: $(\alpha,m)\mathcal{C}(\beta,n)$ if $m = n\ \&\ (\alpha = \beta \text{ or } \alpha,\beta \in \mathbb{Z})$ and $(\alpha,m)\mathcal{K}(\beta,n)$ if $m = n$. Note $\mathcal{C} \subsetneq \mathcal{K}$ since $(\alpha_1,0)\mathcal{K}(\alpha_2,0)$ but $(\alpha_1,0) \not\mathcal{C}\ (\alpha_2,0)$.

1.6. *TRANSITIVE ACTIONS.*

Recapping Theorem 1.5.2 in the transitive setting:

THEOREM 1.6.1. *Let (G,Ω) be a transitive action and $\alpha \in \Omega$. There is a one-to-one containment preserving correspondence between the congruences of (G,Ω) and the blocks that contain α, given by $\mathcal{C} \mapsto \alpha\mathcal{C}$.*

Of the same ilk:

THEOREM 1.6.2. *Let (G,Ω) be a transitive ordered (ℓ-)permutation group and $\alpha \in \Omega$. Every block containing α is the orbit of a unique convex (ℓ-)subgroup of G containing G_α. Conversely, if (G,Ω) is an ℓ-permutation group, the orbit of α under any such convex ℓ-subgroup is a block containing α. This one-to-one correspondence preserves order.*

Proof: Let Δ be a block containing α. We show that $\Delta = \alpha G_{(\Delta)}$. Clearly $\alpha G_{(\Delta)} \subseteq \Delta$. Let $\delta \in \Delta$. By transitivity, there is $g \in G$ such that $\alpha g = \delta$. So $\Delta \cap \Delta g \neq \emptyset$. Hence $\Delta g = \Delta$. Therefore $\delta = \alpha g \in \alpha G_{(\Delta)}$; so $\Delta = \alpha G_{(\Delta)}$. Observe that $G_\alpha \subseteq G_{(\Delta)}$. If $\alpha H = \alpha K$ with $G_\alpha \subseteq H,K$, let $h \in H$. Then $\alpha h = \alpha k$ for some $k \in K$. Thus $hk^{-1} \in G_\alpha \subseteq K$, so $h \in K$; i.e., $H \subseteq K$. By symmetry, $K \subseteq H$ and the uniqueness is proved.

Conversely, let H be a convex ℓ-subgroup of G. If $\alpha f_1 \leq \beta \leq \alpha f_2$ for some $f_1, f_2 \in H$, let $g \in G$ be such that $\alpha g = \beta$ (by transitivity). Therefore $\beta = \alpha[(f_1 \vee g) \wedge f_2]$. But $f_1 \wedge f_2 \leq (f_1 \vee g) \wedge f_2 \leq f_2$ and as $f_2, f_1 \wedge f_2 \in H$, $(f_1 \vee g) \wedge f_2 \in H$. Hence $\beta \in \alpha H$, so αH is convex. If $H \supseteq G_\alpha$, let $\beta \in \alpha H \cap \alpha Hf$ for some $f \in G$. Then $\alpha h_1 = \beta = \alpha h_2 f$ for some $h_1, h_2 \in H$. Thus $h_2 f h_1^{-1} \in G_\alpha \subseteq H$; so $f \in H$. Hence $\alpha Hf = \alpha H$. Consequently, αH is a block.

A transitive action (G,Ω) is called *primitive* if there are no non-trivial congruences of (G,Ω)--see Example 1.4.5. We will say that Ω is *primitive* if $(A(\Omega),\Omega)$ is. In Example 1.4.6 we saw that a homogeneous chain need not be primitive. We will see that $\mathbb{R}$ is primitive (Corollary 1.6.6).

If (G,Ω) is an action with lazy subgroup G, we will say (G,Ω) is *static*; i.e., $\alpha g = \alpha$ for all $\alpha \in \Omega$.

EXAMPLE 1.6.3. $(\{e\},\mathbb{R})$ is a static ℓ-permutation group.

EXAMPLE 1.6.4. $(A(\omega),\omega)$ is a static ℓ-permutation group. (Recall $\omega = \{0,1,2,\ldots\}$.)

Note that if (G,Ω) is static and $\mathcal{C}$ is any equivalence relation on Ω whose classes are convex, then $\mathcal{C}$ is a congruence of (G,Ω). So if $|\Omega| \geq 3$, (G,Ω) has proper congruences. On the other hand, static ordered permutation groups are in some sense basic building bricks. Hence when we extend the definition of primitive to the intransitive case (Section 3.2), we will do so in such a way that static ordered permutation groups will be primitive. Thus having no non-trivial congruences will not necessarily be enough to be primitive in the intransitive case.

An action (G,Ω) is said to satisfy the _separation property_ if whenever $\Delta \subseteq \Omega$ is a non-empty bounded interval of Ω and α,β are distinct points of Ω, there is $g \in G$ such that exactly one of $\alpha g, \beta g$ belongs to Δ. If $|\Omega| = 1$, (G,Ω) enjoys the separation property vacuously.

THEOREM 1.6.5. *A transitive action* (G,Ω) *is primitive if and only if it satisfies the separation property.*

Proof: If (G,Ω) is not primitive, there is a bounded block $\Delta \neq \Omega$ with $|\Delta| > 1$. Let $\alpha,\beta \in \Delta$ with $\alpha \neq \beta$. But if $\delta \in \Delta$, $\delta g \in \Delta$ only if $\Delta g = \Delta$. Hence $\alpha g \in \Delta$ precisely when $\beta g \in \Delta$. So α and β cannot be separated by Δ.

Conversely, let $\alpha,\beta \in \Omega$ be distinct points which cannot be separated by the non-empty bounded interval $\Delta \neq \Omega$. Define $\mathcal{C}$ by: $\sigma \mathcal{C} \tau$ if $(\forall g)(\sigma g, \tau g \in \Delta$ or $\sigma g, \tau g \notin \Delta)$. Then clearly $\mathcal{C}$ is a congruence on (G,Ω) and $\alpha \mathcal{C} \beta$. If $\delta \in \Delta$ and $\delta' \in \Omega \setminus \Delta$, then $\delta' \notin \delta\mathcal{C}$. Hence $\mathcal{C}$ is not trivial. Thus (G,Ω) is not primitive.

For further equivalences of "primitive", see Section 4.1.

Let (G,Ω) be an action such that $\alpha_i < \beta_i$ $(i = 1,2)$ implies there is $g \in G$ with $\alpha_1 g = \alpha_2$ and $\beta_1 g = \beta_2$. Then (G,Ω) is said to be a _doubly transitive action_; e.g., Examples 1.1.2 and 1.2.3. Clearly any doubly transitive action satisfies the separation property. Hence, excluding the static case,

COROLLARY 1.6.6. *Any doubly transitive action is primitive.*

EXAMPLE 1.6.7. Let $G = \{g \in A(\mathbb{R})$: g is differentiable (at each $\alpha \in \mathbb{R})\}$. Then $(G,\mathbb{R})$ is a doubly transitive ordered permutation group ($h_0 \in G$ where h_0 is defined in Example 1.1.2).and hence is primitive. However, it is not an ℓ-permutation group since $g_0 \vee e \notin G$ $(g_0: \alpha \mapsto 2\alpha \; (\alpha \in \mathbb{R}))$.

Next we note (if $G \neq \{e\}$):

LEMMA 1.6.8. *If (G,Ω) is a doubly transitive ordered permutation group, then G is not Abelian. Indeed, the centre of G is $\{e\}$.*

Proof: Let $e \neq g \in G$. For some $\alpha \in \Omega$, $\alpha g \neq \alpha$. By double transitivity, there is $f \in G$ such that $\alpha f = \alpha$ and $\alpha g f = \alpha g^2$. Now $\alpha g f = \alpha g^2 \neq \alpha g = \alpha f g$. Hence $fg \neq gf$. Consequently g does not belong to the centre of G.

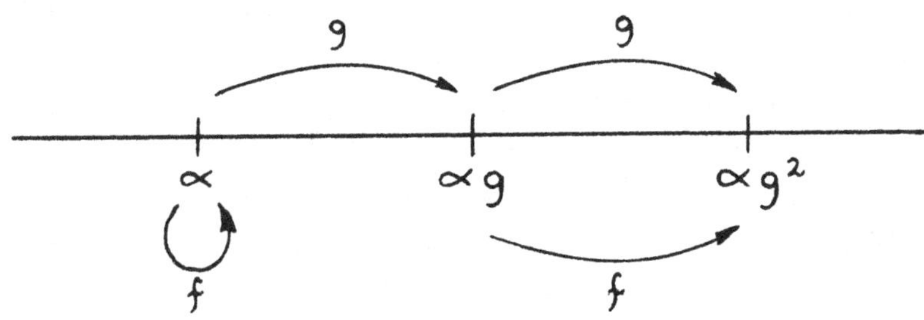

We complete this section with a preliminary study of transitive ℓ-permutation groups. In Theorem 1.6.2, we saw that if (G,Ω) is a transitive ℓ-permutation group and H is a convex ℓ-subgroup containing G_α, then αH is a block. Actually, the first part of the proof of the converse direction of that theorem shows:

LEMMA 1.6.9. *Let (G,Ω) be a transitive ℓ-permutation group, H a convex ℓ-subgroup of G and $\beta \in \Omega$. Then βH is convex. Hence βG_α is convex for all $\alpha,\beta \in \Omega$.*

Hence the orbits of G_α partition Ω into convex subsets and the set of orbits inherits a linear ordering from Ω: $\Delta_1 < \Delta_2$ if $\delta_1 < \delta_2$ for all $\delta_i \in \Delta_i$ $(i = 1,2)$. If Δ is an orbit of G_α, let $\underline{\Delta'} = \{\alpha g\colon \alpha g^{-1} \in \Delta\}$.

LEMMA 1.6.10. *If (G,Ω) is a transitive ℓ-permutation group, $\alpha \in \Omega$ and Δ is a G_α orbit, then so is Δ' and the map $\Delta \mapsto \Delta'$ is an anti-ordermorphism of the set of G_α orbits onto itself.*

Proof: Let $\beta \in \Delta'$. Then $\beta = \alpha f$ for some $f \in G$ with $\alpha f^{-1} \in \Delta$. If $\alpha g \in \Delta'$ with $\alpha g^{-1} \in \Delta$, there is $h \in G_\alpha$ such that $\alpha g^{-1} = \alpha f^{-1}h$. Hence $f^{-1}hg \in G_\alpha$ and $\beta f^{-1}hg = \alpha g$. Thus Δ' is a G_α orbit.

If Δ_1, Δ_2 are distinct G_α orbits, let $\alpha g^{-1} \in \Delta_1 \setminus \Delta_2$. Then $\alpha g \in \Delta_1' \setminus \Delta_2'$ (if $\alpha g = \alpha f \in \Delta_2'$ with $\alpha f^{-1} \in \Delta_2$, then $\alpha g^{-1} = \alpha f^{-1}(fg^{-1}) \in \Delta_2$ since $fg^{-1} \in G_\alpha$). Hence $\Delta_1' \neq \Delta_2'$. Since Δ_1' and Δ_2' are distinct G_α orbits, $\Delta_1' \cap \Delta_2' = \emptyset$. If $\Delta_1 < \Delta_2$, let $\alpha g^{-1} \in \Delta_1$ and $\alpha h^{-1} \in \Delta_2$. Then $g^{-1}h \vee e \in G_\alpha$ so $\max\{\alpha g, \alpha h\} = \alpha g(e \vee g^{-1}h) \in \Delta_1'$. But $\alpha h \in \Delta_2'$ so $\alpha h < \alpha g$. Thus $\Delta_2' < \Delta_1'$.

Since $\Delta'' = \Delta$, the proof is complete.

Note that, as a consequence, $\Delta' = \Delta$ only if $\Delta = \{\alpha\}$.

We partition the set of orbits of G_α into four classes: $|\Delta| = 1 = |\Delta'|$; $|\Delta| = 1 < |\Delta'|$; $|\Delta| > 1 = |\Delta'|$; $|\Delta|, |\Delta'| > 1$. Of course some of these classes may be empty. The set of G_α orbits under the induced order, partitioned as above, is called the *α-configuration* of (G,Ω). We now show:

THEOREM 1.6.11. *Let (G,Ω) be a transitive ℓ-permutation group, $\alpha_1, \alpha_2 \in \Omega$ and $f \in G$ such that $\alpha_1 f = \alpha_2$. Then the map $\Delta \mapsto \Delta f$ is an ordermorphism of the α_1-configuration onto the α_2-configuration of (G,Ω); i.e., $\Delta' f$ is the α_2-paired orbit of Δf and if $|\Delta| = 1 = |\Delta'|$, then $|\Delta f| = 1 = |\Delta' f|$, etc.*

Proof: It is enough to show that Δf is a G_{α_2} orbit and $(\Delta f)' = \Delta' f$. But if $\Delta = \delta G_{\alpha_1}$, $\Delta f = \delta G_{\alpha_1} f = \delta f G_{\alpha_1 f} = \delta f G_{\alpha_2}$ by the fundamental triviality; i.e., Δf is a G_{α_2} orbit. Moreover, if $\delta = \alpha_1 g$, $\delta f = \alpha_1 g f = \alpha_2(f^{-1}gf)$. Thus $(\Delta f)' = \alpha_2 f^{-1}g^{-1}fG_{\alpha_2} = \alpha_1 g^{-1}G_{\alpha_2 f^{-1}} f = \alpha_1 g^{-1}G_{\alpha_1} f = \Delta' f$ and the proof is complete.

Thus the α-configuration is independent of the choice of $\alpha \in \Omega$ and will just speak of the *configuration* of (G,Ω).

If Δ is a G_α orbit, Δ is said to be *positive* if $\Delta > \{\alpha\}$ (*negative* if $\Delta < \{\alpha\}$). If Δ is an orbit of G_α and $|\Delta| > 1$, Δ is called a *long orbit* of G_α. If (G,Ω) is a doubly transitive ℓ-permutation group it has just one positive and one negative G_α orbit and these are long. If whenever Δ is a long orbit of G_α so is Δ', we say that the transitive ℓ-permutation group (G,Ω) is *balanced*. By Theorem 1.6.11, the definition of balanced is independent of the choice of $\alpha \in \Omega$.

THEOREM 1.6.12. *If (G,Ω) is a transitive ℓ-permutation group, $\alpha \in \Omega$ and Δ is a G_α orbit with $|\Delta| > 1 = |\Delta'|$, then Δ is a block of (G,Ω). Hence a primitive transitive ℓ-permutation group is balanced.*

Proof: Let Δ be long with $\Delta' = \{\beta\}$, say. Then $G_\alpha \subseteq G_\beta$ and αG_β is a block by Theorem 1.6.2. Let $g \in G$ be such that $\alpha g \in \Delta$. We show that $\alpha G_\beta g = \Delta$, whence Δ is a block (since αG_β is). Since $\alpha g \in \Delta$, $\alpha g^{-1} \in \Delta' = \{\beta\}$. Hence $\alpha = \beta g$. But $\alpha G_\beta g = \alpha g G_{\beta g} = \alpha g G_\alpha = \Delta$ by the fundamental triviality, and the theorem is proved.

For further results about configurations, see McCleary [72a].

1.7. PRIMITIVE COMPONENTS.

Let (G,Ω) be a transitive action with $\mathcal{C} \subseteq \mathcal{K}$ congruences of (G,Ω). Let Δ be a $\mathcal{K}$ class. Then $(G_{(\Delta)},\Delta)$ is a transitive action in the natural way and $\mathcal{C}' = \mathcal{C}|\Delta \times \Delta$ is a congruence of $(G_{(\Delta)},\Delta)$. Thus $(G_{(\Delta)},\Delta/\mathcal{C}')$ is a transitive action. If L is the lazy subgroup of $(G_{(\Delta)},\Delta/\mathcal{C}')$, let $\underline{\hat{G}_{(\Delta)}} = G_{(\Delta)}/L \cong G_{(\Delta)}|\Delta$. So $(\hat{G}_{(\Delta)},\Delta/\mathcal{C}')$ is a faithful transitive action called the *component* of (G,Ω) relative to $(\mathcal{C},\mathcal{K})$ at Δ.

We first show that the component is independent of the particular $\mathcal{K}$ class chosen.

LEMMA 1.7.1. *Let* (G,Ω) *be a transitive action and* $\mathcal{C} \subseteq \mathcal{K}$ *be congruences of* (G,Ω). *Let* Δ_1, Δ_2 *be* $\mathcal{K}$ *classes. Then the components of* (G,Ω) *relative to* $(\mathcal{C}, \mathcal{K})$ *at* Δ_1 *and* Δ_2 *are o-isomorphic; indeed,* ℓ*-isomorphic if* (G,Ω) *is an* ℓ*-permutation group.*

Proof: There exists $g \in G$ such that $\Delta_1 g = \Delta_2$. By the fundamental triviality, $g^{-1}G_{(\Delta_1)}g = G_{(\Delta_1 g)} = G_{(\Delta_2)}$. Let $\mathcal{C}_i' = \mathcal{C}|\Delta_i \times \Delta_i$ $(i = 1,2)$, and $g^{-1}L_1 g = L_2$ where L_1 is the lazy subgroup of $(G_{(\Delta_1)}, \Delta_1/\mathcal{C}_1')$. Then L_2 is the lazy subgroup of $(G_{(\Delta_2)}, \Delta_2/\mathcal{C}_2')$. For $\delta_1 \in \Delta_1$, let $(\delta_1\mathcal{C}_1')\phi = (\delta_1 g)\mathcal{C}_2'$ and $(L_1 f)\psi = L_2 g^{-1} f g$ $(f \in G_{(\Delta_1)})$. Then $\psi\colon \hat{G}_{(\Delta_1)} \cong \hat{G}_{(\Delta_2)}$ and $\phi\colon \Delta_1/\mathcal{C}_1' \cong \Delta_2/\mathcal{C}_2'$. $[(\delta_1\mathcal{C}_1')\phi](L_1 f)\psi = \delta_1 g\mathcal{C}_2' L_2 g^{-1} f g = \delta_1 f g \mathcal{C}_2'$ $= [(\delta_1\mathcal{C}_1')(L_1 f)]\phi$. Hence $(\hat{G}_{(\Delta_1)}, \Delta_1/\mathcal{C}_1') \cong (\hat{G}_{(\Delta_2)}, \Delta_2/\mathcal{C}_2')$ as desired.

Hence, from now on, we will drop the reference to Δ in the component of a transitive action relative to a pair of congruences. Also, we will write $\underline{\Delta/\mathcal{C}}$ for $\Delta/\mathcal{C} \cap (\Delta \times \Delta)$ if Δ is a union of $\mathcal{C}$ classes.

Note that the definitions make sense in the intransitive case as well, and the proof of Lemma 1.7.1 still works provided $\Delta_2 = \Delta_1 g$ for some $g \in G$. This should be remembered when reading Section 3.2 as should the next theorem.

If $\mathcal{C} \subsetneq \mathcal{K}$ are congruences of a transitive action (G,Ω), we say $\mathcal{K}$ *covers* $\mathcal{C}$ if there is no congruence $\mathcal{D}$ of (G,Ω) with $\mathcal{C} \subsetneq \mathcal{D} \subsetneq \mathcal{K}$.

THEOREM 1.7.2. *If* (G,Ω) *is a transitive action with congruences* $\mathcal{C} \subsetneq \mathcal{K}$, *the component of* (G,Ω) *relative to* $(\mathcal{C}, \mathcal{K})$ *is primitive if and only if* $\mathcal{K}$ *covers* $\mathcal{C}$.

Proof: Let Δ be a $\mathcal{K}$ class. By Theorem 1.4.2, any block of $(\hat{G}_{(\Delta)}, \Delta/\mathcal{C})$ has the form $T/\mathcal{C}$ where $T \subseteq \Delta$ is a block of $(G_{(\Delta)}, \Omega)$. If $g \in G$ and $Tg \cap T \neq \emptyset$, then $\Delta g \cap \Delta \neq \emptyset$ since $T \subseteq \Delta$. Because Δ is a block, $\Delta g = \Delta$. Hence $g \in G_{(\Delta)}$; so

$Tg = T$. Thus T is a block of (G,Ω) and the theorem follows from Theorem 1.6.1.

If (G,Ω) is a transitive action and $\mathcal{K}$ covers $\mathcal{C}$, the component of (G,Ω) relative to $(\mathcal{C},\mathcal{K})$ is said to be a *primitive component* of (G,Ω).

We complete this section with two examples to show how to determine the primitive components of transitive ordered permutation groups and to see how the ordered permutation group is reflected in the set of its primitive components.

EXAMPLE 1.7.3. Let $\Omega = A \overleftarrow{\times} B$ where A and B are primitive, homogeneous chains. Let $\mathcal{S}$ and $\mathcal{U}$ be the trivial equivalence relations (Example 1.4.5). Define $\mathcal{C}$ by: $(\alpha,\beta)\mathcal{C}(\alpha',\beta')$ if $\beta = \beta'$ $(\alpha,\alpha' \in A,\ \beta,\beta' \in B)$. Let $G = \{g \in A(\Omega): \sigma\mathcal{C}\tau \iff \sigma g\mathcal{C}\tau g\}$. So $\mathcal{S}$, $\mathcal{C}$ and $\mathcal{U}$ are congruences of (G,Ω). Note that (G,Ω) is transitive.

$(\mathcal{S},\mathcal{C})$ component: Let Δ be a $\mathcal{C}$ class, say $\Delta = \{(\alpha,\beta_0): \alpha \in A\}$. Note that Δ is ordermorphic to A, so $A(\Delta) \cong A(A)$. Let $s \in A(A)$ and define $g_s \in G_{(\Delta)}$ by:

$$(\alpha,\beta)g_s = \begin{cases} (\alpha s,\beta_0) & \text{if } \beta = \beta_0 \\ (\alpha,\beta) & \text{if } \beta \neq \beta_0 \end{cases}.$$

The map $s \mapsto Lg_s$ is an ℓ-isomorphism between $A(A)$ and $\hat{G}_{(\Delta)}$. Hence the $(\mathcal{S},\mathcal{C})$ component is ℓ-isomorphic to $(A(A),A)$ which is primitive by hypothesis.

$(\mathcal{C},\mathcal{U})$ component: Ω is the $\mathcal{U}$ class and $L = \{g \in G_{(\Omega)}: Tg = T \text{ for all } \mathcal{C} \text{ classes } T\} = \{g \in G: Tg = T \text{ for all } \mathcal{C} \text{ classes } T\}$. Let $s \in A(B)$ and define $\bar{g}_s \in G$ by $(\alpha,\beta)\bar{g}_s = (\alpha,\beta s)$ $(\alpha \in A,\ \beta \in B)$. The map $s \mapsto L\bar{g}_s$ is an ℓ-isomorphism between $A(B)$ and G/L. So the $(\mathcal{C},\mathcal{U})$ component is ℓ-isomorphic to $(A(B),B)$ which is primitive by assumption.

Hence the only congruences of (G,Ω) are $\mathcal{S}$, $\mathcal{C}$ and $\mathcal{U}$.

Let (H_1,A) and (H_2,B) be transitive ordered permutation groups and let $\bar{G} = \{g \in G: g \text{ induces an element of } H_1 \text{ on the}$

$(\mathcal{J},\mathcal{C})$ component and an element of H_2 on the $(\mathcal{C},\mathcal{U})$ component}. Then $(\bar{G},\Omega)$ is a transitive permutation group with $(\mathcal{J},\mathcal{C})$ component (H_1,A) and $(\mathcal{C},\mathcal{U})$ component (H_2,B) as is easily seen by modifying the argument above. Of course, there will be other components unless (H_1,A) and (H_2,B) are primitive.

EXAMPLE 1.7.4 Let $\Omega = A \overleftarrow{\times} B \overleftarrow{\times} \Gamma$ where A,B,Γ are primitive homogeneous chains. Let $\mathcal{J}$ and $\mathcal{U}$ be the trivial equivalence relations. Define $\mathcal{C}$ and $\mathcal{K}$ by: $(\alpha,\beta,\gamma)\mathcal{C}(\alpha',\beta',\gamma')$ if $\gamma = \gamma'$ and $\beta = \beta'$; $(\alpha,\beta,\gamma)\mathcal{K}(\alpha',\beta',\gamma')$ if $\gamma = \gamma'$. So $\mathcal{J} \subseteq \mathcal{C} \subseteq \mathcal{K} \subseteq \mathcal{U}$. Let $G = \{g \in A(\Omega)\colon \sigma\mathcal{C}\tau \iff \sigma g\mathcal{C}\tau g$ and $\sigma\mathcal{K}\tau \iff \sigma g\mathcal{K}\tau g\}$. Then $\mathcal{J}$, $\mathcal{C}$, $\mathcal{K}$ and $\mathcal{U}$ are congruences of (G,Ω) and (G,Ω) is transitive. As in Example 1.7.3, the $(\mathcal{J},\mathcal{C})$ component is $(A(A),A)$ and the $(\mathcal{K},\mathcal{U})$ component is $(A(\Gamma),\Gamma)$ (to within isomorphism). We examine the $(\mathcal{C},\mathcal{K})$ component. Let Δ be a $\mathcal{K}$ class, say $\Delta = \{(\alpha,\beta,\gamma_0)\colon \alpha \in A,\ \beta \in B\}$. Now $\Delta/\mathcal{C} \cong B$. Let $s \in A(B)$ and

define $g_s \in G_{(\Delta)}$ by: $(\alpha,\beta,\gamma)g_s = \begin{cases} (\alpha,\beta s,\gamma_0) & \text{if } \gamma = \gamma_0 \\ (\alpha,\beta,\gamma) & \text{if } \gamma \neq \gamma_0 \end{cases}.$

Here $L = \{f \in G_{(\Delta)}\colon (\forall\alpha \in A)(\forall\beta \in B)(\exists\alpha' \in A)((\alpha,\beta,\gamma_0)f = (\alpha',\beta,\gamma_0))\}$, and the map $s \mapsto Lg_s$ is an ℓ-isomorphism between $A(B)$ and $G_{(\Delta)}/L = \hat{G}_{(\Delta)}$. The $(\mathcal{C},\mathcal{K})$ component is ℓ-isomorphic to $(A(B),B)$ which is primitive by hypothesis. So the only congruences of (G,Ω) are $\mathcal{J}$, $\mathcal{C}$, $\mathcal{K}$ and $\mathcal{U}$.

We can modify this example as we did in Example 1.7.3.

1.8. DEDEKIND COMPLETION AND CHARACTER.

Let Ω be a chain. We say that (A,B) is a *cut* in Ω if A and B are non-empty disjoint subsets of Ω whose union is Ω such that

(1) $A < B$ (i.e., $\alpha < \beta$ for all $\alpha \in A$, $\beta \in B$), and

either

(2a) A has no greatest element,

or

(2b) for some $\beta \in \Omega$, $A = \{\alpha \in \Omega: \alpha < \beta\}$ and $B = \{\alpha \in \Omega: \alpha \geq \beta\}$.

Let $\bar{\Omega}$ be the set of cuts in Ω, the *Dedekind completion* of Ω. We single out clause (2a) by considering the Dedekind completion $\mathbb{R}$ of $\mathbb{Q}$, and (2b) by considering the case that Ω has an element β with an immediate predecessor (to ensure that the predecessor of β "occurs" as a bona fide element of $\bar{\Omega}$-- see below; e.g., $\Omega = \mathbb{Z}$). Totally order $\bar{\Omega}$ by: $(A,B) \leq (\Gamma,\Delta)$ if $A \subseteq \Gamma$. Order embed Ω in $\bar{\Omega}$ via: $\beta \mapsto (A_\beta,B_\beta)$ where $A_\beta = \{\alpha \in \Omega: \alpha < \beta\}$ and $B_\beta = \{\alpha \in \Omega: \alpha \geq \beta\}$. Each $(A,B) \in \bar{\Omega}$ of type (2a) is the supremum of $\{(A_\alpha,B_\alpha): \alpha \in A\}$ and the infimum of $\{(A_\alpha,B_\alpha): \alpha \in B\}$. Moreover, if $(A,B) < (\Gamma,\Delta)$, there is $\alpha \in \Omega$ such that $(A,B) \leq (A_\alpha,B_\alpha) \leq (\Gamma,\Delta)$. We will identify Ω with its image in $\bar{\Omega}$ and write $\bar{\alpha}$ for the general element of $\bar{\Omega}$ ($\bar{\alpha}$ does *not* signify the image (A_α,B_α) of α). So under this identification, Ω is dense in $\bar{\Omega}\backslash\Omega$ (i.e., if $\bar{\alpha} < \bar{\beta}$ in $\bar{\Omega}\backslash\Omega$ there is $\alpha \in \Omega$ with $\bar{\alpha} < \alpha < \bar{\beta}$). If $\bar{\alpha} \in \bar{\Omega}\backslash\Omega$, we say that $\bar{\alpha}$ is a *hole* in Ω; thus a cut in Ω is a point of Ω or a hole in Ω. Note that if $\{(A_i,B_i): i \in I\}$ is a set of cuts in Ω with $(A_i,B_i) \leq (A,B)$ for all $i \in I$, then $(\bigcup_{i \in I} A_i, \bigcap_{i \in I} B_i)$ is the supremum of $\{(A_i,B_i): i \in I\}$ in $\bar{\Omega}$. Also if Δ is a convex subset of Ω, then $\bar{\Delta} \subseteq \bar{\Omega}$.

Let $g \in A(\Omega)$. There is a unique continuous extension $\bar{g}$ of g to a permutation of $\bar{\Omega}$, namely $\bar{\alpha}\bar{g} = \sup\{\alpha g: \alpha \in \Omega \;\&\; \alpha \leq \bar{\alpha}\}$. Now $\bar{g}$ preserves the order on $\bar{\Omega}$ and so $\bar{g} \in A(\bar{\Omega})$. We will identify g with its unique continuous extension $\bar{g}$ and just write $g \in A(\bar{\Omega})$ if $g \in A(\Omega)$.

EXAMPLE 1.8.1. $(A(\mathbb{Q}),\mathbb{Q})$ is transitive but $(A(\mathbb{Q}),\mathbb{R}) = (A(\mathbb{Q}),\bar{\mathbb{Q}})$ is not.

EXAMPLE 1.8.2. Let $\Omega = \mathbb{Z} \overleftarrow{\times} \mathbb{R}$. Then Ω is homogeneous. (If $(n_i,r_i) \in \Omega$ $(i = 1,2)$, let $(m,s)g = (m + n_2 - n_1, s + r_2 - r_1)$ $(m \in \mathbb{Z},\ s \in \mathbb{R})$ (i.e., "local" translation by $n_2 - n_1$, and

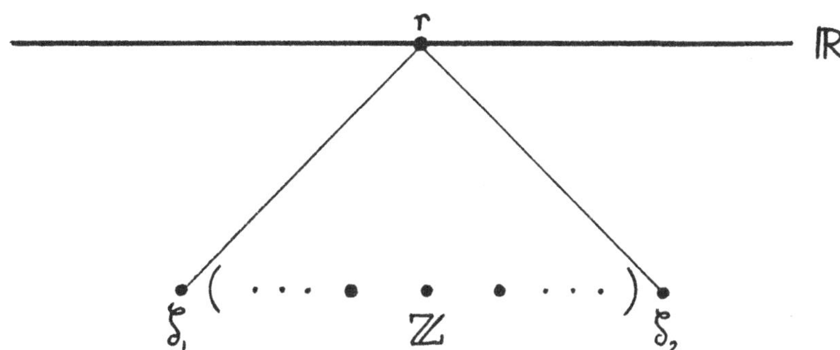

"global" translation by $r_2 - r_1$). So $g \in A(\Omega)$ and $(n_1,r_1)g = (n_2,r_2)$.) However, we now show $\bar{\Omega}$ is not homogeneous. Let ζ_1,ζ_2 be two new symbols and let $T = \mathbb{Z} \cup \{\zeta_1,\zeta_2\}$ ordered by $\zeta_1 < n < \zeta_2$ for all $n \in \mathbb{Z}$. Then $\bar{\Omega} = T \overleftarrow{\times} \mathbb{R}$. Now $(0,0)$ has an immediate predecessor but $(\zeta_2,0)$ does not. Hence there is no $g \in A(\bar{\Omega})$ such that $(0,0)g = (\zeta_2,0)$; i.e., $\bar{\Omega}$ is not homogeneous.

Let $\Lambda_1 \subseteq \Lambda \subseteq \Omega$. Λ_1 is *cofinal* (*coinitial*) in Λ if for each $\lambda \in \Lambda$, there is $\lambda_1 \in \Lambda_1$ such that $\lambda_1 \geq \lambda$ ($\lambda_1 \leq \lambda$). If Λ_1 is cofinal and coinitial in Λ, it is said to be *coterminal* in Λ. The *cofinality* (*coinitiality*) of Λ is $\inf\{|\Lambda_1|: \Lambda_1 \text{ is cofinal in } \Lambda\}$ ($\inf\{|\Lambda_1|: \Lambda_1 \text{ is coinitial in } \Lambda\}$). We will write *cf(Λ)* for the cofinality of Λ, and *ci(Λ)* for the coinitiality of Λ. If $cf(\Lambda) = ci(\Lambda)$ we will say that Λ has *coterminality* $cf(\Lambda)$. Let $\bar{\alpha} \in \bar{\Omega}$. If $\Lambda = \{\beta \in \Omega: \beta < \bar{\alpha}\}$ and $\sup(\Lambda) = \bar{\alpha}$, we call $cf(\Lambda)$ the *initial character* of $\bar{\alpha}$, and dually for the *final character* of $\bar{\alpha}$. If $\bar{\alpha}$ has initial character $\aleph_\mu$ and final character $\aleph_\nu$, we write $c_{\mu\nu}$ for the *character* of $\bar{\alpha}$. If $\bar{\alpha}$ has initial character $\aleph_\mu$ but $\inf\{\beta \in \Omega: \beta > \bar{\alpha}\} \neq \bar{\alpha}$, we write $c_{\mu *}$ for the character of $\bar{\alpha}$. Similarly we may have $c_{*\nu}$ or c_{**} as characters of an element of $\bar{\Omega}$. Note that a hole in Ω must have character $c_{\mu\nu}$ for some ordinals μ,ν since $\inf\{\beta \in \Omega: \beta > \bar{\alpha}\} \neq \bar{\alpha}$ only if $\bar{\alpha}$ has an immediate successor, etc. Observe that $\bar{\alpha}g$ has the same character as $\bar{\alpha}$ for all $g \in A(\bar{\Omega})$. Hence if $\bar{\Omega}$ is homogeneous, all its elements have the same character (cf. Example 1.8.2). For example, $\mathbb{R}$, all of whose points have character c_{00} in Ω if Ω is any chain with $\bar{\Omega} = \mathbb{R}$. Also note that if $\bar{\Omega} = \bar{T}$, the character of all points of $\bar{\Omega}$ (with respect to Ω) is the same as the character of all points of $\bar{\Omega} = \bar{T}$ (with respect to T). This is because Ω and T are dense in $\bar{\Omega}$ and $\bar{T} = \bar{\Omega}$ respectively.

1.9. BUMPS AND SUPPORTS.

Let (G,Ω) be an ordered permutation group and $g \in G$. The *support* of g is $\{\alpha \in \Omega: \alpha g \neq \alpha\} =$ *supp(g)*. If $f,g \in G$ and $\mathrm{supp}(f) \cap \mathrm{supp}(g) = \emptyset$, then we say that f and g are *disjoint*.

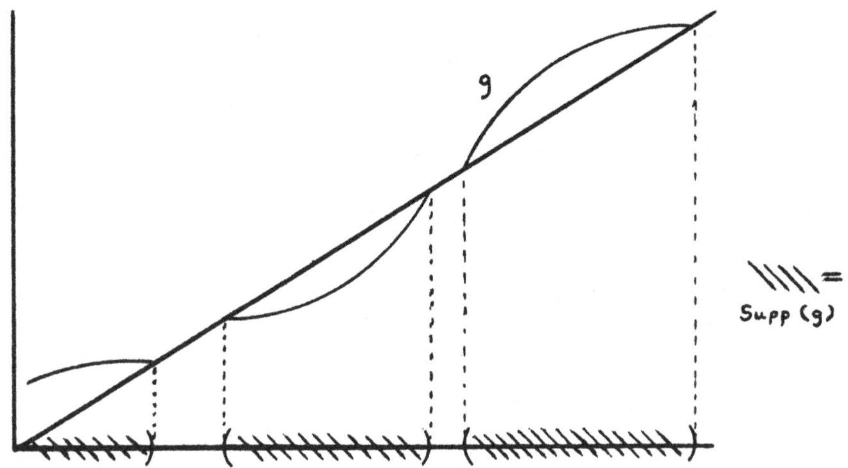

So if $e \leq f,g \in G$, f and g are disjoint precisely when $f \wedge g = e$.

The following trivial lemma is very important.

LEMMA 1.9.1. *Disjoint elements of an ordered permutation group commute.*

Proof: Let (G,Ω) be an ordered permutation group and $f,g \in G$ be disjoint. Let $\alpha \in \Omega$. Note that if $\alpha \neq \alpha f$, then $\alpha f \neq \alpha f^2$; hence $\alpha fg = \alpha f$. Thus

$$\alpha(fg) = \begin{Bmatrix} \alpha f & \text{if } \alpha \in \text{supp}(f) \\ \\ \alpha g & \text{if } \alpha \notin \text{supp}(f) \end{Bmatrix} = \begin{Bmatrix} \alpha f & \text{if } \alpha \notin \text{supp}(g) \\ \\ \alpha g & \text{if } \alpha \in \text{supp}(g) \end{Bmatrix} = \alpha(gf).$$

Therefore $fg = gf$. //

If $\bar{\Delta} \subseteq \bar{\Omega}$, let $\underline{Conv_{\Omega}(\bar{\Delta})} = \{\alpha \in \Omega: (\exists \bar{\delta}_1 \in \bar{\Delta})(\exists \bar{\delta}_2 \in \bar{\Delta})(\bar{\delta}_1 \leq \alpha \leq \bar{\delta}_2)\}$, the *convexification* of $\bar{\Delta}$ in Ω. If for some $\alpha \in \Omega$, $\text{supp}(g) = \text{Conv}_{\Omega}\{\alpha g^n : n \in \mathbb{Z}\}$, we say that g is a *bump* or g has *one bump*. This definition is independent of the choice of $\alpha \in \text{supp}(g)$. For if $\beta \in \text{supp}(g)$ and $\alpha < \alpha g$, $\alpha g^m \leq \beta < \alpha g^{m+1}$ for some $m \in \mathbb{Z}$. Hence $\beta g^{-(m+1)} < \alpha \leq \beta g^{-m}$ and $\beta < \alpha g^{m+1} \leq \beta g$. So $\text{Conv}_{\Omega}\{\beta g^n : n \in \mathbb{Z}\} = \text{supp}(g)$ and $\beta < \beta g$. A similar argument works if $\alpha g < \alpha$ but in this case $\beta g < \beta$. So, if g has one bump, it moves all points in its support up, or it moves them all down. In the former case, we say that g has *positive parity*; in the latter, *negative parity*. Note that if g is a bump, $\text{supp}(g)$ has countable coterminality and is an interval of Ω. Caution: For $g \in G$, $\text{supp}(g)$ may be convex without g being a bump. For example, take $(A(\mathbb{Q}),\mathbb{Q})$ and $g \in A(\mathbb{Q})$ with $\bar{\alpha} g = \bar{\alpha}$ only if $\bar{\alpha} = \sqrt{2}$. $\text{supp}(g) = \mathbb{Q} = \Omega$ which is convex. However $\{0g^n : n \in \mathbb{Z}\} < 2$ so g is not a bump. What is true, of course, is that g has one bump if and only if $\overline{\text{supp}}(g)$ is convex in $\bar{\Omega}$, where $\overline{\textit{supp}}\textit{(g)} = \{\bar{\alpha} \in \bar{\Omega}: \bar{\alpha} g \neq \bar{\alpha}\}$ (regarding $g \in A(\Omega)$ as an element of $A(\bar{\Omega})$ in the natural way).

Let (G,Ω) be an ordered permutation group and $e \neq g \in G$. We say that g is *bounded* if there exist $\alpha,\beta \in \Omega$ such that $\alpha < \text{supp}(g) < \beta$; *bounded above* (*below*) if there exists $\alpha \in \Omega$ such that $\text{supp}(g) < \alpha$ ($\text{supp}(g) > \alpha$); *unbounded* if $\text{supp}(g)$ is coterminal in Ω. A bump can have any of these four boundities (if Ω is countably coterminal):

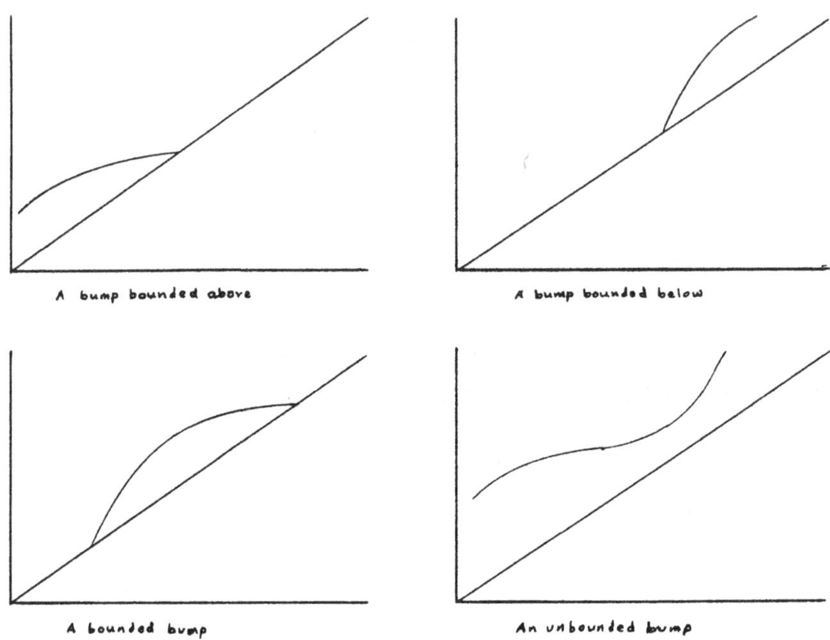

If (G,Ω) is an ordered permutation group, $f \in A(\Omega)$ and $g \in G$, then f is a *bump of g* if f is a bump and $f|\mathrm{supp}(f) = g|\mathrm{supp}(f)$. If f is a bump of g, then its support is an interval and is called a *supporting interval* of g. Indeed, $\mathrm{supp}(g)$ is uniquely a union of its supporting intervals, and so is open in the interval topology on Ω.

We next give an example to show that the bumps of $g \in G$ need not necessarily belong to G even when (G,Ω) is a transitive ℓ-permutation group.

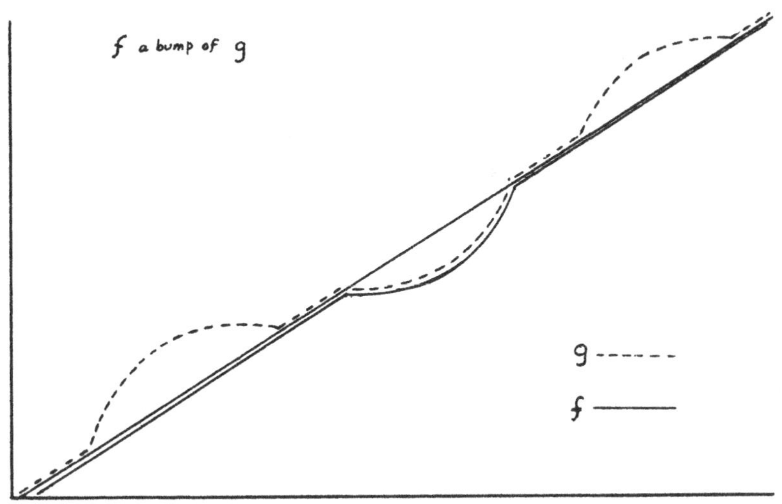

EXAMPLE 1.9.2. Let $G = \{g \in A(\mathbb{R}): (\alpha + 1)g = \alpha g + 1$ for all $\alpha \in \mathbb{R}\}$. So $(G,\mathbb{R})$ is a transitive ℓ-permutation group. Let $g_0 \in G$ with $\text{supp}(g_0) = \mathbb{R}\setminus\mathbb{Z}$. Then no bump of g belongs to G. An example of such a g_0 is: $\alpha g_0 = (\alpha - n)^{\frac{1}{2}} + n$ if $\alpha \in [n,n + 1]$ $(n \in \mathbb{Z})$.

Let (G,Ω) be an ordered permutation group. In view of Lemma 1.9.1, each $g \in G$ is the "product" (in $A(\Omega)$) of its bumps. Moreover, if g has only finitely many bumps, this is accurate. If $g > e$, then $g = \bigvee\{f: f$ is a bump of $g\}$ (This $\bigvee$ (supremum) is possibly infinite and is taken in $A(\Omega)$).

LEMMA 1.9.3. *Let (G,Ω) be an ordered permutation group and $e < f,g \in G$. If $f \wedge gf^{-1} = e$, then each bump of f is a bump of g.*

Proof: Since $\alpha = \alpha(f \wedge gf^{-1}) = \min\{\alpha f, \alpha gf^{-1}\}$, $\alpha = \alpha gf^{-1}$ for all $\alpha \in \text{supp}(f)$. Hence $\alpha f = \alpha g$ for all $\alpha \in \text{supp}(f)$ and the lemma is proved.

LEMMA 1.9.4. *Let (G,Ω) be an ordered permutation group and $e < f,g \in G$. If $f \wedge g^{-1}fg = e$, then $f^n \leq g$ for all $n \in \mathbb{Z}^+$.*

Hence if $supp(f) \subseteq (\alpha,\beta)$ and $\alpha g \geq \beta$, then $f \leq g$.

<u>*Proof:*</u> By the fundamental triviality, $supp(g^{-1}fg) = supp(f)g$. Hence if h is a bump of f, $supp(h) < supp(h)g$. Thus if $\alpha \in supp(h)$, $\alpha f^n \in supp(h) < \alpha g$ for all $n \in \mathbb{Z}^+$. Since this holds for all bumps h of f, $\alpha f^n < \alpha g$ for all $\alpha \in \Omega$ and $n \in \mathbb{Z}^+$.

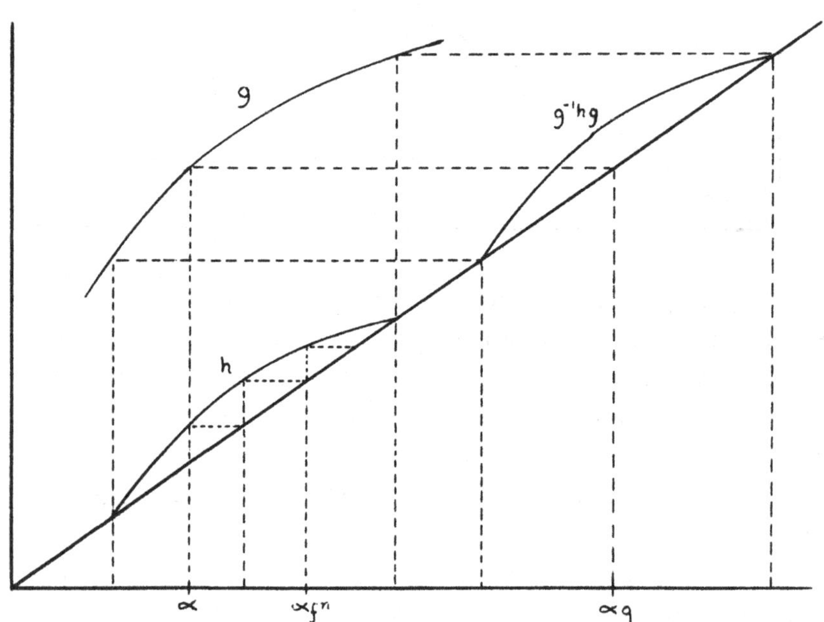

<u>*1.10.*</u> <u>*MULTIPLE TRANSITIVITY.*</u>

Let $m \in \mathbb{Z}^+$. We say that an ordered permutation group (G,Ω) is <u>*m-transitive*</u> if whenever $\alpha_1 < \ldots < \alpha_m$ and $\beta_1 < \ldots < \beta_m$ in Ω, there is $g \in G$ such that $\alpha_i g = \beta_i$ $(1 \leq i \leq m)$. For example, we will see that $(A(\mathbb{R}),\mathbb{R})$ is m-transitive for all $m \in \mathbb{Z}^+$. Indeed, if F is a totally ordered field, $(A(F),F)$ is doubly transitive and hence m-transitive for all $m \in \mathbb{Z}^+$ as we will show.

Throughout the book we will assume that

IF (G,Ω) IS m-TRANSITIVE, THEN Ω IS INFINITE.

This avoids the trivial case that $G = \{e\}$ and $|\Omega| \leq m$ (cf. transitive = 1-transitive).

If $(A(\Omega),\Omega)$ is m-transitive, we will simply say that Ω is *m-homogeneous*. This conforms to the usual model theory usage.

Note that if (G,Ω) is m-transitive, it is n-transitive for all $n \leq m$.

LEMMA 1.10.1. *If (G,Ω) is a doubly transitive ℓ-permutation group, then it is m-transitive for all $m \geq 2$.*

Proof: We will only prove double transitivity implies triple transitivity (the same proof works to spread the m-transitivity disease by induction). Let $\alpha_1 < \alpha_2 < \alpha_3$ and $\beta_1 < \beta_2 < \beta_3$. By double transitivity, there is $g \in G$ such that $\alpha_1 g = \beta_1$ and

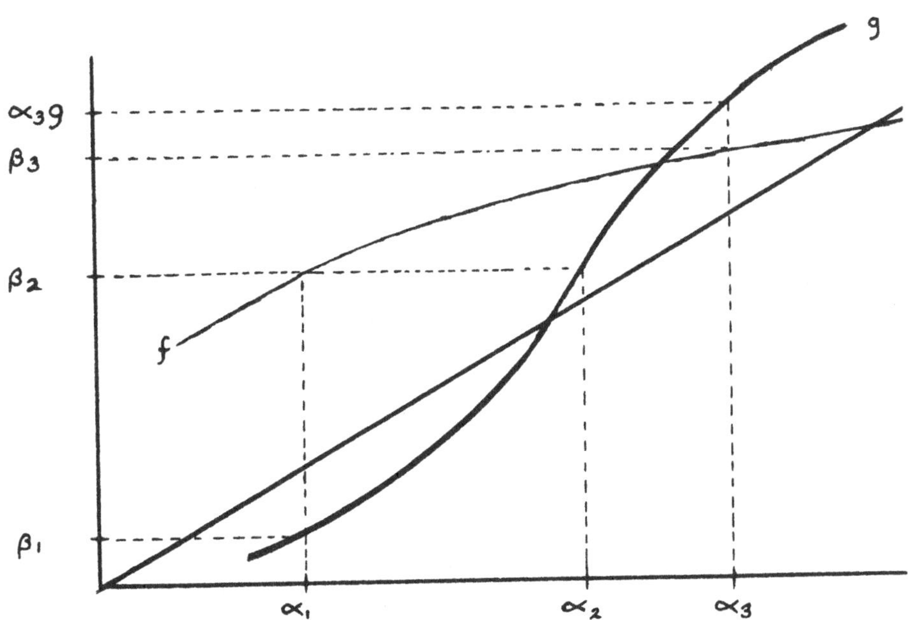

$\alpha_2 g = \beta_2$. If $\beta_3 \leq \alpha_3 g$, let $f \in G$ be such that $\alpha_1 f = \beta_2$ and $\alpha_3 f = \beta_3$. Now $\alpha_i(g \wedge f) = \min\{\alpha_i g, \alpha_i f\} = \beta_i$ $(i = 1,2,3)$. If $\beta_3 > \alpha_3 g$, let $h \in G$ be such that $\alpha_2 h = \beta_1$ and $\alpha_3 h = \beta_3$. $\alpha_i(g \vee h) = \beta_i$ $(i = 1,2,3)$.

As a useful simplification in checking double transitivity for ordered permutation groups, we have:

LEMMA 1.10.2. *Let (G,Ω) be a transitive ordered permutation group. Then (G,Ω) is doubly transitive if whenever $\alpha_1 < \alpha_2 < \alpha_3$, there is $g \in G_{\alpha_1}$ such that $\alpha_2 g = \alpha_3$.*

Proof: Let $\alpha_1 < \alpha_2$ and $\beta_1 < \beta_2$. We must find $g \in G$ such that $\alpha_i g = \beta_i$ $(i = 1,2)$. By transitivity, there is $f \in G$ such that $\alpha_1 f = \beta_1$. If $\alpha_2 f = \beta_2$, let $g = f$. If $\alpha_2 f < \beta_2$, then $\beta_1 = \alpha_1 f < \alpha_2 f < \beta_2$; so there is $h \in G_{\beta_1}$ such that $\alpha_2 fh = \beta_2$. Let $g = fh$. Then $\alpha_i g = \beta_i$ $(i = 1,2)$. Similarly if $\beta_2 < \alpha_2 f$. //

Note that if (G,Ω) is a doubly transitive ordered permutation group, the condition holds. (If $\alpha_1 < \alpha_2 < \alpha_3$, then as $\alpha_1 < \alpha_2$ and $\alpha_1 < \alpha_3$, there is $g \in G$ such that $\alpha_1 g = \alpha_1$ and $\alpha_2 g = \alpha_3$.)

Let $B(\Omega) = \{f \in A(\Omega)$: f is bounded$\}$. If (G,Ω) is a doubly transitive ℓ-permutation group and $G \cap B(\Omega) = \{e\}$, we say that (G,Ω) is a *pathological* ℓ-permutation group. In later chapters, we will see that the name is well chosen.

EXAMPLE 1.10.3. Let $G = \{g \in A(\mathbb{R})$: $(\exists m \in \mathbb{Z}^+)(\forall \alpha \in \mathbb{R})[(\alpha + m)g = \alpha g + m]\}$. (If $g \in G$, the least $m \in \mathbb{Z}^+$ for which $(\alpha + m)g = \alpha g + m$ for all $\alpha \in \mathbb{R}$ is called the *period* of g.) $(G,\mathbb{R})$ is a pathological ℓ-permutation group: If $\alpha \in \mathrm{supp}(g)$, then $(\alpha + km)g = \alpha g + km \neq \alpha + km$ for all $k \in \mathbb{Z}$ where m is the period of g; hence $G \cap B(\Omega) = \{e\}$. Also if $\alpha_1 < \alpha_2 < \alpha_3$, let $m > \alpha_3 - \alpha_1$. There is $e < \hat{g} \in A([\alpha_1, \alpha_1 + m])$ such that $\alpha_2 \hat{g} = \alpha_3$. Clearly $\hat{g}$ uniquely extends to an element of G of period m, say g. Now

$e < g \in G_{\alpha_1}$ and $\alpha_2 g = \alpha_3$. By Lemma 1.10.2, $(G,\mathbb{R})$ is doubly transitive.

We now reexamine Lemma 1.10.1. Unfortunately, there are doubly transitive ordered permutation groups that are not triply transitive so the next lemma is optimal.

LEMMA 1.10.4. *If (G,Ω) is a triply transitive ordered permutation group and $G \cap B(\Omega) \neq \{e\}$, then (G,Ω) is m-transitive for all $m \in \mathbb{Z}^+$.*

Proof: Assume that (G,Ω) is m-transitive for some $m \geq 3$. Let $\alpha_1 < \ldots < \alpha_{m+1}$ and $\beta_1 < \ldots < \beta_{m+1}$. By hypothesis, there is $g \in G$ such that $\alpha_i g = \beta_i$ $(1 \leq i \leq m)$. If $\alpha_{m+1}g = \beta_{m+1}$, we are home, so assume not. Let $e \neq g_0 \in G \cap B(\Omega)$, say $\sigma < \mathrm{supp}(g_0)$ and $\tau \in \mathrm{supp}(g_0)$. Without loss of generality, $\tau < \tau g_0$. If $\alpha_{m+1}g < \beta_{m+1}$, let $f \in G$ be such that $\sigma f = \beta_m$, $\tau f = \alpha_{m+1}g$ and $\tau g_0 f = \beta_{m+1}$. Then $\beta_m < \mathrm{supp}(f^{-1}g_0 f)$ and $\alpha_{m+1}gf^{-1}g_0 f = \beta_{m+1}$. Hence $\alpha_i h = \beta_i$ $(1 \leq i \leq m+1)$ where $h = gf^{-1}g_0 f$. Similarly if $\beta_{m+1} < \alpha_{m+1}g$. //

Note that the above proof shows that the conclusion holds provided (G,Ω) is triply transitive and contains an element whose support is bounded above or below.

LEMMA 1.10.5. *Let (G,Ω) be a triply transitive ordered permutation group and $N \lhd G$ with $N \cap B(\Omega) \neq \{e\}$. Then $(N \cap B(\Omega),\Omega)$ is m-transitive for all $m \in \mathbb{Z}^+$.*

Proof: By induction on m. For $m = 1$, let $\alpha < \beta$ in Ω. Let $e \neq g_0 \in N \cap B(\Omega)$. Let $\gamma \in \mathrm{supp}(g_0)$. Without loss of generality, $\gamma < \gamma g_0$. There is $f \in G$ such that $\gamma f = \alpha$ and $\gamma g_0 f = \beta$ since (G,Ω) is triply transitive. Now $f^{-1}g_0 f \in N \cap B(\Omega)$ and $\alpha f^{-1}g_0 f = \beta$. Hence $(N \cap B(\Omega),\Omega)$ is 1-transitive.

Assume that $(N \cap B(\Omega),\Omega)$ is m-transitive. The proof of the preceding lemma shows that $(N \cap B(\Omega),\Omega)$ is (m+1)-transitive (take $g, g_0 \in N \cap B(\Omega)$). //

An ordered permutation group (G,Ω) is said to have the <u>*interval support property*</u> if for each interval $\Delta \subseteq \Omega$, there is $e \neq g \in G$ such that $\mathrm{supp}(g) \subseteq \Delta$. For example, $(A(\mathbb{R}),\mathbb{R})$ satisfies the interval support property but Examples 1.9.2 and 1.10.3 do not.

<u>COROLLARY</u> <u>1.10.6</u>. *If (G,Ω) is a triply transitive ordered permutation group with $G \cap B(\Omega) \neq \{e\}$, then $(G \cap B(\Omega),\Omega)$ satisfies the interval support property. Indeed, if $\bar{\alpha}_1 < \beta_1 < \beta_2 < \bar{\alpha}_3$, there is $g \in G$ such that $\mathrm{supp}(g) \subseteq (\bar{\alpha}_1,\bar{\alpha}_3)$ and $\beta_1 g = \beta_2$. Hence if $\bar{\alpha}_1 < \bar{\alpha}_2 < \bar{\alpha}_3$, there is $g \in G$ such that $\bar{\alpha}_2 \in \mathrm{supp}(g) \subseteq (\bar{\alpha}_1,\bar{\alpha}_3)$. The above conclusions hold if (G,Ω) is a non-pathological doubly transitive ℓ-permutation group.*

<u>*Proof:*</u> Let $\beta_0,\beta_3 \in \Omega$ with $\bar{\alpha}_1 \leq \beta_0 < \beta_1 < \beta_2 < \beta_3 \leq \bar{\alpha}_3$. By Lemma 1.10.5, there is $g_0 \in G \cap B(\Omega)$ with $\beta_1 g_0 = \beta_2$. Say $\mathrm{supp}(g_0) \subseteq (\gamma,\delta)$. By Lemma 1.10.4, there is $f \in G$ such that $\gamma f = \beta_0$, $\beta_1 f = \beta_1$, $\beta_2 f = \beta_2$ and $\delta f = \beta_3$. Then $\mathrm{supp}(f^{-1}g_0 f) \subseteq (\beta_0,\beta_3) \subseteq (\bar{\alpha}_1,\bar{\alpha}_3)$ and $\beta_1 f^{-1} g_0 f = \beta_2$.

If $\bar{\alpha}_1 < \bar{\alpha}_2 < \bar{\alpha}_3$, let $\beta_1,\beta_2 \in \Omega$ with $\bar{\alpha}_1 < \beta_1 \leq \bar{\alpha}_2 < \beta_2 < \bar{\alpha}_3$ and apply the above. //

We now prove a result (due to McCleary [78b]) which we will use heavily in the next chapter. It shows that in the triply transitive case the order is (almost) definable from the group action. Write <u>$[x,y]$</u> for $x^{-1}y^{-1}xy$, the commutator of x and y.

<u>THEOREM</u> <u>1.10.7</u>. *Let (G,Ω) be a triply transitive ordered permutation group containing a strictly positive element of bounded support. Let $e \neq f,g,p \in G$. Then $[f,(h^{-1}ph)^{-1}g(h^{-1}ph)] = e$ for all $h \in G$ if and only if (1) $e < p$ & $\mathrm{supp}(f) < \mathrm{supp}(g)$ or (2) $p < e$ & $\mathrm{supp}(g) < \mathrm{supp}(f)$.*

<u>*Proof:*</u> Assume (1). Then $\mathrm{supp}((h^{-1}ph)^{-1}g(h^{-1}ph)) = [\mathrm{supp}(g)](h^{-1}ph)$. Since $e < p$,

$e < h^{-1}ph$ for all $h \in G$. Therefore, $\text{supp}(f) < \text{supp}(g)$ implies $\text{supp}(f) < \text{supp}((h^{-1}ph)^{-1}g(h^{-1}ph))$. So f and $(h^{-1}ph)^{-1}g(h^{-1}ph)$ are disjoint. By Lemma 1.9.1, $[f,(h^{-1}ph)^{-1}g(h^{-1}ph)] = e$. Similarly, (2) implies $[f,(h^{-1}ph)^{-1}g(h^{-1}ph)] = e$ for all $h \in G$.

Conversely, assume $[f,(h^{-1}ph)^{-1}g(h^{-1}ph)] = e$ for all $h \in G$. Suppose $p \not< e$. If there are $\alpha \in \text{supp}(f)$ and $\beta \in \text{supp}(g)$ with $\beta < \alpha$, then $\beta g^{-2} < \beta g^{-1} < \beta < \alpha < \alpha f < \alpha f^2$ without loss of generality (replacing f by f^{-1} or g by g^{-1} if necessary). Let $\sigma \in \Omega$ with $\sigma < \sigma p$. Choose $\tau_1, \tau_2 \in \Omega$ with $\sigma < \tau_1 < \tau_2 < \sigma p$. By Lemma 1.10.4, there is $h \in G$ such that $\sigma h = \beta g^{-2}$, $\tau_1 h = \beta g^{-1}$, $\tau_2 h = \beta$, $\sigma ph = \alpha$, $\tau_1 ph = \alpha f$ and $\tau_2 ph \neq \alpha f^2$. Now $\alpha f(h^{-1}ph)^{-1}g(h^{-1}ph) = \tau_2 ph \neq \alpha f^2 = \alpha(h^{-1}ph)^{-1}g(h^{-1}ph)f$. Thus $[f,(h^{-1}ph)^{-1}g(h^{-1}ph)] \neq e$, a contradiction. Hence $p \not< e$ implies $\text{supp}(f) < \text{supp}(g)$. Similarly

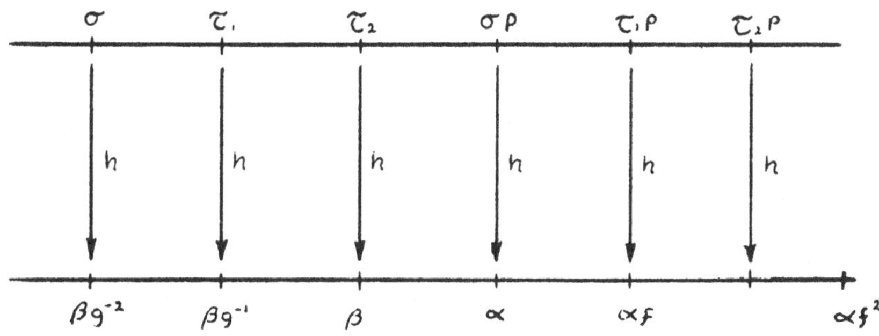

$p \not> e$ implies $\text{supp}(f) > \text{supp}(g)$. So $(p \not< e \ \& \ p \not> e)$ is impossible. Therefore $p > e$ or $p < e$ and the theorem follows.

<u>COROLLARY</u> <u>1.10.8</u>. *Let* (G,Ω) *be a triply transitive ordered permutation group containing a strictly positive element of bounded support, and let* $p \in G$. *Then* $(p > e$ *or* $p < e)$ *precisely when there are* $f,g \in G$ *such that* $[f,(h^{-1}ph)^{-1}g(h^{-1}ph)] = e$ *for all* $h \in G$. *Furthermore,* $p,q > e$ *or* $p,q < e$ *if and only if* $(\forall h)([f,(h^{-1}ph)^{-1}g(h^{-1}ph)] = e \leftrightarrow [f,(h^{-1}qh)^{-1}g(h^{-1}qh)] = e)$, *provided* $p > e$ *or* $p < e$.

Finally, we note an obvious result about homogeneous chains which we will use repeatedly.

LEMMA 1.10.9 *(The Patching Lemma). Let* Ω *be a chain and* $\{A_i: i \in I\}$, $\{B_i: i \in I\}$ *be covers of* Ω. *Suppose that for each* $i \in I$, *there is an ordermorphism* $f_i: A_i \cong B_i$ *such that* $\alpha f_i = \alpha f_j$ *for all* $\alpha \in A_i \cap A_j$ *and* $\alpha_i f_i < \alpha_j f_j$ *whenever* $\alpha_i \in A_i$, $\alpha_j \in A_j$ *and* $\alpha_i < \alpha_j$. *Then* $f = \bigcup_{i \in I} f_i \in A(\Omega)$. *In particular, let* A *and* B *be intervals of a chain* Ω, *and* $A = \bigcup\{A_i: i \in I\}$ *and* $B = \bigcup\{B_i: i \in I\}$ *with* $sup(A)$, $sup(B)$ *belonging to the same orbit of* $A(\Omega)$, *or* A *and* B *both final intervals of* Ω; *and* $inf(A)$, $inf(B)$ *belonging to the same orbit of* $A(\Omega)$, *or* A *and* B *both initial intervals of* Ω. *If* $f_i: A_i \cong B_i$ $(i \in I)$ *are ordermorphisms with* $\alpha f_i = \alpha f_j$ *whenever* $\alpha \in A_i \cap A_j$ *and* $\alpha_i f_i < \alpha_j f_j$ *whenever* $\alpha_i \in A_i$, $\alpha_j \in A_j$ *and* $\alpha_i < \alpha_j$, *then there exists* $f \in A(\Omega)$ *such that* $f|A_i = f_i$ $(i \in I)$.

We give one application of the Patching Lemma now (others appear in later chapters).

LEMMA 1.10.10. *Let* Ω *be a doubly homogeneous chain. Then any two holes in* $\bar{\Omega}$ *of character* c_{00} *belong to the same* $A(\Omega)$ *orbit.*

Proof: Let $\bar{\alpha}, \bar{\beta} \in \bar{\Omega}\setminus\Omega$ have character c_{00}. Let $\{\alpha_n: n \in \omega\}$ and $\{\beta_n: n \in \omega\}$ be increasing sequences in Ω, and $\{\gamma_n: n \in \omega\}$ and $\{\delta_n: n \in \omega\}$ be decreasing sequences in Ω such that $\sup\{\alpha_n: n \in \omega\} = \bar{\alpha} = \inf\{\gamma_n: n \in \omega\}$ and $\sup\{\beta_n: n \in \omega\} = \bar{\beta} = \inf\{\delta_n: n \in \omega\}$. Since any two bounded intervals of Ω (with endpoints in Ω) are ordermorphic (Ω is doubly homogeneous), there exist ordermorphisms $f_n: [\alpha_n, \alpha_{n+1}] \cong [\beta_n, \beta_{n+1}]$ and $g_n: [\gamma_{n+1}, \gamma_n] \cong [\delta_{n+1}, \delta_n]$ $(n \in \omega)$. Also any two final (initial) intervals of Ω with endpoint in Ω are ordermorphic since Ω is doubly homogeneous. Hence there are ordermorphisms $h_0: (-\infty, \alpha_0) \cong (-\infty, \beta_0)$ and $h_1: (\gamma_0, \infty) \cong (\delta_0, \infty)$.

Let $f = \bigcup_{n\in\omega} f_n \cup h_0 \cup h_1 \cup \bigcup_{n\in\omega} g_n$. Then $f \in A(\Omega)$ by the Patching Lemma. Moreover, $\bar{\alpha}f = \sup\{\alpha_n f: n \in \omega\} = \sup\{\beta_n: n \in \omega\} = \bar{\beta}$, as desired.

1.11. *IDENTITIES.*

We now use Holland's Theorem to obtain some identities about ℓ-groups.

LEMMA 1.11.1. *Every ℓ-group is torsion-free.*

Proof: Let (G,Ω) be an ℓ-permutation group and $g \in G$. If $g^n = e$ for some $n \in \mathbb{Z}^+$ and $g \neq e$, there is $\alpha \in \Omega$ such that $\alpha g \neq \alpha$. Without loss of generality, $\alpha < \alpha g$. Hence $\alpha g < \alpha g^2$ so $\alpha < \alpha g < \alpha g^2 < \ldots < \alpha g^n = \alpha e = \alpha$, the desired contradiction. Thus $g^n = e$ implies $g = e$; i.e., G is torsion-free.

LEMMA 1.11.2. *Every ℓ-group is a distributive lattice.*

Proof: Let (G,Ω) be an ℓ-permutation group and $f,g,h \in G$. Let $\alpha \in \Omega$. $\alpha[(f \vee g) \wedge h] = \min\{\max\{\alpha f,\alpha g\},\alpha h\} = \max\{\min\{\alpha f,\alpha h\},\min\{\alpha g,\alpha h\}\}$ since Ω is a chain. Hence $\alpha[(f \vee g) \wedge h] = \alpha[(f \wedge h) \vee (g \wedge h)]$. Since this holds for all $\alpha \in \Omega$, $(f \vee g) \wedge h = (f \wedge h) \vee (g \wedge h)$; i.e., G is a distributive lattice.

Since $f(g \vee h) = fg \vee fh$, $(g \vee h)f = gf \vee hf$ and dually, Lemma 1.11.2 gives:

COROLLARY 1.11.3. *Every ℓ-group word has a normal form $\bigvee_{i\in I} \bigwedge_{j\in J} w_{ij}$ where w_{ij} are group words, and I,J are finite.*

Warning: This normal form is not necessarily unique as we now show:

LEMMA 1.11.4. *If G is an ℓ-group and $g \in G$, then $g \vee g^{-1} \geq e$. Hence $(g \vee e) \wedge (g^{-1} \vee e) = e$.*

Proof: Let (G,Ω) be an ℓ-permutation group and $\alpha \in \Omega$. Either $\alpha g \leq \alpha$ or $\alpha \leq \alpha g$. In the first case $\alpha \leq \alpha g^{-1}$; so, in either case, $\alpha \leq \alpha(g \vee g^{-1})$. Thus $e \leq g \vee g^{-1}$. Therefore $(g^{-1} \wedge g) = (g \vee g^{-1})^{-1} \leq e$. Hence $(g \vee e) \wedge (g^{-1} \vee e) = (g \wedge g^{-1}) \vee e = e$ by Lemma 1.11.2.

We will write $\underline{|g|}$ for $g \vee g^{-1}$. So $|g| \geq e$.

We close with a less trivial identity. Recall that $[x,y] = x^{-1}y^{-1}xy$.

<u>LEMMA</u> <u>1.11.5</u>. *In any ℓ-group G, if $f_i, g_i \in G$ $(i = 1,2)$ and $f_1 \wedge f_2 = e = g_1 \wedge g_2$ then $[(f_1f_2^{-1})^{-1}g_1(f_1f_2^{-1}), f_1^{-1}g_2f_1[g_2,f_2^{-1}]] = e$.*

<u>Proof</u>: By Lemma 1.9.1, it is enough to show that $(f_1f_2^{-1})^{-1}g_1(f_1f_2^{-1})$ is disjoint from $f_1^{-1}g_2f_1[g_2,f_2^{-1}]$. Since f_1 and f_2 commute (by Lemma 1.9.1), conjugating both sides by $(f_1f_2^{-1})^{-1}$ reduces the proof to showing g_1 is disjoint from $f_2^{-1}g_2f_1g_2^{-1}f_2g_2f_1^{-1}$. So suppose $\alpha < \alpha g_1$ (whence $\alpha g_2 = \alpha$). We must show $\alpha f_2^{-1}g_2f_1g_2^{-1}f_2g_2f_1^{-1} = \alpha$.

<u>Case</u> <u>1</u>. $\alpha f_1 = \alpha = \alpha f_2$. Then $\alpha f_2^{-1}g_2f_1g_2^{-1}f_2g_2f_1^{-1} = \alpha$ since $\alpha g_2 = \alpha$.

<u>Case</u> <u>2</u>. $\alpha f_1 \neq \alpha$ (so $\alpha f_2 = \alpha$). Now $\alpha < \alpha f_1$ so

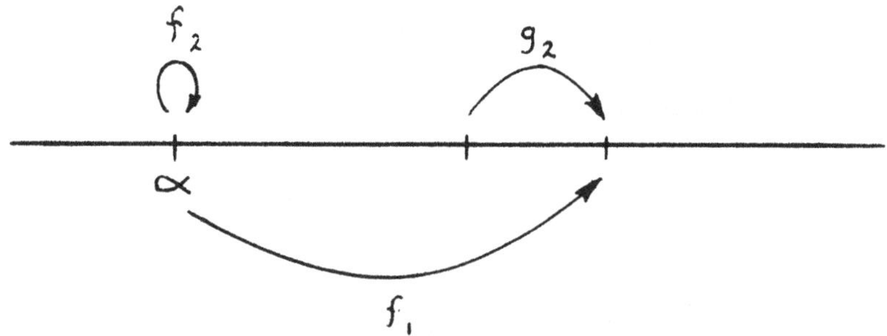

$\alpha = \alpha g_2^{-1} < \alpha f_1 g_2^{-1} \leq \alpha f_1$. Since $f_1 \wedge f_2 = e$, $\alpha f_1 g_2^{-1} f_2 = \alpha f_1 g_2^{-1}$. Hence $\alpha f_2^{-1} g_2 f_1 g_2^{-1} f_2 g_2 f_1^{-1} = \alpha f_1 g_2^{-1} f_2 g_2 f_1^{-1} = \alpha f_1 g_2^{-1} g_2 f_1^{-1} = \alpha$.

<u>Case 3</u>. $\alpha f_2 \neq \alpha$ (so $\alpha f_1 = \alpha$). Now $\alpha f_2^{-1} \leq \alpha f_2^{-1} g_2 < \alpha g_2 = \alpha$; so $\alpha f_2^{-1} g_2 \in \text{supp}(f_2)$. Hence $\alpha f_2^{-1} g_2 f_1 = \alpha f_2^{-1} g_2$. Thus $\alpha f_2^{-1} g_2 f_1 g_2^{-1} f_2 g_2 f_1^{-1} = \alpha f_2^{-1} g_2 g_2^{-1} f_2 g_2 f_1^{-1} = \alpha$.

Q.E.D.

Note that Lemma 1.11.5 gives an identity as well as an implication (Replace f_1 by $f \vee e$, f_2 by $f^{-1} \vee e$, g_1 by $g \vee e$ and g_2 by $g^{-1} \vee e$ and use Lemma 1.11.4). Lemma 1.11.5 is due to Holland and Scrimger [72]. For further "disguised" identities, see Holland [77] and Franchello [77].

CHAPTER 2
DOUBLY TRANSITIVE $A(\Omega)$

This chapter is devoted to the study of $A(\Omega)$ when it is well-behaved and has a lot of additional properties. The assumption that gives us nice theorems is that Ω is doubly homogeneous. Under this extra hypothesis, we can apply the Patching Lemma (Lemma 1.10.9). This helps us give algebraic conditions equivalent to geometric ones. Indeed, we begin Section 2.1 with a series of lemmas in which we capture various geometric properties by algebraic statements. These we will use repeatedly in this chapter (as well as later in the book). Although they may appear quite technical at first, their use in proving non-technical results emphasises the need to examine them early on. We will prove the following theorem due to Michèle Jambu-Giraudet [81a] (Recall Ω^* is Ω with the order reversed).

<u>THEOREM</u> <u>2A</u>. *Let* Ω_1 *and* Ω_2 *be doubly homogeneous chains, and* $A(\Omega_1)$ *and* $A(\Omega_2)$ *satisfy the same group-theoretic sentences. Then* $A(\Omega_1)$ *and either* $A(\Omega_2)$ *or* $A(\Omega_2^*)$ *satisfy the same* ℓ*-group sentences.*

We will also use the technique to prove two theorems of Gurevich and Holland [81]:

<u>THEOREM</u> <u>2B</u>. *If* Ω *is doubly homogeneous and* $A(\Omega)$ *satisfies the same* ℓ*-group sentences as* $A(\mathbb{R})$, *then* $\Omega \cong \mathbb{R}$. *In particular, if* $A(\Omega) \cong A(\mathbb{R})$ *as* ℓ*-groups, then* $\Omega \cong \mathbb{R}$.

<u>THEOREM</u> <u>2C</u>. *If* Ω *is doubly homogeneous and* $A(\Omega)$ *satisfies the same* ℓ*-group sentences as* $A(\mathbb{Q})$, *then* $\Omega \cong \mathbb{Q}$ *or* $\mathbb{R}\backslash\mathbb{Q}$. *In*

particular, if $A(\Omega) \cong A(\mathbb{Q})$ *as* ℓ*-groups, then* $\Omega \cong \mathbb{Q}$ *or* $\mathbb{R}$ *.*

As a consequence we obtain (by Theorem 2A):

THEOREM 2B* (2C*). *Same as Theorem 2B (2C) with "*ℓ*-group" replaced by "group."*

Another application of our ideas is:

THEOREM 2D. *If* Ω *and* T *are doubly homogeneous chains and* $\psi: A(\Omega) \cong A(T)$ *as* ℓ*-groups (groups), then* Ω *is ordermorphic (or antiordermorphic) to an orbit of* $A(T)$ *in* $\overline{T}$*. If* $\hat{\psi}$ *denotes this (anti)ordermorphism, then* $g\psi = \hat{\psi}^{-1}g\hat{\psi}$ $(g \in A(\Omega))$*.*

The ℓ-group part of Theorem 2D is due to Holland [65a], the group part to McCleary [78b] and Rabinovich [75]. In view of Theorem 2A, we can give essentially the same proof for both parts; it is quite different from any of the original proofs and is due to Jambu-Giraudet [81a].

We next wish to examine closely the lattice of $A(\Omega)$ for Ω a doubly homogeneous chain. If G is a group, we will use $G^{\dagger}$ for the set G with the group operation $f \dagger g = gf = (f^{-1}g^{-1})^{-1}$. We obtain the following analogues (due to Jambu-Giraudet [81a]) of the preceding theorems.

THEOREM 2A†. *Let* Ω_1 *and* Ω_2 *be doubly homogeneous chains, and* $A(\Omega_1)$ *and* $A(\Omega_2)$ *satisfy the same lattice-theoretic sentences with constant* e*. Then* $A(\Omega_1)$ *and either* $A(\Omega_2)$ *or* $A(\Omega_2)^{\dagger}$ *satisfy the same* ℓ*-group sentences.*

THEOREM 2B† (2C†). *Same as Theorem 2B (2C) with "*ℓ*-group" replaced by "lattice with* e*."*

THEOREM 2D†. *If* Ω *and* T *are doubly homogeneous chains and* $\psi: A(\Omega) \cong A(T)$ *as lattices with* e*, then* Ω *is ordermorphic* *(or antiordermorphic) to an orbit of* $A(T)$ *in* $\overline{T}$. *If* $\hat{\psi}$ *denotes this (anti)ordermorphism, then* $g\psi = \hat{\psi}^{-1}g\hat{\psi}$ $(\hat{\psi}^{-1}g^{-1}\hat{\psi})$ $(g \in A(\Omega))$.

We will next turn our attention to some immediate applications of the Patching Lemma (Lemma 1.10.9). We will give a geometric criterion equivalent to the algebraic notion that two elements of doubly transitive $A(\Omega)$ are conjugate. We will also prove:

THEOREM 2E. *If* $A(\Omega)$ *is doubly transitive, then it is divisible.*

and

THEOREM 2F. *If* Ω *is doubly homogeneous, every element of* $A(\Omega)$ *is a commutator.*

Recall that $B(\Omega) = \{g \in A(\Omega)$: g has bounded support$\}$. Let $L(\Omega) = \{g \in A(\Omega)$: supp(g) is bounded above$\}$ (i.e., $L(\Omega)$ is the set of all elements of $A(\Omega)$ that live on the left), and $R(\Omega) = \{g \in A(\Omega)$: supp(g) is bounded below$\}$. By the fundamental triviality, these are normal subgroups of $A(\Omega)$. We will use our conjugacy criterion to prove, in particular:

THEOREM 2G. *If* Ω *is doubly homogeneous,* $B(\Omega)$ *is a simple group and is contained in every normal subgroup of* $A(\Omega)$ $(L(\Omega)$, $R(\Omega))$ *except* $\{e\}$.

THEOREM 2H. *The only proper normal subgroups of* $A(\mathbb{R})$ *are* $B(\mathbb{R})$, $L(\mathbb{R})$ *and* $R(\mathbb{R})$.

Theorem 2H fails if we replace $\mathbb{R}$ by an arbitrary doubly homogeneous chain as we will show by example.

We next consider the ℓ-automorphisms and group automorphisms of doubly transitive $A(\Omega)$. As a special case of our work, we will obtain

THEOREM 2I. *The only ℓ-automorphisms of $A(\mathbb{R})$ are inner-- i.e., conjugation by elements of $A(\mathbb{R})$. The only automorphisms of the group $A(\mathbb{R})$ are ℓ-automorphisms possibly followed by ϕ_- where $\alpha(g\phi_-) = -((-\alpha)g)$ $(\alpha \in \mathbb{R};\ g \in A(\mathbb{R}))$.*

The proof we give here relies on our work in Section 2.1. An alternative proof will be given in Chapter 9 when Ω is not necessarily doubly homogeneous.

Finally, we prove that doubly homogeneous chains are universal. Specifically:

THEOREM 2J. *For any chain T, there exist a doubly homogeneous chain Ω and an ℓ-embedding of $A(T)$ in $A(\Omega)$ such that*

(1) the restriction maps $B(T)$ into $B(\Omega)$ and

(2) $|\Omega| = \max\{|T|, \aleph_0\}$.

As an immediate consequence of Theorems 2E, 2J and Holland's Theorem we have

COROLLARY 2K. *Every ℓ-group can be ℓ-embedded in a divisible ℓ-group.*

and

COROLLARY 2L. *Every ℓ-group G can be ℓ-embedded in $B(\Omega)$ for some doubly homogeneous chain Ω with $|\Omega| \leq \max\{|G|, \aleph_0\}$.*

(If G is an ℓ-group, it can be ℓ-embedded in $A(T)$ for some doubly homogeneous chain T by Holland's Theorem and Theorem 2J. We can ℓ-embed $A(T)$ in $B(T \overleftarrow{\times} \mathbb{Z})$ via: $f \mapsto \hat{f}$ where

$$(\tau,n)\hat{f} = \begin{cases} (\tau f, 0) & \text{if } n = 0 \\ (\tau, n) & \text{if } n \neq 0 \end{cases}$$

. By Theorem 2J, we can ℓ-embed $B(T \overleftarrow{\times} \mathbb{Z})$ in $B(\Omega)$ for some doubly homogeneous chain Ω. Clearly the composition of these maps ℓ-embeds G in $B(\Omega)$.)

This whole chapter can probably best be expressed in the caption "Geometry versus Algebra." The interplay of these two is of central importance. The techniques are as valuable as the theorems they prove.

2.1. *GEOMETRY VERSUS ALGEBRA.*

This section is devoted to establishing the machinery used frequently when dealing with doubly transitive $A(\Omega)$. Specifically, we will use it in this section to prove Theorems 2A-2D and their analogues.

Typically, we will take a geometric (pictorial) concept and write down an algebraic statement which will be equivalent for doubly transitive $A(\Omega)$.

Recall the definition of a bump (p. 25), a most geometric notion.

We begin our catalogue with an algebraic description of bumps.

Write *bump(g)* for $(\forall f_1)(\forall f_2)(f_1 \wedge f_2 = e \;\&\; f_1 \vee f_2 = g \rightarrow$ $\rightarrow f_1 = e \text{ or } f_2 = e)$.

LEMMA 2.1.1. *Let* Ω *be any chain and* $e < g \in A(\Omega)$. g *has one bump if and only if* $bump(g)$.

Proof: If g has more than one bump, let h be a bump of g. Then $h \in A(\Omega)$ and as $h \wedge gh^{-1} = e$, $h \vee gh^{-1} = g$ and $gh^{-1} \neq e$, bump(g) is false (cf. Lemma 1.9.3). Conversely, if g has one bump and $f_1 \vee f_2 = g$ with $f_1 \wedge f_2 = e$, then $\overline{supp}(g)$ is the disjoint union of the open sets $\overline{supp}(f_1)$ and $\overline{supp}(f_2)$. Since $\overline{supp}(g)$ is an interval of $\bar{\Omega}$, it is connected, Hence $\overline{supp}(f_i) = \emptyset$ for $i = 1$ or 2; i.e., $f_1 = e$ or $f_2 = e$.

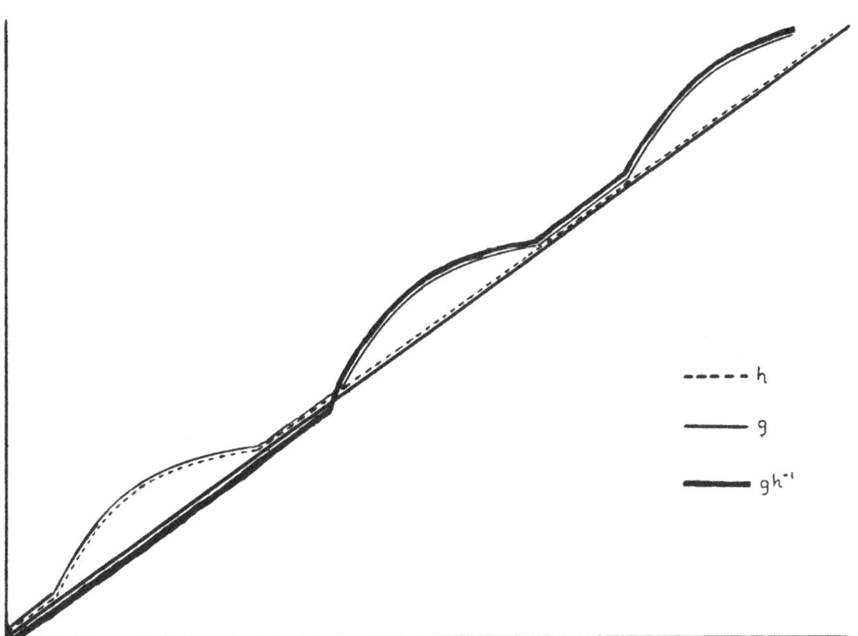

Write *bump(h,g)* for bump(h) & $gh^{-1} \wedge h = e$. By Lemmas 1.9.3 and 2.1.1:

COROLLARY 2.1.2. *Let* Ω *be any chain and* $e < g,h \in A(\Omega)$. *Then bump(h,g) if and only if* h *is a bump of* g.

Let (G,Ω) be an ℓ-permutation group and $g_1, g_2 \in G$. We say that g_1 is to the *left* of g_2 if for all $\alpha_i \in \mathrm{supp}(g_i)$ $(i = 1,2)$, $\alpha_1 < \alpha_2$. Write $g_1 \mathcal{L} g_2$ if $(\forall f \geq e)(|g_1| \wedge |f^{-1}g_2f| = e)$.

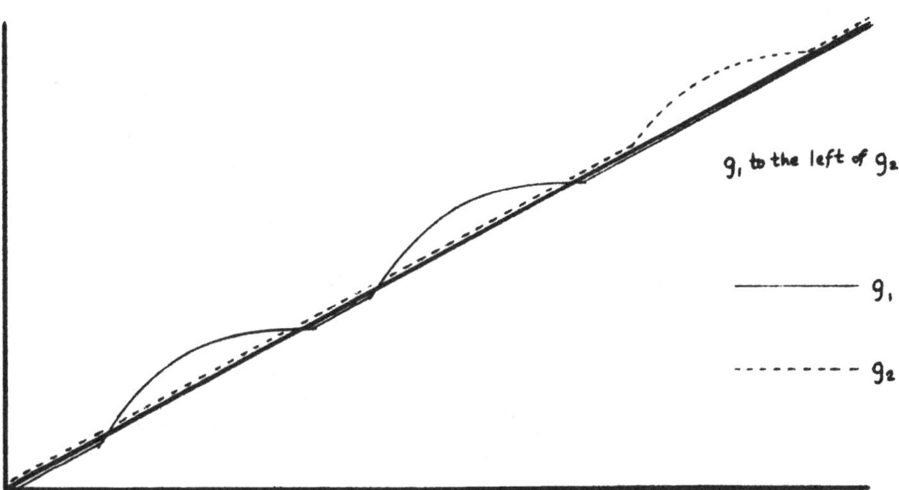

<u>LEMMA</u> <u>2.1.3</u>. *If (G,Ω) is a transitive ℓ-permutation group and $e \neq g_1, g_2 \in G$, then $g_1 \mathcal{L} g_2$ if and only if g_1 is to the left of g_2.*

Proof: If g_1 is to the left of g_2, then $g_1 \mathcal{L} g_2$ by the fundamental triviality. Conversely, if g_1 is not to the left of g_2, there exist $\alpha_i \in \operatorname{supp}(g_i)$ $(i = 1,2)$ such that $\alpha_2 \leq \alpha_1$. Let $e \leq f \in G$ be such that $\alpha_2 f = \alpha_1$. Then $\alpha_1 \in \operatorname{supp}(g_1) \cap \operatorname{supp}(f^{-1} g_2 f)$, so not $g_1 \mathcal{L} g_2$.

Let <u>$R(g)$</u> be the formula $(\exists y \neq e)(y \mathcal{L} g)$, <u>$L(g)$</u> the formula $(\exists y \neq e)(g \mathcal{L} y)$, and <u>$B(g)$</u> the formula $L(g)$ & $R(g)$.

<u>COROLLARY</u> <u>2.1.4</u>. *If Ω is doubly homogeneous and $g \in A(\Omega)$, then $g \in B(\Omega)$ $(g \in L(\Omega), g \in R(\Omega))$ if and only if $B(g)$ $(L(g), R(g))$. Hence $B(\Omega)$, $L(\Omega)$, and $R(\Omega)$ do not satisfy the same ℓ-group sentences (and so are not ℓ-isomorphic).*

Let $e \neq g_1, g_2 \in A(\Omega)$. We say that g_1 is <u>adjacent</u> to g_2 if g_1 is to the left of g_2 and $\sup(\operatorname{supp}(g_1)) = \inf(\operatorname{supp}(g_2))$ $(\in \bar{\Omega})$. We write <u>$Adj(g_1, g_2)$</u> for: $g_1 \mathcal{L} g_2$ & $\neg(\exists f \neq e)(g_1 \mathcal{L} f$ & $f \mathcal{L} g_2)$.

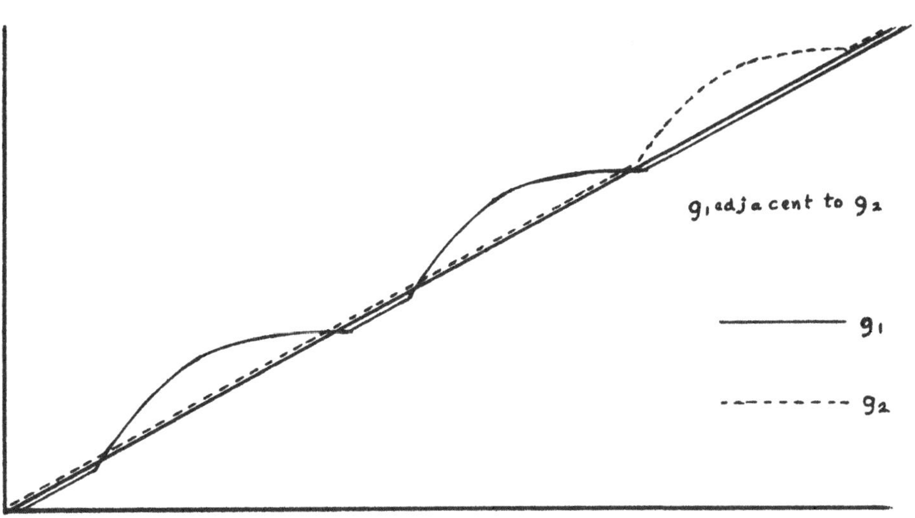

CORLLARY 2.1.5. *If Ω is doubly homogeneous and $e \neq g_1, g_2 \in A(\Omega)$, then g_1 is adjacent to g_2 if and only if $Adj(g_1,g_2)$.*

Proof: By Lemma 2.1.3, if g_1 is adjacent to g_2, then $\mathrm{Adj}(g_1,g_2)$. Conversely, if g_1 is not adjacent to g_2 but g_1 is to the left of g_2, let $\bar{\alpha} = \sup(\mathrm{supp}(g_1))$ and $\bar{\beta} = \inf(\mathrm{supp}(g_2))$. Since g_1 is to the left of (but not adjacent to) g_2, $\bar{\alpha} < \bar{\beta}$. By Corollary 1.10.6, there exists $e \neq f \in A(\Omega)$ such that $\mathrm{supp}(f) \subseteq (\bar{\alpha},\bar{\beta})$. Hence $g_1 \mathcal{L} f$ & $f \mathcal{L} g_2$, so not $\mathrm{Adj}(g_1,g_2)$.

LEMMA 2.1.6. *If Ω is doubly homogeneous and $\bar{\alpha} < \bar{\beta}$ (in $\bar{\Omega}$), there is $e < g \in B(\Omega)$ such that $\bar{\alpha} = \inf(\mathrm{supp}(g))$ and $\bar{\beta} = \sup(\mathrm{supp}(g))$.*

Proof: By Corollary 1.10.6, there is $e < g_1 \in B(\Omega)$ with $\mathrm{supp}(g_1) \subseteq (\bar{\alpha},\bar{\beta})$. If $\inf(\mathrm{supp}(g_1)) = \bar{\alpha}_1 > \bar{\alpha}$, let $e < g_2 \in B(\Omega)$ with $\mathrm{supp}(g_2) \subseteq (\bar{\alpha},\bar{\alpha}_1)$; and if $\sup(\mathrm{supp}(g_1)) = \bar{\beta}_1 < \bar{\beta}$, let $e < g_3 \in B(\Omega)$ with $\mathrm{supp}(g_3) \subseteq (\bar{\beta}_1,\bar{\beta})$. Continuing in this way (possibly transfinitely) and taking the supremum (in $A(\Omega)$) of the resulting set of positive elements of $A(\Omega)$, we obtain $g \in B(\Omega)$ having the desired properties.

LEMMA 2.1.7. *If* Ω *is doubly homogeneous, then it is Dedekind complete if and only if* $A(\Omega)$ *satisfies:* $(\forall g \neq e)(B(g) \to (\exists h)Adj(g,h^{-1}gh))$.

Proof: If Ω is not Dedekind complete, let $\bar{\alpha} \in \Omega$ and $\bar{\beta} \in \bar{\Omega}\setminus\Omega$ with $\bar{\alpha} < \bar{\beta}$. Let $e < g \in B(\Omega)$ satisfy Lemma 2.1.6. If $h \in A(\Omega)$ and g is to the left of $h^{-1}gh$, then $\bar{\beta} \leq \bar{\alpha}h = \inf(supp(g))h = \inf(supp(h^{-1}gh))$. Since $\bar{\beta} \notin \Omega$ and $\bar{\alpha}h \in \Omega$, $\bar{\beta} < \bar{\alpha}h$. By Corollary 1.10.6, there is $e < f_0 \in A(\Omega)$ such that $supp(f_0) \subseteq (\bar{\beta},\bar{\alpha}h)$. Hence not $Adj(g,h^{-1}gh)$.

Conversely, if Ω is Dedekind complete, let $e \neq g \in B(\Omega)$. Let $\bar{\alpha} = \inf(supp(g))$ and $\bar{\beta} = \sup(supp(g))$. Since Ω is Dedekind complete, $\bar{\alpha},\bar{\beta} \in \Omega$ and there is $h \in A(\Omega)$ such that $\bar{\alpha}h = \bar{\beta}$. Clearly g is adjacent to $h^{-1}gh$.

Let $e \neq f,g \in B(\Omega)$. Define $\underline{f \approx g}$ if $\forall h(f\mathcal{L}h \leftrightarrow g\mathcal{L}h)$. Then $\approx$ is an equivalence relation on $B(\Omega)$. By Lemma 2.1.6:

THEOREM 2.1.8. *Let* Ω *be a doubly homogeneous chain. For each* $e \neq f \in B(\Omega)$, *let* $\bar{\alpha}_f = \sup(supp(f)) \in \bar{\Omega}$. *Then* $f \mapsto \bar{\alpha}_f$ *maps* $B(\Omega)\setminus\{e\}$ *onto* $\bar{\Omega}$, *and* $\bar{\alpha}_f = \bar{\alpha}_g$ *if and only if* $f \approx g$. *Moreover,* $\bar{\alpha}_f < \bar{\alpha}_g$ *if and only if, for some* $e \neq h \in B(\Omega)$, $h \approx g \ \& \ f\mathcal{L}h$.

Since $B(\Omega)$ is characterised by Corollary 2.1.4, this "interpretation" of $\bar{\Omega}$ (together with its order) in $A(\Omega)$ is in the first order ℓ-group language. Can we distinguish Ω in $\bar{\Omega}$ in this way? If $e \neq f,g \in B(\Omega)$ interpret $\bar{\alpha}_f$ and $\bar{\alpha}_g$ respectively, and $\bar{\alpha}_f \in \Omega$, then $\{h^{-1}fh\colon h \in A(\Omega)\}$ interprets the points of Ω. So if $\bar{\alpha}_g \in \Omega$, $h^{-1}fh \approx g$ for some $h \in A(\Omega)$. Indeed:

COROLLARY 2.1.9. *Let* Ω *be doubly homogeneous and* $e \neq f,g \in B(\Omega)$. *Then* $\bar{\alpha}_f$ *and* $\bar{\alpha}_g$ *belong to the same* $A(\Omega)$ *orbit if and only if* $h^{-1}fh \approx g$ *for some* $h \in A(\Omega)$.

Note that this last statement is a first order condition (in the language of ℓ-groups).

Now if $A(\Omega)$ cannot be distinguished (in the ℓ-group language) from $A(\mathbb{R})$, then Ω is Dedekind complete by Lemma 2.1.7, so there is only one orbit and no problem arises. However, if we replace $\mathbb{R}$ by $\mathbb{Q}$, then $A(\Omega)$ has just two orbits in $\bar{\Omega}$ since $A(\mathbb{Q})$ does in $\mathbb{R}$. $\lceil A(\mathbb{Q})$ satisfies $(\exists f \neq e)(\exists g \neq e)(B(f)\ \&\ B(g)\ \&\ (\forall h)(h^{-1}fh \not\approx g)\ \&\ (\forall k_1 \neq e)(B(k_1) \rightarrow (\exists k_2)(k_2^{-1}k_1k_2 \approx f$ or $k_2^{-1}k_1k_2 \approx g))).]$

In order to prove Theorems 2B and 2C we must still capture countability. We first need a technical lemma.

Let G be a group and $X \subseteq G$. The <u>*centraliser*</u> of X in G, <u>$C_G(X)$</u> $= \{f \in G: (\forall x \in X)(fx = xf)\}$. We will often write <u>$C(X)$</u> for $C_G(X)$ if G is clear from context. If $g \in G$, write <u>$C(g)$</u> for $C(\{g\})$, and <u>$\langle g\rangle$</u> $= \{g^n : n \in \mathbb{Z}\}$. Note that $C(\langle g\rangle) = C(g)$. So $\langle g\rangle \subseteq C(C(g)) \subseteq C(g)$.

<u>LEMMA</u> <u>2.1.10</u>. *Let* Ω *be doubly homogeneous and* $e < g \in A(\Omega)$ *have one bump. Then* $\langle g\rangle = C(C(g))$.

<u>*Proof:*</u> For reductio ad absurdum, assume $f \in C(C(g))\backslash\langle g\rangle$. Let $\Delta = \mathrm{supp}(g)$. For each $n \in \mathbb{Z}$, the sets $\{\bar{\delta} \in \bar{\Delta}: \bar{\delta}f \leq \bar{\delta}g^n\}$ and $\{\bar{\delta} \in \bar{\Delta}: \bar{\delta}f \geq \bar{\delta}g^{n+1}\}$ are closed (in $\bar{\Delta}$) and disjoint. Since $\bar{\Delta}$ is connected, there exists $\bar{\alpha} \in \bar{\Delta}$ such that $\bar{\alpha}f \notin \{\bar{\alpha}g^n : n \in \mathbb{Z}\}$.

<u>Case</u> <u>1</u>. $\bar{\alpha}f \notin \bar{\Delta}$. By Corollary 1.10.6, there exists $e < h \in A(\Omega)$ such that $\bar{\alpha}f \in \mathrm{supp}(h)$ and $\mathrm{supp}(h) \cap \Delta = \emptyset$. Now $h \wedge g = e$ so $h \in C(g)$. Hence $f \in C(h)$. But $\bar{\alpha}hf = \bar{\alpha}f < \bar{\alpha}fh$, a contradiction.

<u>Case</u> <u>2</u>. $\bar{\alpha}f \in \bar{\Delta}$. So for some $m \in \mathbb{Z}$, $\bar{\alpha}g^m < \bar{\alpha}f < \bar{\alpha}g^{m+1}$. Let $e < h_0 \in A(\Omega)$ be such that $\bar{\alpha}f \in \mathrm{supp}(h_0) \subseteq (\bar{\alpha}g^m, \bar{\alpha}g^{m+1})$. We take h_0 and from it construct $h \in C(g)$ by taking the pointwise supremum of the disjoint set of elements $\{g^{-n}h_0g^n : n \in \mathbb{Z}\}$ of $A(\Omega)$; i.e., $h|[\bar{\alpha}g^n, \bar{\alpha}g^{n+1}] = g^{-n+m}h_0g^{n-m}|[\bar{\alpha}g^n, \bar{\alpha}g^{n+1}]$ and

$\text{supp}(h) \subseteq \Delta$. Then $h \in C(g)$ so $f \in C(h)$. But $\bar{\alpha}fh = \bar{\alpha}fh_0 > \bar{\alpha}f = \bar{\alpha}g^m h_0 g^{-m} f = \bar{\alpha}hf$, a contradiction.

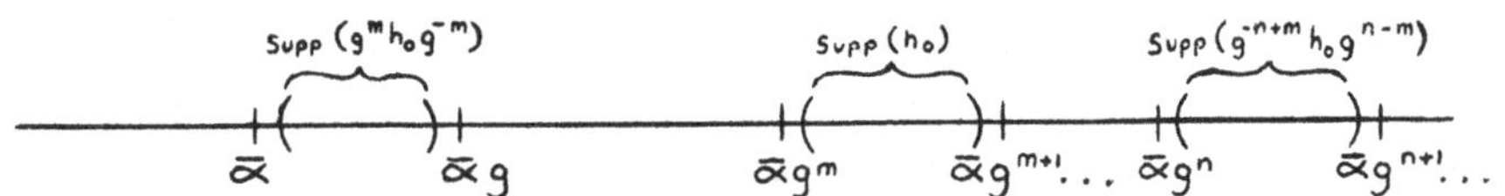

Note that the condition $f \in C(C(g))$ can be expressed by a sentence $(\forall h)(hg = gh \rightarrow hf = fh)$, but the condition $f \in \langle g \rangle$ usually cannot ($f = g^0$ or g^1 or g^{-1} or g^2 or g^{-2}... is not permissible).

We can now turn to countability.

LEMMA 2.1.11. *Let Ω be a doubly homogeneous chain and $\Delta \subseteq \Omega$ with $\sigma < \inf(\Delta) < \sup(\Delta) < \tau$. Δ is countable if and only if there exist $\alpha \in \Omega$ and $f, g \in B(\Omega)$ such that $\Delta = \{\alpha f^m g^{-m} : m \in \omega\}$ and $\text{supp}(f) = \text{supp}(g) \subseteq (\sigma, \tau)$.*

Proof: Clearly $\{\alpha f^m g^{-m} : m \in \omega\}$ is a countable subset. Conversely, enumerate $\Delta = \{\delta_n : n \in \omega\}$. Let $\alpha = \delta_0$ and $\sigma < \beta < \Delta$. There exists $e < g \in A(\Omega)$ having one bump such that $\Delta < \beta g$ and $\text{supp}(g) \subseteq (\sigma, \tau)$. By double transitivity and Corollary 1.10.6, there is $e < f \in A(\Omega)$ having one bump such that $\text{supp}(f) = \text{supp}(g) \subseteq (\sigma, \tau)$, $\beta f^m = \beta g^m$ and $\alpha f^m = \delta_m g^m$ $(m \in \omega)$. Hence $\Delta = \{\alpha f^m g^{-m} : m \in \omega\}$.

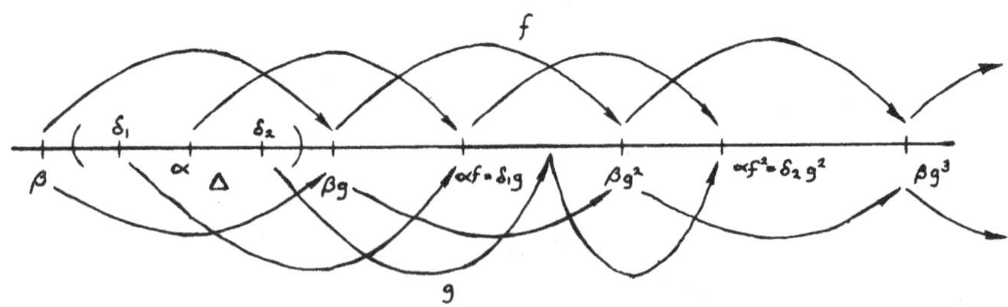

Note that $\Delta = \{\alpha ab^{-1}: e < a \in \langle f \rangle, e < b \in \langle g \rangle\} \cap (\beta, \beta g)$.

LEMMA 2.1.12. *Let* Ω *be doubly homogeneous and* $\Delta \subseteq \Omega$. *Then* Δ *is countable if and only if there are* $\alpha_0, \alpha_1 \in \Omega$ *and* $h, f_0, f_1, g_0, g_1 \in A(\Omega)$ *such that* $\Delta = \{\alpha_0 h^{2n} f_0^m g_0^{-m}: m \in \omega, n \in \mathbb{Z}\} \cup \{\alpha_1 h^{2n} f_1^m g_1^{-m}: m \ \omega, n \ \mathbb{Z}\}$.

Proof: Clearly any such set is countable, so assume Δ is countable. If Δ is bounded, the lemma follows at once from Lemma 2.1.11. Assume Δ is unbounded and let $\{\alpha_n: n \in \mathbb{Z}\}$, $\{\beta_n: n \in \mathbb{Z}\}$, $\{\sigma_n: n \in \mathbb{Z}\}$ and $\{\tau_n: n \in \mathbb{Z}\}$ be coterminal in Δ (and hence in Ω) with $\alpha_n < \sigma_n < \beta_n < \tau_n < \alpha_{n+1}$ $(n \in \mathbb{Z})$. Let $\Delta_{0,n} = \Delta \cap [\beta_{2n-1}, \beta_{2n}]$

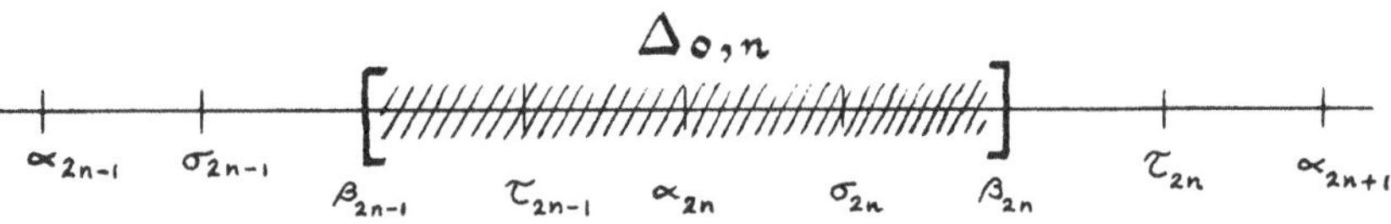

and $\Delta_{1,n} = \Delta \cap (\beta_{2n}, \beta_{2n+1})$. By Lemma 2.1.11, there are $e < f_{i,n}, g_{i,n} \in A(\Omega)$ $(i = 0,1)$ having one bump with $\text{supp}(f_{0,n}) = \text{supp}(g_{0,n}) \subseteq (\alpha_{2n-1}, \alpha_{2n+1})$, $\sigma_{2n-1} g_{0,n} = \tau_{2n}$, $\Delta_{0,n} = \{\alpha_{2n} f_{0,n}^m g_{0,n}^{-m}: m \in \omega\} = \{\alpha_{2n} a_{0,n} b_{0,n}^{-1}: e < a_{0,n} \in \langle f_{0,n} \rangle$, $e < b_{0,n} \in \langle g_{0,n} \rangle\} \cap (\sigma_{2n-1}, \tau_{2n})$ and $\text{supp}(f_{1,n}) = \text{supp}(g_{1,n}) \subseteq (\alpha_{2n}, \alpha_{2n+2})$, $\sigma_{2n} g_{1,n} = \tau_{2n+1}$, $\Delta_{1,n} = \{\alpha_{2n+1} f_{1,n}^m g_{1,n}^{-m}: m \in \omega\} = \{\alpha_{2n+1} a_{1,n} b_{1,n}^{-1}: e < a_{1,n} \in \langle f_{1,n} \rangle$, $e < b_{1,n} \in \langle g_{1,n} \rangle\} \cap (\sigma_{2n}, \tau_{2n+1})$ $(n \in \mathbb{Z})$. Let f_i and g_i be the supremum of the disjoint elements $\{f_{i,n}: n \in \mathbb{Z}\}$ and $\{g_{i,n}: n \in \mathbb{Z}\}$ respectively $(i = 0,1)$. Then $\Delta = \bigcup_{n \in \mathbb{Z}} \{\alpha_{2n} f_0^m g_0^{-m}: m \in \omega\} \cup \bigcup_{n \in \mathbb{Z}} \{\alpha_{2n+1} f_1^m g_1^{-m}: m \in \omega\}$. Let $h_n: [\alpha_n, \alpha_{n+1}] \cong [\alpha_{n+1}, \alpha_{n+2}]$ with $\sigma_n h_n = \sigma_{n+1}$ and $\tau_n h_n = \tau_{n+1}$ $(n \in \mathbb{Z})$. Let $h = \bigcup_{n \in \mathbb{Z}} h_n$. Thus

$\Delta = \{\alpha_0 h^{2n} f_0^m g_0^{-m} : m \in \omega, n \in \mathbb{Z}\} \cup \{\alpha_1 h^{2n} f_1^m g_1^{-m} : m \in \omega, n \in \mathbb{Z}\}$, as desired. Similarly if Δ is bounded above or below.

Note that the proof works for Δ a subset of any orbit of $A(\Omega)$ (in $\bar{\Omega}$). It also gives that if Δ is countable and unbounded, there are $e < h \in A(\Omega)$ having one (unbounded) bump, $e < f_i, g_i \in A(\Omega)$ and $\bar{\alpha}_i, \bar{\sigma}_i, \bar{\tau}_i \in \bar{\Omega}$ belonging to the same $A(\Omega)$ orbit $(i = 0,1)$ such that $\bar{\alpha}_0 < \bar{\sigma}_0 < \bar{\tau}_0 < \bar{\alpha}_1 = \bar{\alpha}_0 h$, $\bar{\sigma}_1 = \bar{\sigma}_0 h$, $\bar{\tau}_1 = \bar{\tau}_0 h$, and each interval $(\bar{\alpha}_i kh^{-1}, \bar{\alpha}_i kh)$ contains the support of a unique bump c_i (d_i) of f_i (g_i) with $\bar{\sigma}_i kh^{-1}, \bar{\tau}_i k \in \text{supp}(c_i) \cap \text{supp}(d_i)$ $(i = 0,1;\ k \in \langle h^2 \rangle)$; and $\bar{\delta} \in \Delta$ if and only if $\bar{\delta} = \bar{\alpha}_i k a_i b_i^{-1}$ for some $i \in \{0,1\}$, $k \in \langle h^2 \rangle$, $e < a_i \in \langle c_i \rangle$ and $e < b_i \in \langle d_i \rangle$ with c_i (d_i) that bump of f_i (g_i) containing $\bar{\alpha}_i k$ in its support, and $\bar{\sigma}_i kh^{-1} < \bar{\alpha}_i k a_i b_i^{-1} < \bar{\tau}_i k$. So we can code countable subsets of an orbit of $A(\Omega)$ by tuples $(\bar{\alpha}_0, \bar{\sigma}_0, \bar{\tau}_0, h, f_0, g_0, f_1, g_1)$ where the coordinates satisfy the above conditions. If $f_{\bar{\alpha}_0}, f_{\bar{\sigma}_0}, f_{\bar{\tau}_0} \in B(\Omega)$ with $\sup(\text{supp}(f_{\bar{\rho}_0})) = \bar{\rho}_0$ $(\rho \in \{\alpha, \sigma, \tau\})$, then the octuple $(f_{\bar{\alpha}_0}, f_{\bar{\sigma}_0}, f_{\bar{\tau}_0}, h, f_0, g_0, f_1, g_1)$ of elements of $A(\Omega)$ codes Δ. Moreover, the above conditions on the coordinates of this tuple can be expressed in the language of ℓ-groups by Corollaries 2.1.2, 2.1.9, Lemma 2.1.10 and the remarks following Theorem 2.1.8. What is more, we can determine (in the ℓ-group language of $A(\Omega)$) if two such octuples code the same countable subset Δ of an orbit of $A(\Omega)$, and if $\bar{\beta} \in \Delta$. Hence we can use these octuples to detect $=$ and $\subseteq$ between countable subsets of an orbit of $A(\Omega)$. Furthermore, "there exists a countable Δ..." can be translated into our language by "there exists such an octuple...", etc. From now on we will assume that this translation has been performed.

We now prove Theorems 2B and 2C.

By the remarks following Corollary 2.1.9, $A(\mathbb{Q})$ has two orbits one of which is countable. By the above, this is expressible in the language of ℓ-groups and so holds for any doubly transitive $A(\Omega) \equiv A(\mathbb{Q})$. Thus Ω or $\bar{\Omega}\backslash\Omega$ is countable proving Theorem 2C. To prove Theorem 2B, note that if doubly transitive $A(\Omega) \equiv A(\mathbb{R})$, then Ω is Dedekind complete (Lemma 2.1.7) and (by the previous paragraph) has a countable dense subset (Δ is dense in $\bar{\Omega}$ if $(\forall\bar{\alpha})(\forall\bar{\beta})[\bar{\alpha} < \bar{\beta} \rightarrow (\exists\bar{\delta} \in \Delta)(\bar{\alpha} < \bar{\delta} < \bar{\beta})]$). Hence $\Omega \cong \mathbb{R}$.

Actually, we have shown that there are sentences $\theta_{\mathbb{R}}$ and $\theta_{\mathbb{Q}}$ of the language of ℓ-groups such that doubly transitive $A(\Omega)$ satisfies $\theta_{\mathbb{R}}$ ($\theta_{\mathbb{Q}}$) if and only if $\Omega \cong \mathbb{R}$ ($\mathbb{Q}$ or $\mathbb{R}\backslash\mathbb{Q}$). In Chapter 4, we will show that "doubly transitive" can be replaced by "transitive" in Theorems 2B and 2C.

We now wish to consider the group language instead of the ℓ-group language. Recall that Corollary 1.10.8 (in conjunction with Lemma 1.10.1) shows: There is a first order formula $\phi_0(p)$ of the language of groups which holds in a doubly transitive $A(\Omega)$ if and only if p or p^{-1} exceeds e. Moreover, by Corollary 1.10.8, there is a first order formula $\phi_p(y,z)$ (of the group language) which holds in doubly transitive $A(\Omega)$ if and only if $y \leq z$ (when $p \geq e$) or $y \geq z$ (when $p^{-1} \geq e$). Now given a sentence θ in the ℓ-group language (using $\leq$ in place of the lattice operations), replace each occurrence of $y \leq z$ by $\phi_p(y,z)$ and call the resulting sentence θ_p. Let θ' be $(\exists p)(\phi_0(p)\ \&\ \theta_p)$. Clearly, the ℓ-group $A(\Omega)$ satisfies θ if and only if $A(\Omega)$ or $A(\Omega^*)$ satisfies the group-theoretic sentence θ'. This proves Theorem 2A.

Since $\mathbb{R}^* = \mathbb{R}$ and $\mathbb{Q}^* = \mathbb{Q}$, we obtain Theorems 2B' and 2C' from Theorems 2A, 2B and 2C. Again, in Chapter 4, we will show that "doubly transitive" can be replaced by "transitive" in Theorems 2B' and 2C'.

We now prove Theorem 2D.

So suppose Ω and T are doubly homogeneous, and $A(\Omega) \cong A(T)$ (as groups). Let ψ be the isomorphism. Let $\bar{\alpha} \in \bar{\Omega}$. By Lemma 2.1.6, there are $e \neq f_1, f_2 \in B(\Omega)$ such that $\sup(\mathrm{supp}(f_1)) = \bar{\alpha} = \inf(\mathrm{supp}(f_2))$. Such a pair (f_1,f_2) satisfies: $(\forall p)(\phi_0(p)\ \&\ f_1 \mathcal{L}_p f_2 \rightarrow \mathrm{Adj}_p(f_1,f_2))$. Moreover, if

$\sup(\mathrm{supp}(g_1)) = \bar{\beta} = \inf(\mathrm{supp}(g_2))$, then $\bar{\alpha} = \bar{\beta}$ if and only if $(\forall p)(\phi_0(p) \;\&\; f_1 \mathcal{L}_p f_2 \;\&\; g_1 \mathcal{L}_p g_2 \rightarrow (\forall h)[f_1 \mathcal{L}_p h \leftrightarrow g_1 \mathcal{L}_p h])$. Hence, by Theorem 2.1.8 and Corollaries 2.1.4 and 2.1.9, ψ induces an ordermorphism $\hat{\psi}$ from $\bar{\Omega}$ onto $\bar{T}$ or $\bar{T}^*$ such that the restriction of $\hat{\psi}$ to Ω gives an orbit of $A(T)$ in $\bar{T}$ (or $\bar{T}^*$). Hence Ω is ordermorphic (or anti-ordermorphic) to an orbit of $A(T)$ in $\bar{T}$. Moreover, if $\hat{\psi}$ is an ordermorphism and $\bar{\beta} \in \bar{T}$, let $f_1\psi \in B(T)$ with $\sup(\mathrm{supp}(f_1\psi)) = \bar{\beta}$. Since $\sup(\mathrm{supp}(h_1))\hat{\psi} = \sup(\mathrm{supp}(h_1\psi))$ $(h_1 \in A(\Omega))$, $\bar{\beta}\hat{\psi}^{-1}g\hat{\psi} = \sup(\mathrm{supp}(f_1))g\hat{\psi} = \sup(\mathrm{supp}((g^{-1}f_1g)\psi)) = \sup(\mathrm{supp}(f_1\psi))(g\psi) = \bar{\beta}(g\psi)$ $(g \in A(\Omega))$. Hence $\hat{\psi}^{-1}g\hat{\psi} = g\psi$ as desired. If $\hat{\psi}$ is an antiordermorphism, the proof is the same except $\sup(\mathrm{supp}(h_1))\hat{\psi} = \sup(\mathrm{supp}(h_2\psi))$ where (h_1,h_2) is an adjacent pair.

Finally, we wish to consider the lattice language. Observe that the formula we gave for $\mathrm{bump}(g)$ was in the lattice language and did not involve the group operation. If $g,h \in A(\Omega)$ have one bump and $g \wedge h = e$ we say that $f \in A(\Omega)$ *lies between* g and h if $\mathrm{supp}(g) < \mathrm{supp}(f) < \mathrm{supp}(h)$ or

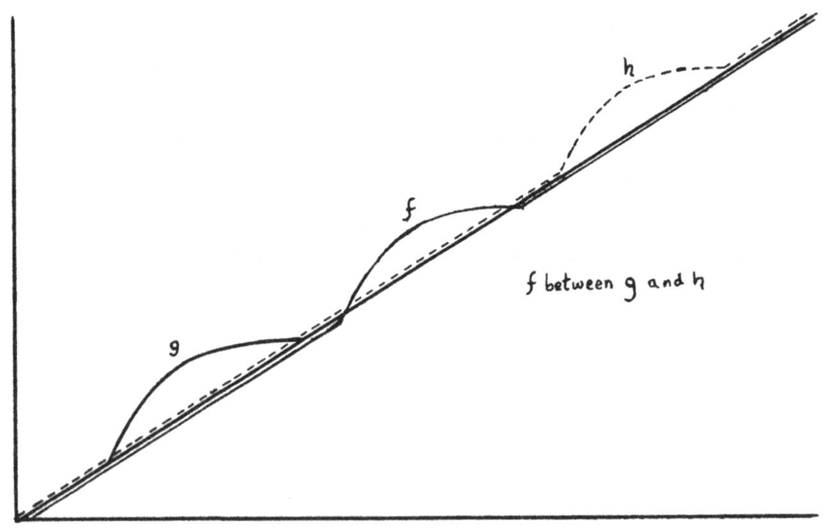

$\mathrm{supp}(h) < \mathrm{supp}(f) < \mathrm{supp}(g)$. Let $\underline{B(f,g,h)}$ be the lattice formula: $e < f$ & bump(g) & bump(h) & $g \wedge h = e$ & $f \wedge g = e$ & $f \wedge h = e$ & $(\forall k > e)(\mathrm{bump}(k)$ & $k \wedge g \neq e$ & $k \wedge h \neq e \rightarrow$ $(\forall f' > e)(\mathrm{bump}(f')$ & $f' \leq f \rightarrow k \wedge f' \neq e))$.

LEMMA 2.1.13. *Let* Ω *be doubly homogeneous and* $g,h \in A(\Omega)$ *have one bump with* $g \wedge h = e$. *Let* $e < f \in A(\Omega)$. *Then* $B(f,g,h)$ *if and only if* f *lies between* g *and* h.

Proof: Without loss of generality, $\bar{\alpha} = \sup(\mathrm{supp}(g)) \leq \inf(\mathrm{supp}(h)) = \bar{\beta}$. If f lies between g and h and $e < k \in A(\Omega)$ has one bump with $k \wedge g \neq e \neq k \wedge h$, let $\alpha \in \mathrm{supp}(f)$. Then $\bar{\alpha} < \alpha < \bar{\beta}$ and since k has one bump and $\bar{\alpha},\bar{\beta} \in \mathrm{supp}(k)$, $\bar{\alpha}k^n \geq \bar{\beta}$ for some $n \in \mathbb{Z}^+$. Hence $\alpha < \bar{\beta} \leq \bar{\alpha}k^n < \alpha k^n$ so $\alpha \in \mathrm{supp}(k^n) = \mathrm{supp}(k)$. Thus $\mathrm{supp}(f) \subseteq \mathrm{supp}(k)$, so $B(f,g,h)$.

Conversely, we assume $B(f,g,h)$. If f does not lie between g and h, there exists $\alpha \in \mathrm{supp}(f)$ such that $\alpha \notin (\bar{\alpha},\bar{\beta})$. Let $\sigma \in \mathrm{supp}(g)$ and $\tau \in \mathrm{supp}(h)$. Since $g \wedge f = e = h \wedge f$, and g and h each have one bump, $\alpha \notin (\sigma,\tau)$. There exists $e < k \in A(\Omega)$ such that $\bar{\beta} < \bar{\alpha}k$ and $\sigma k = \sigma$, $\tau k = \tau$. Let $k_0 \in A(\Omega)$ be the bump of k whose support contains $\bar{\alpha}$. Then $k_0 \wedge g \neq e \neq k_0 \wedge h$, but $k_0 \wedge f' = e$ where f' is the bump of f whose support contains α. So $B(f,g,h)$ fails, the desired contradiction.

Let $\underline{\bar{B}(f)}$ be the formula: $(\exists g > e)(\exists h > e)B(f,g,h)$. Then:

COROLLARY 2.1.14. *If* Ω *is doubly homogeneous and* $e < f \in A(\Omega)$, *then* $\bar{B}(f)$ *if and only if* $f \in B(\Omega)$.

We next seek a left-right orientation on $A(\Omega)$. Suppose $e < a,b \in B(\Omega)$ have one bump and are disjoint. Let $e < f,g \in B(\Omega)$ has one bump with $\{a,b,f,g\}$ pairwise disjoint. Then we can use betweenness to express in the lattice language: a is to the left of b if and only if f is to the left of g. This is done by taking the disjunction of six possibilities each of the form $B(\ ,\ ,\ )$ & $B(\ ,\ ,\ )$. Let $\underline{L(a,b,f,g)}$ be the conjunction of this formula and the hypotheses on a,b,f,g. Note that $L(a,b,f,g)$ is a formula of the lattice language.

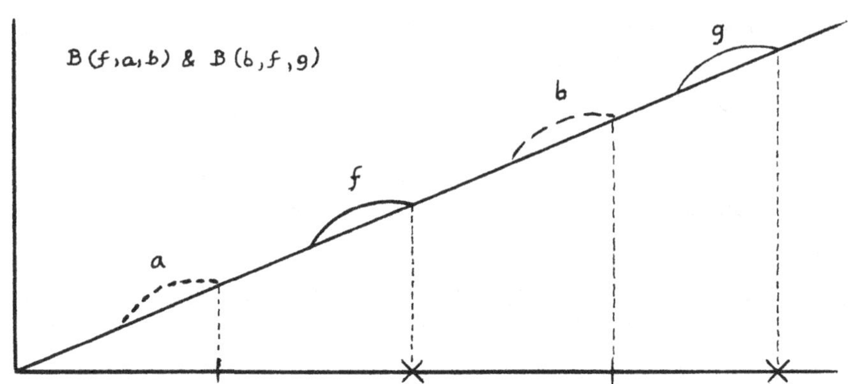

If $e < f,g \in B(\Omega)$ have one bump and are disjoint (but not necessarily disjoint from a and b), we can still express in the lattice language: a is to the left of b if and only if f is to the left of g by: $(\exists c)(\exists d)[L(a,b,c,d) \,\&\, L(c,d,f,g)]$. Let $\underline{L'(a,b,f,g)}$ be this formula. More generally, if $e < f_1,f_2 \in B(\Omega)$ are disjoint, write $\underline{\mathcal{L}(a,b,f_1,f_2)}$ for $(\forall e < f_1' \leq f_1)(\forall e < f_2' \leq f_2)(\text{bump}(f_1') \,\&\, \text{bump}(f_2') \rightarrow L'(a,b,f_1',f_2'))$.

LEMMA 2.1.15. *Let Ω be doubly homogeneous, and $e < a,b,f_1,f_2 \in B(\Omega)$ with a,b having one bump and $a \wedge b = e$. Then $A(\Omega)$ satisfies $\mathcal{L}(a,b,f_1,f_2)$ if and only if (a is to the left of b and f_1 is to the left of f_2) or (b is to the left of a and f_2 is to the left of f_1).*

Let $e < f_1,f_2 \in B(\Omega)$. We say that f_1,f_2 is an *adjacent pair* if $\sup(\text{supp}(f_1)) = \inf(\text{supp}(f_2))$ or $\sup(\text{supp}(f_2)) = \inf(\text{supp}(f_1))$.

LEMMA 2.1.16. *Let Ω be doubly homogeneous and $e < f_1,f_2 \in B(\Omega)$. Then f_1,f_2 is an adjacent pair precisely when: $(\forall a,b > e$ one bumps$)(a \wedge b = e \rightarrow \mathcal{L}(a,b,f_1,f_2)$ or $\mathcal{L}(a,b,f_2,f_1))$ & $(\forall h > e)(\exists e < f_1' \leq f_1)(\exists e < f_2' \leq f_2)[\text{bump}(f_1') \,\&\, \text{bump}(f_2') \rightarrow \neg B(h,f_1',f_2')]$.*

Let $\bar{\alpha},\bar{\beta} \in \bar{\Omega}$. Choose $e < f_1, f_2 \in B(\Omega)$ so that $\sup(\text{supp}(f_1)) = \bar{\alpha} = \inf(\text{supp}(f_2))$, and $e < g_1, g_2 \in B(\Omega)$ so that $\sup(\text{supp}(g_1)) = \bar{\beta} = \inf(\text{supp}(g_2))$. Then $\bar{\alpha} = \bar{\beta}$ if and only if $(\forall \text{ bump } h > e)(h \wedge f_1 \neq e \ \& \ h \wedge f_2 \neq e \leftrightarrow h \wedge g_1 \neq e \ \& \ h \wedge g_2 \neq e)$. By Corollary 2.1.14 and Lemmas 2.1.15 and 2.1.16, we can interpret the set $\bar{\Omega}$ together with equality in the lattice language of $A(\Omega)$ by adjacent pairs of elements of $B(\Omega)$. Indeed, using parameters $e \neq a,b \in B(\Omega)$ with $a \wedge b = e$, we can put an ordering on $\bar{\Omega}$. If a is to the left of b, this coincides with the original ordering on $\bar{\Omega}$; otherwise it is the reverse ordering $\bar{\Omega}^*$.

We now wish to obtain the action of $A(\Omega)$ on $\bar{\Omega}$. Note that $\text{supp}(f) \subseteq \text{supp}(g)$ is equivalent to $(\forall h)(h \wedge g = e \rightarrow h \wedge f = e)$ if Ω is doubly homogeneous and $e < f,g \in A(\Omega)$.

<u>LEMMA 2.1.17</u>. *Let Ω be doubly homogeneous, $\bar{\alpha} \leq \bar{\beta}$ in $\bar{\Omega}$ and $e < h \in A(\Omega)$. Then $\bar{\alpha}h \leq \bar{\beta}$ if and only if for all $e < x \in B(\Omega)$ having one bump, if $\bar{\alpha},\bar{\beta} \in \text{supp}(x)$, there exists $e < y \not\leq h$ with $\text{supp}(y) \subseteq \text{supp}(x)$.*

<u>*Proof:*</u> If $\bar{\alpha}h > \bar{\beta}$, let $\sigma \in \Omega$ be such that $\bar{\beta} < \sigma h < \bar{\alpha}h$. Let x have one bump with $\bar{\alpha},\bar{\beta} \in \text{supp}(x) \subseteq (\sigma,\sigma h)$. Now if $\text{supp}(y) \subseteq \text{supp}(x)$ and $\tau \in \text{supp}(y)$, $\tau y < \sigma h < \tau h$; so $y < h$.

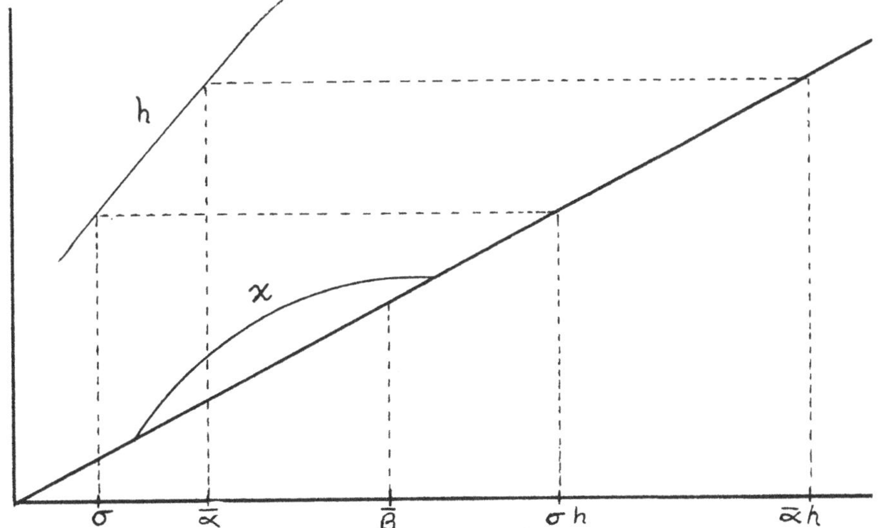

Conversely, if $\bar{\alpha}h \leq \bar{\beta}$, let $e < x \in B(\Omega)$ have one bump and assume $\bar{\alpha}, \bar{\beta} \in \mathrm{supp}(x)$. Then for some $m \in \mathbb{Z}^+$, $\bar{\beta} < \bar{\alpha}x^m$. Let $y = x^m \not\leq h$. Then $\mathrm{supp}(y) = \mathrm{supp}(x)$. //

By Lemma 2.1.17 and the remarks preceding it, we can express $\bar{\alpha}h = \bar{\beta}$ in the lattice language of $A(\Omega)$ (with the aid of an appropriate pair of parameters a,b) if $h > e$. Clearly we can prove an analogue of the lemma if $h < e$; so for an arbitrary $h \in A(\Omega)$ since $\bar{\alpha}h = \bar{\beta}$ $(\geq \bar{\alpha})$ if and only if $\bar{\alpha}(h \vee e) = \bar{\beta}$, and $\bar{\alpha}h = \bar{\beta}$ $(\leq \bar{\alpha})$ if and only if $\bar{\alpha}(h \wedge e) = \bar{\beta}$. Finally, we can define $f \cdot g = h$ in the lattice language by: $(\forall\bar{\alpha})(\forall\bar{\beta})(\bar{\alpha}f = \bar{\beta} \rightarrow \bar{\alpha}h = \bar{\beta}g)$. The only problem is that we cannot express "a is to the left of b." However, had we chosen a to the right of b (we can say $a,b > e$ are disjoint, bounded and have one bump; so a is to the left or right of b), the expression we used to capture $\bar{\alpha}h = \bar{\beta}$ would give $\bar{\alpha} = \bar{\beta}h$. Hence we have proved Theorem $2A^\dagger$. Theorems $2B^\dagger$ and $2C^\dagger$ now follow.

To prove Theorem $2D^\dagger$, we note that either $A(\Omega) \cong A(T)$ as ℓ-groups or $A(\Omega) \cong A(T)^\dagger$ as ℓ-groups. In the first case, the result follows from Theorem 2D. In the latter case, if $\psi\colon A(\Omega) \cong A(T)^\dagger$, define ψ' by $g\psi' = (g\psi)^{-1}$. Then $(fg)\psi' = ((fg)\psi)^{-1} = (f\psi + g\psi)^{-1} = [(g\psi)(f\psi)]^{-1} = f\psi' g\psi'$. Also $(f \vee g)\psi' = f\psi' \wedge g\psi'$ and dually. Hence ψ' is a group isomorphism between $A(\Omega)$ and $A(T)$ that reverses the order. Theorem $2D^\dagger$ now follows from Theorem 2D.

2.2. *DIVISIBILITY AND CONJUGACY IN $A(\Omega)$.*

This section could equally well be entitled "Applications of the Patching Lemma." Indeed, the technique is just as important as the results. Throughout this section, therefore, Ω will be doubly homogeneous. Our first application will be to prove that $A(\Omega)$ is divisible (Theorem 2E). Our next will be to give a completely geometric description of when two elements of $A(\mathbb{R})$ are conjugate. This will give a much simpler proof than that of the previous section to show that $A(\mathbb{Q}) \not\cong A(\mathbb{R})$ (actually, $A(\mathbb{Q}) \not\equiv A(\mathbb{R})$, in the language of groups, lattices or ℓ-groups). We generalise our results to give a geometric description of when two elements of

$A(\Omega)$ are conjugate (Ω any doubly homogeneous chain). All of these results are due to Holland [63] or folklore. We give one more example of the technique to simultaneously make several elements of $A(\mathbb{R})$ conjugate by the same conjugator (under certain hypotheses). This particular example can be found in Glass and Gurevich [81] and illustrates the power of the technique. Further examples and applications can then be left to the reader's imagination.

We first prove Theorem 2E:

Suppose that Ω is doubly homogeneous, $g \in A(\Omega)$ and $m \in \mathbb{Z}^+$. First assume $g > e$ has one bump. Let $\alpha \in \Delta = \text{supp}(g)$. So $\Delta = \bigcup_{n \in \mathbb{Z}} [\alpha g^n, \alpha g^{n+1}]$. There are $\beta_0, \beta_1, \ldots, \beta_m \in \Delta$ such that $\alpha = \beta_0 < \beta_1 < \ldots < \beta_m = \alpha g$. Since Ω is doubly homogeneous, there are ordermorphisms $f_n\colon [\beta_{n-1}, \beta_n] \cong [\beta_n, \beta_{n+1}]$ $(1 \leq n < m)$. Let

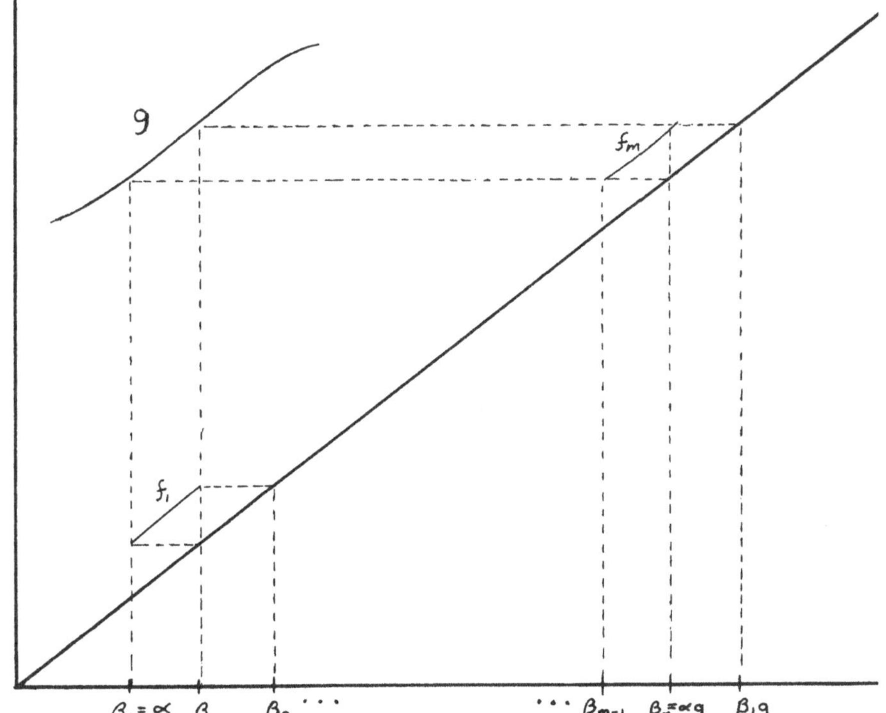

$f_m = (f_1 \dots f_{m-1})^{-1} g$. Now $\beta_{m-1} f_m = \alpha g = \beta_m$ and $\beta_m f_m = \beta_1 g$. Define $h \in A(\Omega)$ by: $\gamma h = \gamma g^{-k} f_n g^k$ if $\gamma \in [\beta_{n-1} g^k, \beta_n g^k]$ for some $k \in \mathbb{Z}$ and $1 \leq n \leq m$ and $\gamma h = \gamma$ otherwise. Now $\gamma h^m = \gamma g^{-k} f_n f_{n+1} \dots f_m g^{-1} f_1 \dots f_{n-1} g^{k+1}$ if $\gamma \in [\beta_{n-1} g^k, \beta_n g^k]$ for some $k \in \mathbb{Z}$ and $1 \leq n \leq m$ and $\gamma h = \gamma$ otherwise. By the definition of f_m, $\gamma h^m = \left\{\begin{matrix} \gamma g \text{ if } \gamma \in \Delta \\ \\ \gamma \text{ if } \gamma \notin \Delta \end{matrix}\right\} = \gamma g$. Hence $h^m = g$.

In the general case, for each bump g_i of g, there exists $h_i \in A(\Omega)$ with $h_i^m = g_i$. Note $\mathrm{supp}(h_i) = \mathrm{supp}(g_i)$. By the patching lemma, there exists $h \in A(\Omega)$ such that $h|\mathrm{supp}(g_i) = h_i|\mathrm{supp}(g_i)$ and $h|\Lambda = e|\Lambda$, where $\Lambda = \Omega \backslash \mathrm{supp}(g)$. Then $h^m = g$ and Theorem 2E is proved.

Note that the proof yields, for example: If $g \in A(\Omega)$, $\alpha \in \mathrm{supp}(g)$ and $\alpha < \beta_1 < \beta_2 < \alpha g$, there exists $h \in A(\Omega)$ such that $h^3 = g$ and $\alpha h = \beta_1$, $\beta_1 h = \beta_2$ and $\beta_2 h = \alpha g$.

We now give another application of the patching lemma. Again the technique is at least as important as the result.

Recall that a bump is either bounded, bounded only above, bounded only below, or unbounded. We call this the *boundity* of the bump.

LEMMA 2.2.1. *Let* $e < g,h \in A(\mathbb{R})$ *each have one bump and the same boundity. Then they are conjugate in* $A(\mathbb{R})$.

Proof: Choose $\alpha \in \mathrm{supp}(g)$ and $\beta \in \mathrm{supp}(h)$. Since $A(\mathbb{R})$ is doubly transitive, there is an ordermorphism $f_0: [\alpha, \alpha g] \cong [\beta, \beta h]$. For each $n \in \mathbb{Z}$, let $f_n = g^{-n} f_0 h^n$. So $f_n: [\alpha g^n, \alpha g^{n+1}] \cong [\beta h^n, \beta h^{n+1}]$. By the patching lemma (since g and h have the same boundity), there exists $f \in A(\mathbb{R})$ of the same boundity as g and h such that $f|[\alpha g^n, \alpha g^{n+1}] = f_n$ for all $n \in \mathbb{Z}$. We now show that $f^{-1} g f = h$. $\mathrm{supp}(g) f = \mathrm{supp}(h)$ since $\mathrm{supp}(g) = \bigcup_{n \in \mathbb{Z}} [\alpha g^n, \alpha g^{n+1}]$ and $\mathrm{supp}(h) = \bigcup_{n \in \mathbb{Z}} [\beta h^n, \beta h^{n+1}]$.

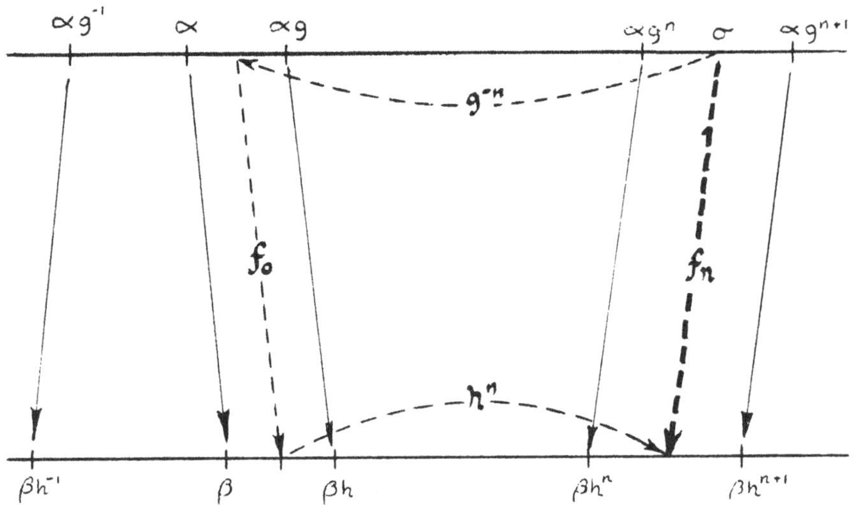

Let $\gamma \notin \text{supp}(h)$. Then $\gamma h = \gamma$ and $\gamma f^{-1} \notin \text{supp}(g)$. Hence $\gamma f^{-1}gf = \gamma f^{-1}f = \gamma = \gamma h$. Let $\gamma \in \text{supp}(h)$, say $\gamma \in [\beta h^m, \beta h^{m+1}]$. Then $\gamma f^{-1}gf = \gamma f_m^{-1}gf_{m+1} = \gamma h^{-m}f_0^{-1}g^m gg^{-(m+1)}f_0 h^{m+1} = \gamma h$. Thus $\gamma f^{-1}gf = \gamma h$ for all $\gamma \in \mathbb{R}$; i.e., $f^{-1}gf = h$.

As a consequence of the proof:

COROLLARY 2.2.2. *Let* $e < g,h \in B(\mathbb{R})$ $(L(\mathbb{R}),R(\mathbb{R}))$ *each have one bump and the same boundity. Then they are conjugate in* $B(\mathbb{R})$ $(L(\mathbb{R}),R(\mathbb{R}))$.

We can now give an easier proof of the fact that $A(\mathbb{Q}) \not\cong A(\mathbb{R})$:

COROLLARY 2.2.3. $A(\mathbb{Q}) \not\cong A(\mathbb{R})$ *as* ℓ*-groups. Indeed, as* ℓ*-groups,* $A(\mathbb{Q}) \not\equiv A(\mathbb{R})$.

Proof: Let ϕ be the sentence which says that any two strictly positive bounded one bump elements are conjugate. (This is expressible as a sentence by Lemma 2.1.1 and Corollary 2.1.4.) $A(\mathbb{R})$ satisfies ϕ but $A(\mathbb{Q})$ does not since there are

$e < g_0, h_0 \in B(\mathbb{Q})$ such that g_0 and h_0 each have one bump, $\text{supp}(g_0) = (0,1)$ and $\text{supp}(h_0) = (0,\sqrt{2})$. (If $f^{-1}g_0f = h_0$, then $1f = \sqrt{2}$; so no such $f \in A(\mathbb{Q})$ exists.)

If $X \subseteq G$ and G is a group, $\underline{\langle X \rangle}$ will denote the subgroup of G generated by X. If $X = \{g\}$, we write $\langle g \rangle$ for $\langle \{g\} \rangle$, conforming to our previous notation.

Let $g \in G$ where (G,Ω) is an ordered permutation group. The *orbitals of g* are the convexifications of the orbits of $\langle g \rangle$. The non-singleton orbitals of g are just the orbitals of the bumps of g and are the supporting intervals of g (see page 26). The *parity of an orbital* is that of the bump of g that it supports. The set of orbitals of g is totally ordered by left-right; i.e., if Δ_1 and Δ_2 are orbitals of g, either $\Delta_1 < \Delta_2$ (for all $\delta_i \in \Delta_i$ $(i = 1,2)$, $\delta_1 < \delta_2$) or $\Delta_2 < \Delta_1$ or $\Delta_1 = \Delta_2$.

THEOREM 2.2.4. *Let $g,h \in A(\mathbb{R})$. Then g and h are conjugate in $A(\mathbb{R})$ if and only if there exists a one-to-one left-right preserving correspondence between the orbitals of g and those of h such that corresponding orbitals have the same parity.*

Proof: If $h = f^{-1}gf$, then the map $I \mapsto If$ is such a map by the fundamental triviality, where I ranges over the orbitals of g.

Conversely, if ϕ is such a map between the orbitals of g and those of h, let I be an orbital of g. Let g_I be the bump of g whose support is I. $I\phi$ is an orbital of h; let $h_{I\phi}$ be the bump of h whose support is $I\phi$. Now g_I and $h_{I\phi}$ have the same parity and boundity so are conjugate by f_I, say (by Lemma 2.2.1) if I is either a positive or negative orbital of g. If $|I| = 1$, let Δ be maximal closed interval of $\mathbb{R} \cup \{\pm\infty\}$ containing only fixed points of g. For each $\delta \in \Delta$, there exists $\delta' \in \mathbb{R}$ such that $\{\delta\}\phi = \{\delta'\}$, where $\delta'h = \delta'$. Then ϕ induces a one-to-one order-preserving map f_Δ of Δ onto $\Delta\phi$ defined by: $\{\delta f_\Delta\} = \{\delta\}\phi$. By the patching lemma, there exists $f \in A(\mathbb{R})$ such that $f|I = f_I|I$ for all orbitals I of g with

$|I| > 1$, and $\delta f = \delta f_\Delta$ if $\delta \in \Delta$ and Δ is a maximal closed interval of $\mathbb{R} \cup \{\pm\infty\}$ containing only fixed points of g. Hence $f^{-1}gf|I = f_I^{-1}g_I f_I|I\phi = h_{I\phi}|I\phi = h|I\phi$ for all orbitals I of g (even if $|I| = 1$). Thus $f^{-1}gf = h$.

We observe that the above proof can easily be modified to prove Theorem 2.2.4 with $L(\mathbb{R})$, $R(\mathbb{R})$ or $B(\mathbb{R})$ in place of $A(\mathbb{R})$.

As examples of applications of Theorem 2.2.4, we see that the following are conjugate

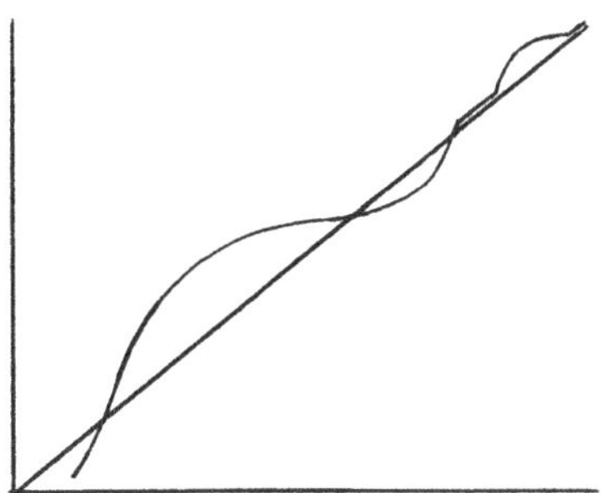

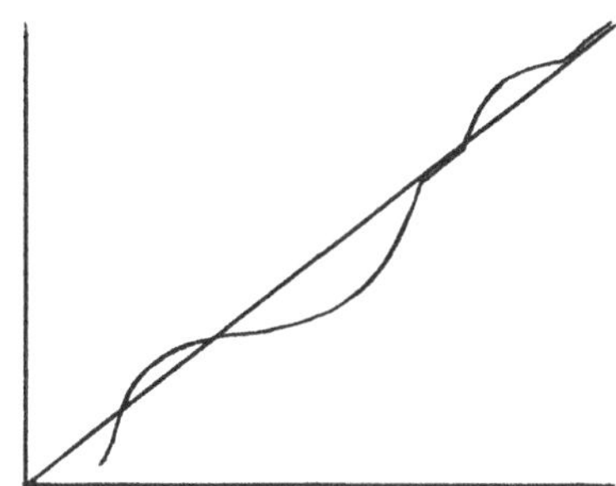

but none of the following are:

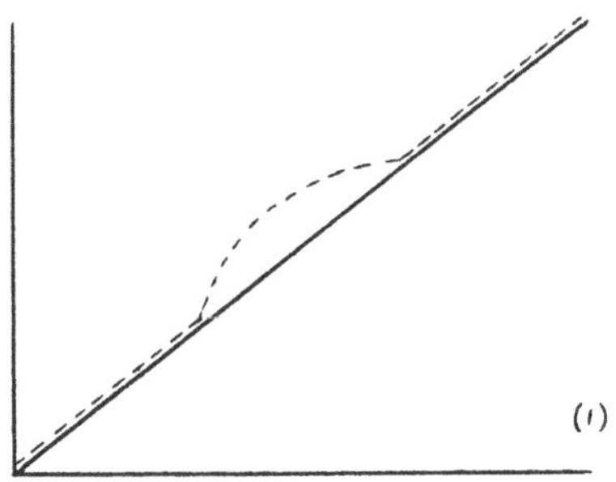

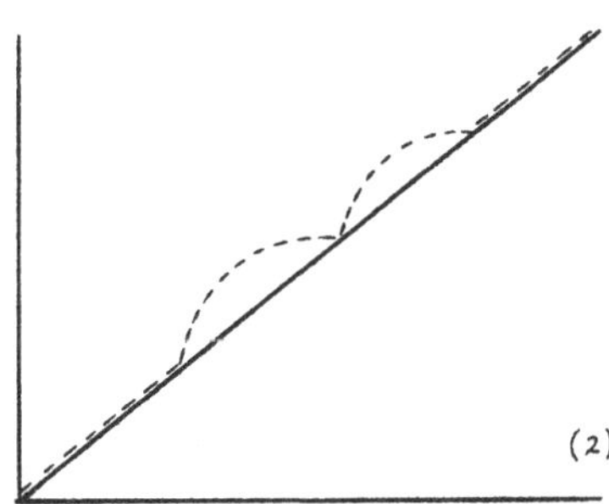

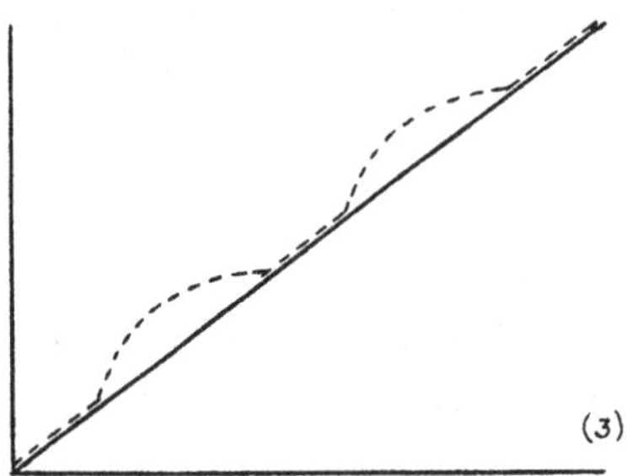

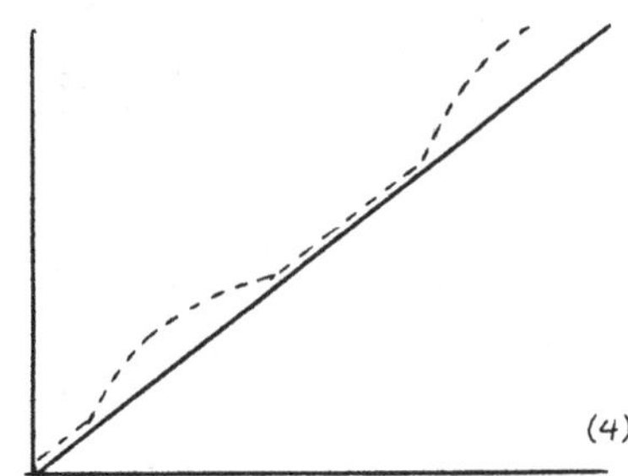

What happens if we replace $\mathbb{R}$ by an arbitrary doubly homogeneous chain Ω? As we saw in the proof of Corollary 2.2.3, Lemma 2.2.1 fails if we replace $\mathbb{R}$ by $\mathbb{Q}$. However, when we consider Theorem 2.2.4, there is no one-to-one map of the orbitals of g_0 to those of h_0 which preserves left-right and maps orbitals of g_0 to those of h_0 of the same parity and boundity, where $g_0, h_0 \in A(\mathbb{Q})$ are given in the proof of Corollary 2.2.3. (There is a least orbital of g_0 to the right of its positive orbital--viz.: $\{1\}$--but no least orbital of h_0 to the right of its positive orbital.) Indeed, since we are taking orbitals of points of Ω (i.e., $\text{Conv}_\Omega\{\alpha g^n : n \in \mathbb{Z}\}$ $(\underline{\alpha \in \Omega})$), a moment's reflection shows that the proof of Theorem 2.2.4 generalises to give:

<u>THEOREM</u> <u>2.2.5</u>. *Let* Ω *be doubly homogeneous and* $g,h \in A(\Omega)$. *Then* g *and* h *are conjugate in* $A(\Omega)$ *if and only if there exists a one-to-one left-right preserving map from the orbitals of* g *onto those of* h *such that corresponding orbitals have the same parity. Moreover, the conjugator in this case can be taken to belong to* $B(\Omega)$ $(L(\Omega), R(\Omega))$ *if* g *and* h *do.*

The following corollary implies Theorem 2F.

COROLLARY 2.2.6. *If* Ω *is doubly homogeneous and* $g \in A(\Omega)$, *there exists* $f \in A(\Omega)$ *such that* $g^2 = f^{-1}gf$; *i.e.,* $g = [g,f]$.

COROLLARY 2.2.7. *If* Ω *is doubly homogeneous and* $g \in B(\Omega)$, *there exists* $f \in B(\Omega)$ *such that* $g = [f,g]$. *Similarly with* $L(\Omega)$ *or* $R(\Omega)$ *in place of* $B(\Omega)$.

Note that the fact that $g \in A(\Omega)$ is conjugate to g^m ($m \in \mathbb{Z}^+$) in $A(\Omega)$ if Ω is doubly homogeneous gives an alternative proof of Theorem 2E.

THEOREM 2.2.8. *The group theory and lattice theory of lattice-ordered groups are both undecidable. That is, there is no algorithm which determines whether or not an arbitrary group (lattice) statement is a consequence of the* ℓ*-group axioms.*

Proof: By standard arguments in logic (see Tarski-Mostowski-Robinson [53]), it is enough to find an ℓ-group G and an element $g_0 \in G$ such that we can code into G (using g_0) the integers together with addition and multiplication. The key results that make this possible are 2.1.10 and 2.2.6. Let $G = A(\mathbb{R})$ and $e < g_0$ have one bump. Now $f \in G$ will be declared an integer if and only if $f \in C(C(g_0))$ (i.e., $(\forall h)(hg_0 = g_0h \rightarrow hf = fh)$. By Lemma 2.1.10, any integer is g_0^n for some $n \in \mathbb{Z}$. Define $g_0^n \oplus g_0^m = g_0^p$ if $g_0^n \cdot g_0^m = g_0^p$. This is standard addition since $g_0^n \cdot g_0^m = g_0^{n+m}$. By Corollary 2.2.6, for each $n \in \mathbb{Z}^+$, there is $x \in A(\Omega)$ such that $x^{-1}g_0x = g_0^n$. For this x, $x^{-1}g_0^mx = (x^{-1}g_0x)^m = g_0^{nm}$. Hence define $g_0^n \otimes g_0^m = g_0^p$ if $(\exists x)(x^{-1}g_0x = g_0^n \;\&\; x^{-1}g_0^mx = g_0^p)$ if $n > 0$, and extend in the natural way for $n \leq 0$. Thus we obtain the standard multiplication. The proof of the theorem is completed by applying Theorems 2A and $2A^\dagger$ respectively.

The group part of Theorem 2.2.8 has been proved independently by Mal'cev [71; Chapter 15] (torsion-free nilpotent groups are orderable, see Fuchs [63, p. 48]--the translator of Mal'cev uses "metabelian" for "nilpotent class 2"), Jambu-Giraudet [81b], Buszkowski [79], and Glass and Pierce [80c]. The proof just given is the simplest. The lattice part of the theorem is due to Jambu-Giraudet [81b]. For another proof of the lattice part, see Gurevich [67].

In Chapter 13 we will give an example of a finitely presented ℓ-group with insoluble group word problem.

We give one more illustration of the technique to obtain a stronger result than Lemma 2.2.1.

LEMMA 2.2.9. *Let* $e < g_i \in B(\mathbb{R})$ *each have one bump and* $\alpha_i \in supp(g_i) = (\sigma_i, \tau_i)$ $(i = 0,1,2,3)$. *Suppose that* $\alpha_{i+1} < \sigma_i < \tau_i < \alpha_{i+1} g_{i+1}$ $(i = 0,1,2)$. *Then there is* $h \in B(\mathbb{R})$ *such that* $h^{-1} g_0 h = g_0$, $h^{-1} g_1 h = g_2$ *and* $h^{-1} g_3 h = g_3$.

Proof: Let f_1, f_2 be ordermorphisms

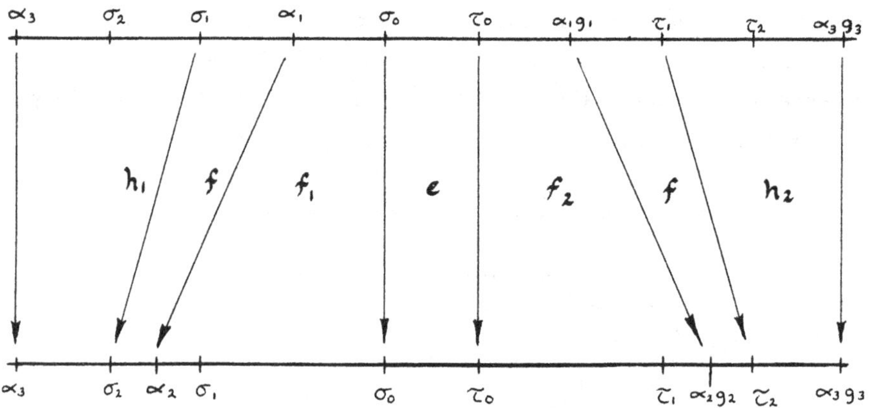

such that $f_1 \colon [\alpha_1, \sigma_0] \cong [\alpha_2, \sigma_0]$ and $f_2 \colon [\tau_0, \alpha_1 g_1] \cong [\tau_0, \alpha_2 g_2]$. Let $f_0 = f_1 \cup f_2 \cup e|[\sigma_0, \tau_0]$; so $f_0 \colon [\alpha_1, \alpha_1 g_1] \cong [\alpha_2, \alpha_2 g_2]$. As in the proof of Lemma 2.2.1, we can extend f_0 to

$f: [\sigma_1,\tau_1] \cong [\sigma_2,\tau_2]$ so that $f^{-1}g_1f = g_2$. Let h_1,h_2 be ordermorphisms with $h_1: [\alpha_3,\sigma_1] \cong [\alpha_3,\sigma_2]$ and $h_2: [\tau_1,\alpha_3g_3] \cong [\tau_2,\alpha_3g_3]$. Let $h_0 = h_1 \cup f \cup h_2$; so $h_0: [\alpha_3,\alpha_3g_3] \cong [\alpha_3,\alpha_3g_3]$. As in the proof of Lemma 2.2.1, we can extend h_0 to $h \in B(\mathbb{R})$ so that $h^{-1}g_3h = g_3$. Since $h|[\sigma_1,\tau_1] = f$, $h^{-1}g_1h = f^{-1}g_1f = g_2$. Also since $h|[\sigma_0,\tau_0] = f|[\sigma_0,\tau_0] = f_0|[\sigma_0,\tau_0] = e|[\sigma_0,\tau_0]$, $h^{-1}g_0h = e^{-1}g_0e = g_0$.

Once the technique is fully grasped, other applications can be handled as needed. See, for example, Chapter 13 which depends only on this section.

2.3. THE NORMAL SUBGROUPS OF $A(\Omega)$.

In this section we wish to consider the normal subgroups of $A(\Omega)$ for Ω a doubly homogeneous chain.

A normal convex ℓ-subgroup of an ℓ-group is called an *ℓ-ideal*. We simplify our problem by first proving the following generalisation of Lloyd [64]:

THEOREM 2.3.1. *If Ω is doubly homogeneous, every normal subgroup of $A(\Omega)$ is an ℓ-ideal.*

From this we will be able to easily prove Theorem 2G: If Ω is doubly homogeneous, $B(\Omega)$ is simple.

We further use Theorem 2.3.1 to give a quick proof of the following generalisation of Theorem 2H:

THEOREM 2.3.2. *If Ω is doubly homogeneous and has countable coterminality, the only non-trivial normal subgroups of $A(\Omega)$ are $L(\Omega)$, $R(\Omega)$, and $B(\Omega)$. Moreover, $B(\Omega)$ is the only proper normal subgroup of either $L(\Omega)$ or $R(\Omega)$.*

In contrast, we give an example of a doubly homogeneous chain Ω such that $A(\Omega)$ has a proper normal subgroup other than $B(\Omega)$, $L(\Omega)$ or $R(\Omega)$. This will illustrate the need to impose "countable coterminality" in the hypotheses of Theorem 2.3.2.

We begin by proving Theorem 2.3.1.

Let Ω be doubly homogeneous and N be a normal subgroup of $A(\Omega)$. We first show that N is convex. So assume $e \leq g \leq f \in N$. Now fg and f have exactly the same intervals of support. Since both fg and f are positive ($\geq e$), there exists $h \in A(\Omega)$ such that $h^{-1}fh = fg$ (by Theorem 2.2.5). Since $f \in N$, $g = f^{-1}(h^{-1}fh) \in N$. Hence N is convex.

Next suppose that $f \in N$ and consider $f \vee e$. By Theorem 2E, there exists $h \in A(\Omega)$ such that $h^2 = f \vee e$. Define $g \in A(\Omega)$

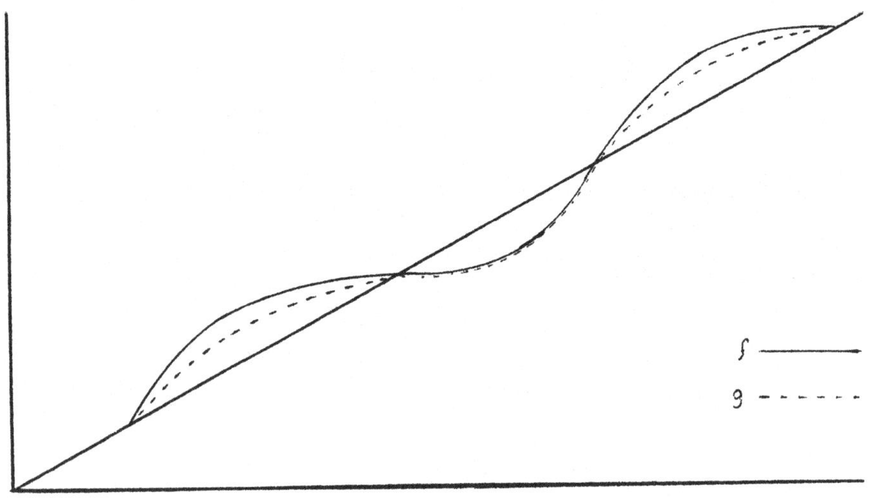

by: $\alpha g = \begin{cases} \alpha f & \text{if } \alpha f \leq \alpha \\ \alpha h & \text{if } \alpha f > \alpha \end{cases}$. Then g and f have exactly the same intervals of support with corresponding intervals having the same parity. Hence g and f are conjugate, so $g \in N$. Thus $fg^{-1} \in N$. But fg^{-1} and $f \vee e$ have the same (positive) intervals of support, and so are conjugate. Hence $f \vee e \in N$ so N is an ℓ-subgroup of $A(\Omega)$. This establishes the theorem.

Note that the proof shows (using the last part of Theorem 2.2.5):

COROLLARY 2.3.3. *Let* Ω *be doubly homogeneous. Every normal subgroup of* $B(\Omega)$ $(L(\Omega), R(\Omega))$ *is an* ℓ*-ideal of* $B(\Omega)$ $(L(\Omega), R(\Omega))$.

Theorem 2G follows immediately from Corollary 2.3.3 and the following lemma:

LEMMA 2.3.4. *Let* Ω *be doubly homogeneous,* $e \leq f \in B(\Omega)$ *and* $e < g \in A(\Omega)$. *Then there exists* $h \in B(\Omega)$ *such that* $f \leq h^{-1}gh$.

Proof: Let $\mathrm{supp}(f) \subseteq (\alpha,\beta)$ and $\gamma \in \mathrm{supp}(g)$. There exists $h \in B(\Omega)$ such that $\gamma h = \alpha$ and $\gamma gh = \beta$. Thus $\alpha h^{-1}gh = \beta$ so $h^{-1}gh \geq f$ by Lemma 1.9.4.

In order to prove Theorem 2.3.2 we need only prove the following lemma similar to Lemma 2.3.4.

LEMMA 2.3.5. *Let* Ω *be doubly homogeneous of countable coterminality. If either*

(i) $e \leq f \in R(\Omega)$ *and* $e < g \notin L(\Omega)$,

(ii) $e \leq f \in L(\Omega)$ *and* $e < g \notin R(\Omega)$,

or (iii) $e \leq f$ *and* $e < g$ *is unbounded,*

there exist $h_1, h_2 \in A(\Omega)$ *such that* $f \leq h_1^{-1}gh_1 \vee h_2^{-1}gh_2$, *and* $h_1, h_2 \in L(\Omega)$ $(R(\Omega))$ *if* g *does.*

Proof: (i) Let $\sigma \leq \inf(\mathrm{supp}(f))$.

CASE 1. There exists $\tau \in \Omega$ such that $\bar{\alpha}g > \bar{\alpha}$ for all $\bar{\alpha} \in \bar{\Omega}$ with $\bar{\alpha} > \tau$.

There exists $h \in B(\Omega)$ such that $\tau h = \sigma$. Now $g' = h^{-1}gh$ fixes no element of $\bar{\Omega}$ above σ. Hence $g' \vee f$ and g' have the same intervals of support. Thus there is $h_1 \in A(\Omega)$ of the same boundity as g' (and so g) such that $h_1^{-1}g'h_1 = g \vee f$; i.e., $f \leq h_1^{-1}h^{-1}ghh_1$.

<u>Case 2</u>. g fixes arbitrarily large elements of $\bar{\Omega}$.

We can find $\alpha_1 < \beta_1 < \alpha_2 < \beta_2 < \dots$ cofinal in Ω such that $[\alpha_n, \beta_n] \subseteq \text{supp}(g)$ but there are $\bar{\alpha}_n \in \bar{\Omega}$ such that $\beta_n < \bar{\alpha}_n < \alpha_{n+1}$ and $\bar{\alpha}_n g = \bar{\alpha}_n (n \in \mathbb{Z}^+)$. By double transitivity and the patching lemma, there exists $h \in R(\Omega)$ such that $\alpha_n h = \beta_n$ and $\beta_n h = \alpha_{n+1}$ $(n \in \mathbb{Z}^+)$. Now $g \vee h^{-1}gh$ fixes no point of $\bar{\Omega}$ to the right of α_1. By Case 1, there exists $h_1 \in A(\Omega)$ of the same boundity as g such that $f \leq h_1^{-1}(g \vee h^{-1}gh)h_1$.

(ii) The dual of case (i).

(iii) If $f \in L(\Omega) \cup R(\Omega)$, use (i) or (ii). So assume f is unbounded. Let $\beta < \sigma < \tau$. Define $e < f_1, f_2 \in A(\Omega)$ such that: $\alpha f_1 = \begin{cases} \alpha f & \text{if } \alpha \geq \sigma \\ \alpha & \text{if } \alpha \leq \beta \\ \leq \alpha f & \text{if } \alpha \in (\beta, \sigma) \end{cases}$ and $\alpha f_2 = \begin{cases} \alpha f & \text{if } \alpha \leq \sigma \\ \alpha & \text{if } \alpha \geq \tau \\ \leq \alpha f & \text{if } \alpha \in (\sigma, \tau) \end{cases}$.

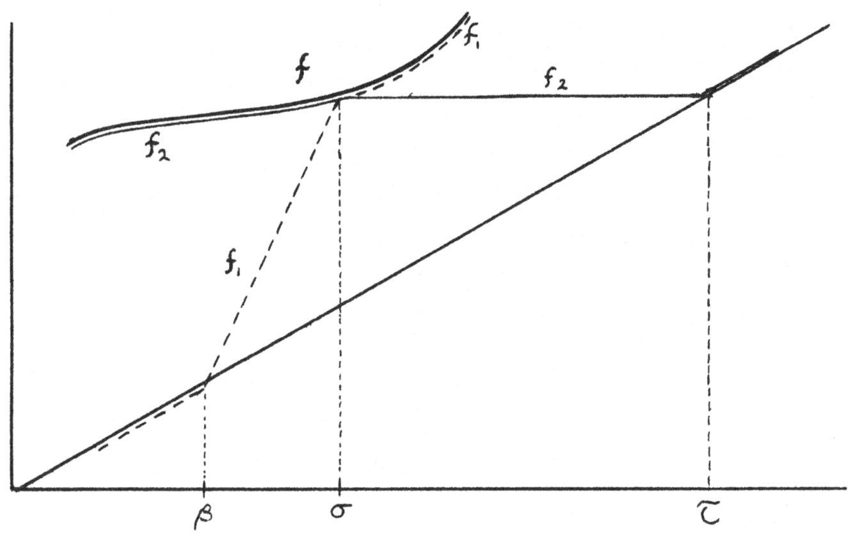

Then $f = f_1 \vee f_2$ and by (i) and (ii) $f_i \leq g_i \vee g_i'$ $(i = 1,2)$ where g_i, g_i' are conjugates of g. We may clearly arrange for $g_1 = g_2$ and $g_1' = g_2'$. Hence $f \leq g_1 \vee g_2$.

Note that as a consequence of the proof of 2.3.1 and 2.3.5 we obtain $f = \prod_{i=1}^{16} g_i$ whenever Ω is doubly homogeneous of countable coterminality and f belongs to the normal subgroup of $A(\Omega)$ generated by g, where each g_i is a conjugate of g or g^{-1}. Indeed, M. Droste [81] has recently shown that the conclusion holds even if the hypothesis of countable coterminality is dropped. He has also proved that if Ω is doubly homogeneous, then any normal subgroup of $A(\Omega)$ is divisible and any subnormal subgroup of $A(\Omega)$ is actually a normal subgroup of $A(\Omega)$.

We complete this section with an example of a doubly homogeneous chain Ω such that $A(\Omega)$ has a proper normal subgroup other than $B(\Omega)$, $L(\Omega)$ or $R(\Omega)$. This shows that the "countable coterminality" hypothesis cannot be removed from Theorem 2.3.2. The example is due to Ball [74].

For the rest of this section only, let I_0 be the half-open interval $[0,1)$ of the real line. Let $\underline{\mathbb{L}'} = I_0 \overleftarrow{\times} \omega_1$; i.e., $(r,\mu) \in \mathbb{L}'$ if and only if $r \in I_0$ and μ is a countable ordinal. $\mathbb{L}'$ has a least element $(0,0)$. Let $\underline{\mathbb{L}} = \mathbb{L}' \backslash \{(0,0)\}$, a *long line*. Any initial segment of $\mathbb{L}$ is ordermorphic to $\mathbb{R}$. Consequently, $\mathbb{L}$ is Dedekind complete, doubly homogeneous, has coinitiality $\aleph_0$ and cofinality $\aleph_1$, and every point of $\mathbb{L}$ has countable character c_{00}. Let $e < g_0 \in A(\mathbb{L})$ be such that $(r,\mu)g_0 = (r,\mu)$ if and only if $(\mu > 0$ and $0 \leq r \leq \frac{1}{2})$. Note that $g_0 \notin R(\mathbb{L}) \cup L(\mathbb{L})$. We will show that the normal subgroup generated by g_0 is not $A(\mathbb{L})$, whence it is a proper normal subgroup of $A(\mathbb{L})$ other than $B(\mathbb{L})$, $L(\mathbb{L})$ or $R(\mathbb{L})$. To prove this we first need a lemma. For $g \in A(\Omega)$, let $\underline{Fx(g)} = \{\alpha \in \Omega: \alpha g = \alpha\}$.

LEMMA 2.3.6.

(1) If $g \in A(\mathbb{L})$, $Fx(g)$ is closed and cofinal in $\mathbb{L}$.

(2) The intersection of a countable collection of closed cofinal subsets of $\mathbb{L}$ is closed and cofinal in $\mathbb{L}$.

Proof: Since any supporting interval is open and has cofinality $\aleph_0$, (1) follows since $\mathbb{L}$ has cofinality $\aleph_1$.

(2) Let $\{\Delta_n : n \in \omega\}$ be a collection of closed cofinal subsets of $\mathbb{L}$. Then $\bigcap_{n \in \omega} \Delta_n$ is closed. Let $n_0, n_1, n_2, \ldots$ be any sequence of natural numbers in which each natural number occurs infinitely often. Let $\beta \in \mathbb{L}$. Form a sequence $\beta < \delta_0 < \delta_1 < \ldots$ of elements of $\mathbb{L}$ with $\delta_j \in \Delta_{n_j}$ $(j \in \omega)$. Since $\mathbb{L}$ has cofinality $\aleph_1$ and each Δ_n is cofinal in $\mathbb{L}$, such a sequence exists. Moreover, so does $\sup\{\delta_j : j \in \omega\} = \delta \in \mathbb{L}$. Since each m occurs as infinitely many n_j, $\delta = \sup\{\delta_j : n_j = m\}$. But $\{\delta_j : n_j = m\} \subseteq \Delta_m$ and as Δ_m is closed, $\delta \in \bigcap_{m \in \omega} \Delta_m$. Thus $\bigcap_{n \in \omega} \Delta_n$ is cofinal in $\mathbb{L}$.

We can now prove:

<u>THEOREM 2.3.7</u>. *$A(\mathbb{L})$ contains a proper normal subgroup other than $B(\mathbb{L})$, $L(\mathbb{L})$ or $R(\mathbb{L})$.*

<u>*Proof:*</u> We show that the normal subgroup of $A(\mathbb{L})$ generated by g_0 is proper. Let $f_1, \ldots, f_n \in A(\mathbb{L})$ and consider $g = f_1^{-1} g_0^{m_1} f_1 \ldots f_n^{-1} g_0^{m_n} f_n$, a typical element of N the normal subgroup of $A(\mathbb{L})$ generated by g_0. $Fx(g) \supseteq \bigcap_{i=1}^{n} Fx(g_0) f_i$. Consider $\Delta = \{(0,\mu) : 0 \neq \mu \in \omega_1\} \subseteq Fx(g_0)$. Let $f_0 = e$ and consider $\bigcap_{i=0}^{n} \Delta f_i$, the intersection of a finite number of closed cofinal subsets of $\mathbb{L}$. By Lemma 2.3.6(2), this intersection is non-empty. Hence there exists $\lambda \in \omega_1$ such that $(0,\lambda) \in \bigcap_{i=0}^{n} \Delta f_i$. Let $(0,\lambda_i) \in \Delta$ be such that $(0,\lambda_i) f_i = (0,\lambda)$ and let $r \in I_0$ be defined by: $(r,\lambda) = \min\{(\frac{1}{2}, \lambda_i) f_i : 0 \leq i \leq n\}$. Since $r > 0$, $Fx(g) \supseteq [(0,\lambda),(r,\lambda)]$ a non-degenerate interval in $\mathbb{L}$. But there exists $e < h_0 \in A(\mathbb{L})$ such that $Fx(h_0) = \Delta$. Hence $h_0 \notin N$, so $N \neq A(\mathbb{L})$ as desired.

2.4. *THE AUTOMORPHISMS OF A(Ω).*

For any group G, there are obvious automorphisms obtained by conjugation by an element of G; i.e., if $f \in G$, the map $g \mapsto f^{-1}gf$ is an automorphism of G. Such automorphisms are called *inner*. *I(G)*, the set of all inner automorphisms of G, forms a normal subgroup of *Aut(G)*, the group of all automorphisms of the group G. The *group of outer automorphisms* of G is the group Aut(G)/I(G).

For any set Ω, if $|\Omega| \neq 6$, every automorphism of the symmetric group on Ω is inner (even if Ω is infinite). If Ω is a doubly homogeneous chain, then A(Ω) may have an automorphism that is not inner; e.g., let ϕ_- be defined by: $\alpha(g\phi_-) = -((-\alpha)g)$ $(\alpha \in \mathbb{R};\ g \in A(\mathbb{R}))$. Now $\phi_- \in \mathrm{Aut}(A(\mathbb{R}))$ but ϕ_- is an involution (its square is the identity automorphism). The crux here is that $\mathbb{R}^*$ ($\mathbb{R}$ with the reverse order) is ordermorphic to $\mathbb{R}$.

Let Ω be a doubly homogeneous chain. Suppose that Ω is ordermorphic to Ω^*, say by i. Then ϕ_i is an automorphism of the group A(Ω) where: $\alpha(g\phi_i) = ((\alpha i)g)\overline{i}$ $(\alpha \in \Omega,\ g \in A(\Omega))$. Of course, ϕ_i does not preserve order (If $e < g$, $e > g\phi_i$). We will prove the following generalisation of Theorem 2I:

THEOREM 2.4.1. *If Ω is doubly homogeneous, then every ℓ-automorphism of A(Ω) is conjugation by an element of $A(\bar{\Omega})$, and every automorphism of A(Ω) is an ℓ-automorphism of A(Ω) followed, possibly, by ϕ_i (if there is $i: \Omega \cong \Omega^*$).*

Schreier conjectured that if G is a finite simple group, the outer automorphism group of G is soluble. We will give an example due to McCleary [78a] to show:

THEOREM 2.4.2. *There is an infinite simple group whose outer automorphism group is insoluble.*

The proofs we give rely heavily on the results of Section 2.1. For an alternative method, see Chapter 9.

We begin by proving Theorem 2.4.1. By Corollary 2.1.4, if $e < f,g \in B(\Omega)$ and ψ is an ℓ-automorphism of $A(\Omega)$, then $e < f\psi, g\psi \in B(\Omega)$ and $f \approx g$ implies $f\psi \approx g\psi$. By Theorem 2.1.8, ψ induces an ordermorphism $\hat{\psi}$ of $\bar{\Omega}$ onto $\bar{\Omega}$; i.e., $\hat{\psi} \in A(\bar{\Omega})$. By Theorem 2D, $g\psi = \hat{\psi}^{-1}g\hat{\psi}$ for all $g \in A(\Omega)$. So ψ is induced by an element of $A(\bar{\Omega})$; i.e., ψ is conjugation by an element of $A(\bar{\Omega})$.

If ψ were just an automorphism of $A(\Omega)$, let $e < g_0 \in B(\Omega)$. If $e < g_0\psi$, then by Theorems 2A and 2.1.8, ψ is an ℓ-automorphism; and if $e > g_0\psi$, $\psi\phi_i$ is an ℓ-automorphism. Since $\phi_i^2 = e$, Theorem 2.4.1 now follows from the previous paragraph.

We note as a consequence of the proof:

<u>THEOREM</u> <u>2.4.3</u>. *If Ω is doubly homogeneous, then every ℓ-automorphism of $B(\Omega)$ is conjugation by an element of $A(\bar{\Omega})$, and every automorphism of $B(\Omega)$ is an ℓ-automorphism of $B(\Omega)$ followed possibly by ϕ_i (if $i: \Omega \cong \Omega^*$).*

Caution: We do not claim in Theorems 2.4.1 and 2.4.3 that conjugation by an element of $A(\bar{\Omega})$ is indeed an automorphism of $A(\Omega)$. Often it is not. For example (Lloyd [64]):

<u>COROLLARY</u> <u>2.4.4</u>. (cf. Theorems 2C and 2I). *Every ℓ-automorphism of $A(\mathbb{Q})$ is inner.*

<u>Proof:</u> By Theorem 2.4.1, if ψ is an ℓ-automorphism of $A(\mathbb{Q})$, it is conjugation by an element $\hat{\psi}$ of $A(\mathbb{R})$. Hence $\mathbb{Q}\hat{\psi}$ is an orbit of $A(\mathbb{Q})$ in $\mathbb{R}$, and it clearly must be ordermorphic to $\mathbb{Q}$. But the only two orbits of $A(\mathbb{Q})$ in $\mathbb{R}$ are $\mathbb{Q}$ and $\mathbb{R}\backslash\mathbb{Q}$. Since $\mathbb{R}\backslash\mathbb{Q} \not\cong \mathbb{Q}$, $\mathbb{Q}\hat{\psi} = \mathbb{Q}$ so $\hat{\psi}|\mathbb{Q} \in A(\mathbb{Q})$; i.e., ψ is inner.

For an example of a doubly homogeneous chain Ω such that $A(\Omega)$ has an outer ℓ-automorphism, see Chapter 9. (The interested reader can refer to it now since it requires no results from any intermediate chapters.)

Finally we prove Theorem 2.4.2. By Theorem 2.4.3, $\mathrm{Aut}(B(\mathbb{R})) \cong A(\mathbb{R}) \rtimes \mathbb{Z}_2$ so the group of outer automorphisms of $B(\mathbb{R})$ is $A(\mathbb{R})/B(\mathbb{R}) \rtimes \mathbb{Z}_2$. But by Theorem 2H, the only proper normal subgroups of $A(\mathbb{R})/B(\mathbb{R})$ are $L(\mathbb{R})/B(\mathbb{R})$ and $R(\mathbb{R})/B(\mathbb{R})$. Since these are simple and non-Abelian, the group of outer automorphisms of $B(\mathbb{R})$ is insoluble. Since $B(\mathbb{R})$ is simple (Theorem 2G), Theorem 2.4.2 follows.

For more on automorphisms of $A(\Omega)$--even when Ω is not doubly homogeneous--see Chapter 9. For other examples of infinite simple groups having insoluble outer automorphism groups, see McCleary [78a].

2.5. *EMBEDDING IN DOUBLY TRANSITIVE* $A(\Omega)$.

This section is devoted to the proof of Theorem 2J. The proof we give here is due to Weinberg [67]. For an alternative proof, see Chapter 10. (The proof given there does not depend on the intermediate chapters and can be read straight after this section.)

In order to prove Theorem 2J, it is enough to prove the following two lemmas:

LEMMA 2.5.1. *Any chain* T *can be order embedded in a totally ordered field* Ω *in such a way that each member of* $A(T)$ *can be extended to a member of* $A(\Omega)$ *with* $|\Omega| = \max\{\aleph_0, |T|\}$.

LEMMA 2.5.2. *If* T *is a subchain of* Ω *and every element of* $A(T)$ *can be extended to an element of* $A(\Omega)$*, then* $A(T)$ *can be* ℓ*-embedded in* $A(\Omega)$ *(the restriction mapping* $B(T)$ *into* $B(\Omega)$*).*

Before proving these lemmas, we need some facts from algebra.

Let G be a group and R be a ring. The *group ring* $\underline{R[G]} = \{\sum_{i=1}^{m} r_i x^{g_i} : r_i \in R, g_i \in G\}$ where $+$ is the coordinate addition and $\cdot$ is polynomial multiplication. It is straightforward to show that $R[G]$ is an integral domain if R is and G is an Abelian o-group. If R is totally ordered, we get a natural total order on $R[G]$ in this case--lexicographic from the right. (If $\alpha = \sum_{i=1}^{m} r_i x^{g_i}$ with each $0 \neq r_i \in R$ and $g_1 < \ldots < g_m$, then $\alpha > 0$ if $r_m > 0$.)

We now prove Lemma 2.5.1. Let T be a chain and G be the free Abelian group on $\{\tau : \tau \in T\}$. Order G by: $\sum_{i=1}^{m} n_i \tau_i > 0$ if $n_j > 0$ where $\tau_j = \max\{\tau_i : 1 \leq i \leq m \ \& \ n_i \neq 0\}$. This clearly extends the order on T and G is an Abelian o-group. Form the group ring $\mathbb{Q}[G]$ and let Ω be its field of fractions. We totally order Ω by: $\alpha\beta^{-1} > 0$ if and only if $\alpha\beta > 0$ in $\mathbb{Q}[G]$. This extends the order on $\mathbb{Q}[G]$ and makes Ω a totally ordered field. Note $|\Omega| = \max\{|T|, \aleph_0\}$.

Let $f \in A(T)$. Define $\phi: A(T) \to A(G)$ by: $(\sum_{i=1}^{m} n_i \tau_i)(f\phi) = \sum_{i=1}^{m} n_i(\tau_i f)$. Then $f\phi \in A(G)$ and extends f. Define $\psi: A(T)\phi \to A(\mathbb{Q}[G])$ by $(\sum_{j=1}^{m} q_j x^{g_j})((f\phi)\psi) = \sum_{j=1}^{m} q_j x^{g_j(f\phi)}$. Then $f\phi\psi \in A(\mathbb{Q}[G])$ and extends $f\phi$ (and hence f). We now extend $f\phi\psi$ in the obvious way to a member of $A(\Omega)$; viz: $(\alpha\beta^{-1})(f\phi\psi) = \alpha(f\phi\psi)(\beta(f\phi\psi))^{-1}$ $(\alpha, \beta \in \mathbb{Q}[G])$. This extension is easily seen to be well-defined and $f\phi\psi \in A(\Omega)$ extends f proving the lemma.

We now prove Lemma 2.5.2. Let $\ell(T)$ be the set of all left segments of T in Ω; i.e., $\Delta \in \ell(T)$ if there exists $\alpha \in \Omega$ such that $\Delta = \{\tau \in T : \tau < \alpha\}$. If $\Delta \in \ell(T)$, let $C(\Delta)$ be the cut in Ω determined by Δ; i.e., $C(\Delta) = \{\beta \in \Omega : \Delta < \beta \leq T\setminus\Delta\}$. For a fixed $\alpha_0 \in \Omega$, let $\Delta_0 = \{\tau \in T : \tau < \alpha_0\}$. Thus $\alpha_0 \in C(\Delta_0)$. If $\Delta_1 \in \ell(T)$ and $\Delta_1 \neq \Delta_0$, then $\Delta_1 \subsetneq \Delta_0$ or $\Delta_0 \subsetneq \Delta_1$. In the first instance, there exists $\tau \in \Delta_0\setminus\Delta_1$. Hence

$C(\Delta_1) \leq \tau < C(\Delta_0)$. In the second case, $C(\Delta_0) < C(\Delta_1)$ similarly. So $\{C(\Delta): \Delta \in \ell(T)\}$ is linearly ordered in the natural way.

Define an equivalence relation ξ on $\ell(T)$ by: $\Delta_1 \xi \Delta_2$ if $C(\Delta_1) \cong C(\Delta_2)$ (as totally ordered sets). Choose an element from each equivalence class. If Δ were chosen and $\Delta_i \xi \Delta$ $(i = 1,2)$, define $\pi_{\Delta_1,\Delta_2} = \pi_{\Delta_1,\Delta}\pi_{\Delta,\Delta_2}$ where $\pi_{\Delta,\Delta}$ is the identity on $C(\Delta)$, π_{Δ,Δ_1} is an ordermorphism of $C(\Delta)$ onto $C(\Delta_1)$ and $\pi_{\Delta_1,\Delta} = \pi^{-1}_{\Delta,\Delta_1}$. So π_{Δ_1,Δ_2} is an ordermorphism of $C(\Delta_1)$ onto $C(\Delta_2)$ and if $\Delta_3 \xi \Delta$, $\pi_{\Delta_1,\Delta_3} = \pi_{\Delta_1,\Delta_2}\pi_{\Delta_2,\Delta_3}$.

Let $\Delta \in \ell(T)$ and $\hat{f} \in A(\Omega)$ be an extension of f. If $\alpha \in C(\Delta)$, then for all $\tau_1 \in \Delta$ and $\tau_2 \in T\setminus\Delta$, $\tau_1 < \alpha \leq \tau_2$. Hence for all $\tau_1 f \in \Delta f$ and $\tau_2 f \in T\setminus\Delta f$, $\tau_1 f = \tau_1\hat{f} < \alpha\hat{f} \leq \tau_2\hat{f} = \tau_2 f$ so $\alpha\hat{f} \in C(\Delta f)$. Similarly, $C(\Delta f)\hat{f}^{-1} \subseteq C(\Delta)$. Thus $C(\Delta f) = C(\Delta)\hat{f}$. Thus $\Delta \xi \Delta f$ for all $\Delta \in \ell(T)$; i.e., the $C(\Delta)$ get mapped onto ordermorphic copies under any extension of an element of $A(T)$ to $A(\Omega)$.

Next define $\psi: A(T) \to A(\Omega)$ by: $\alpha(f\psi) = \alpha\pi_{\Delta,\Delta f}$ where Δ is the unique element of $\ell(T)$ such that $\alpha \in C(\Delta)$ $(\alpha \in \Omega;$ $f \in A(T))$. Now for all $\alpha \in \Omega$ and $f,g \in A(T)$, if $\alpha \in C(\Delta)$, then $\alpha(f\psi) \in \Delta f$; so $\alpha(f\psi)(g\psi) = \alpha\pi_{\Delta,\Delta f}\pi_{\Delta f,\Delta fg} = \alpha\pi_{\Delta,\Delta fg} = \alpha(fg)\psi$. Hence ψ is a homomorphism. If $f < g$, there exists $\tau_0 \in T$ such that $\tau_0 f < \tau_0 g$. But for all $\Delta \in \ell(T)$, $\Delta f \subseteq \Delta g$; so $\alpha(f\psi) = \alpha\pi_{\Delta,\Delta f} \leq \alpha\pi_{\Delta,\Delta g} = \alpha(g\psi)$. Let $\Delta_0 = \{\tau \in T: \tau < \tau_0\}$. Then $\tau_0 f \in \Delta_0 g\setminus\Delta_0 f$, so $\Delta_0 f \subsetneqq \Delta_0 g$. Therefore $C(\Delta_0 f) < C(\Delta_0 g)$ and thus $\tau_0(f\psi) < \tau_0(g\psi)$. Hence $f\psi < g\psi$. Consequently, ψ is a one-to-one order-preserving homomorphism of $A(T)$ into $A(\Omega)$. An easy computation shows that $\Delta(f \vee e) = \Delta f \cup \Delta$ for all $\Delta \in \ell(T)$. It follows that $\pi_{\Delta,\Delta(f\vee e)} = \pi_{\Delta,\Delta f} \vee \pi_{\Delta,\Delta}$. Thus $(f \vee e)\psi = f\psi \vee e$ and ψ is an ℓ-embedding as desired. By definition, $B(T)\psi \subseteq B(\Omega)$. Q.E.D.

PART II

THE STRUCTURE THEORY

CHAPTER 3
CONGRUENCES AND BLOCKS

In Sections 1.4-1.7 we obtained results concerning the set of congruences of an ordered permutation group (G,Ω). We now wish to make a far more extensive study of this set. We will first deal with the transitive case as it is both easier and more intuitive. Recall that in this case each class of a congruence is a block and each block is the class of a unique congruence (Theorem 1.6.1). We will first prove:

<u>THEOREM</u> <u>3A</u>. *Let (G,Ω) be a transitive ordered permutation group and $\alpha \in \Omega$. The blocks of (G,Ω) that contain α form a chain (under inclusion). Hence the congruences of (G,Ω) form a chain under inclusion.*

As an immediate consequence we will have:

<u>COROLLARY</u> <u>3B</u>. *If (G,Ω) is a transitive ordered permutation group, then the set of congruences of (G,Ω) is closed under arbitrary unions and intersections.*

Recall that if (G,Ω) is a transitive ordered permutation group and $\mathfrak{C}$ and $\mathfrak{D}$ are congruences of (G,Ω), then $\mathfrak{D}$ covers $\mathfrak{C}$ if $\mathfrak{C} \subsetneqq \mathfrak{D}$ and there is no congruence $\mathfrak{E}$ of (G,Ω) such that $\mathfrak{C} \subsetneqq \mathfrak{E} \subsetneqq \mathfrak{D}$. Let <u>$k(G,\Omega)$</u> denote the set of covering pairs of convex congruences of (G,Ω). We will simply write <u>$k(\Omega)$</u> for $k(A(\Omega),\Omega)$. We write <u>$(\mathfrak{C}_K,\mathfrak{C}^K) = K \in k(G,\Omega)$</u> where $\mathfrak{C}^K$ covers $\mathfrak{C}_K$. By Theorem 3A, $k(G,\Omega)$ is a chain where $K < K'$ if $\mathfrak{C}^K \subseteq \mathfrak{C}_{K'}$ (i.e., $\mathfrak{C}^K \subsetneqq \mathfrak{C}^{K'}$). Let α and β be distinct points of Ω. Let $\mathfrak{C}_K$ be the union of all congruences $\mathfrak{C}$ of (G,Ω) for which

$\beta \notin \alpha\mathcal{C}$, and $\mathcal{C}^K$ be the intersection of all congruences $\mathcal{C}'$ of (G,Ω) for which $\alpha\mathcal{C}'\beta$. By Corollary 3B, $\mathcal{C}_K$ and $\mathcal{C}^K$ are congruences of (G,Ω) and $\mathcal{C}^K$ clearly covers $\mathcal{C}_K$. Hence $K = (\mathcal{C}_K, \mathcal{C}^K) \in k(G,\Omega)$ and is uniquely determined by (α,β). In this case we write $K = Val(\alpha,\beta)$, the *value of* (α,β). We have therefore established half of the following theorem:

THEOREM 3C. *Let* (G,Ω) *be a transitive ordered permutation group. For each* $\alpha,\beta \in \Omega$ *with* $\alpha \neq \beta$, $Val(\alpha,\beta) \in k(G,\Omega)$. *Moreover, each* $K \in k(G,\Omega)$ *is* $Val(\alpha,\beta)$ *for some* $\alpha,\beta \in \Omega$.

We next show that $k(G,\Omega)$ generates the set of congruences of (G,Ω). Specifically:

THEOREM 3D. *Let* (G,Ω) *be a transitive ordered permutation group and* $\mathcal{C}$ *be a congruence of* (G,Ω). *Then* $\bigcap\{\mathcal{C}_K: K \in k \;\&\; \mathcal{C} \subseteq \mathcal{C}_K\} = \mathcal{C} = \bigcup\{\mathcal{C}^K: K \in k \;\&\; \mathcal{C}^K \subseteq \mathcal{C}\}$, *where* $k = k(G,\Omega)$.

Let (G,Ω) be a transitive ordered permutation group, $\alpha \in \Omega$ and $K \in k(G,\Omega)$. Now $G_{(\alpha\mathcal{C}^K)} = \{g \in G: \alpha g\,\mathcal{C}^K\alpha\}$ is a convex subgroup of G (actually, a convex ℓ-subgroup if (G,Ω) is a transitive ℓ-permutation group) as is easily verified. $G_{(\alpha\mathcal{C}^K)}$ acts on the chain $\alpha\mathcal{C}^K/\mathcal{C}_K$ of $\mathcal{C}_K$ classes contained in $\alpha\mathcal{C}^K$ (ordered in the natural way). Of course, $G_{(\alpha\mathcal{C}^K)}$ may not act faithfully on $\alpha\mathcal{C}^K/\mathcal{C}_K$. However, if we factor out the lazy subgroup, we obtain a (ℓ-)homomorphic image of $G_{(\alpha\mathcal{C}^K)}$ which we denote by $\hat{G}_{(\alpha\mathcal{C}^K)}$ as before. (The lazy subgroup is $\bigcap\{G_{(\beta\mathcal{C}_K)}: \beta\mathcal{C}_K \subseteq \alpha\mathcal{C}^K\}$, a convex normal ($\ell$-)subgroup of the ($\ell$-)group $G_{(\alpha\mathcal{C}^K)}$.) $(\hat{G}_{(\alpha\mathcal{C}^K)}, \alpha\mathcal{C}^K/\mathcal{C}_K)$ is a transitive ordered permutation group that is an ℓ-permutation group if (G,Ω) is. By Theorem 1.7.2, $(\hat{G}_{(\alpha\mathcal{C}^K)}, \alpha\mathcal{C}^K/\mathcal{C}_K)$ is primitive. (Recall that a transitive ordered permutation group is primitive if it has no non-trivial congruences.) Moreover, $(\hat{G}_{(\alpha\mathcal{C}^K)}, \alpha\mathcal{C}^K/\mathcal{C}_K)$ is (ℓ-)isomorphic to $(\hat{G}_{(\beta\mathcal{C}^K)}, \beta\mathcal{C}^K/\mathcal{C}_K)$ if $\beta \in \Omega$ by Lemma 1.7.1. Hence:

THEOREM 3E. *If* (G,Ω) *is a transitive ordered (ℓ-)permutation group,* $\alpha \in \Omega$ *and* $K \in k(G,\Omega)$, *then* $(\hat{G}_{(\alpha\mathcal{C}^K)}, \alpha\mathcal{C}^K/\mathcal{C}_K)$ *is a primitive transitive ordered (ℓ-)permutation group independent of the choice of* $\alpha \in \Omega$.

Since $(\hat{G}_{(\alpha\mathcal{C}^K)}, \alpha\mathcal{C}^K/\mathcal{C}_K)$ is independent of the choice of $\alpha \in \Omega$, we will simply write it as (G_K, Ω_K) and get $\{(G_K,\Omega_K): K \in k(G,\Omega)\}$, the set of *primitive components* of (G,Ω). In future chapters we will frequently prove results about (G,Ω) by examining its individual primitive components.

We next wish to consider the intransitive case. Unfortunately, there is a profusion of blocks and congruences some of which are more useful than others. The idea is to select sufficiently many that we get enough insight into (G,Ω), and yet not too many since we want analogues of Theorems 3A-3E to still hold and yield a nice structure theory. That we can indeed do this is somewhat technical and we postpone a close analysis and discussion to Section 3.2. However, the intuition of this paragraph together with the discussion at the beginning of Section 3.2 is adequate for the intransitive applications in later chapters.

3.1. TRANSITIVE ORDERED PERMUTATION GROUPS.

In this section we wish to complete the proofs of Theorems 3A-3E. Specifically, we need to prove Theorems 3A and 3D and half of Theorem 3C. The proofs are quite straightforward and present no difficult. To establish this claim, let us now provide the proofs.

Proof of Theorem 3A: Let $\alpha \in \Omega$ and Δ_1, Δ_2 be blocks with $\alpha \in \Delta_1 \cap \Delta_2$. If $\Delta_1 \not\subseteq \Delta_2$ and $\Delta_2 \not\subseteq \Delta_1$, we have $\inf(\Delta_1) < \inf(\Delta_2) < \sup(\Delta_1) < \sup(\Delta_2)$ without loss of generality. Let $\delta_i \in \Delta_i$ $(i = 1,2)$ with $\delta_1 < \inf(\Delta_2)$ and $\delta_2 > \sup(\Delta_1)$. By the transitivity of (G,Ω), there exists $g \in G$ such that

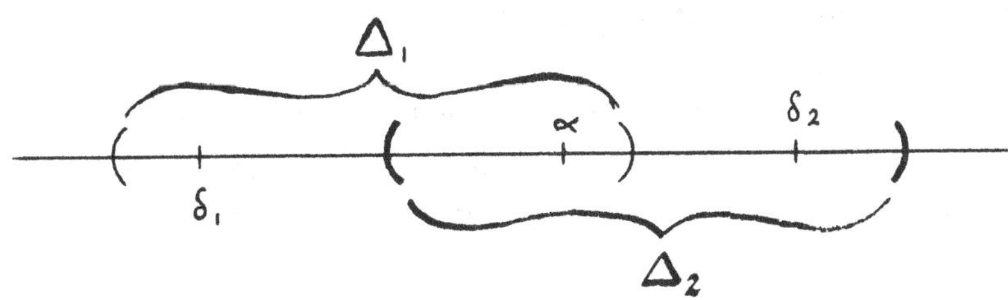

$\alpha g = \delta_2$. So $\Delta_2 g \cap \Delta_2 \neq \emptyset$. Hence $\Delta_2 g = \Delta_2$. Thus $\inf(\Delta_2)g = \inf(\Delta_2)$ and consequently $\Delta_1 g \cap \Delta_1 \neq \emptyset$. Therefore $\Delta_1 g = \Delta_1$. But $\alpha \in \Delta_1$ and $\alpha g = \delta_2 \notin \Delta_1$, the required contradiction.

Proof of Theorem 3C: Let $K \in k(G,\Omega)$. Since $\mathcal{C}_K \subsetneq \mathcal{C}^K$, there exist $\alpha,\beta \in \Omega$ with $\alpha\,\mathcal{C}^K\beta$ but not $\alpha\,\mathcal{C}_K\beta$. Since $\alpha\,\mathcal{C}^K\beta$, $\mathrm{Val}(\alpha,\beta) \leq K$ by Theorem 3A; and since $\beta \notin \alpha\mathcal{C}_K$, $K \leq \mathrm{Val}(\alpha,\beta)$ (again by Theorem 3A). Hence $K = \mathrm{Val}(\alpha,\beta)$.

Proof of Theorem 3D: Clearly $\mathcal{D} = \bigcup\{\mathcal{C}^K : K \in k \;\&\; \mathcal{C}^K \subseteq \mathcal{C}\} \subseteq \mathcal{C}$. If $\mathcal{D} \neq \mathcal{C}$, let $\alpha,\beta \in \Omega$ be such that $\alpha\,\mathcal{C}\,\beta$ but not $\alpha\,\mathcal{D}\,\beta$. Then if $K_0 = \mathrm{Val}(\alpha,\beta)$, we have $\mathcal{D} \subseteq \mathcal{C}_{K_0} \subsetneq \mathcal{C}^{K_0} \subseteq \mathcal{C}$. Since $\mathcal{C}^{K_0} \subseteq \mathcal{C}$, $\mathcal{C}^{K_0} \subseteq \mathcal{D}$ by the definition of $\mathcal{D}$. So $\mathcal{C}^{K_0} \subseteq \mathcal{D} \subsetneq \mathcal{C}^{K_0}$ which is absurd. Similarly, $\mathcal{C} = \bigcap\{\mathcal{C}_K : K \in k \;\&\; \mathcal{C}_K \supseteq \mathcal{C}\}$.

3.2. INTRANSITIVE ORDERED PERMUTATION GROUPS.

We give a very full introduction to this section so that a trusting reader will have all the facts needed for any applications (of the intransitive structure theory) in future chapters. Recall that our goal is to select enough "good" blocks so as to be able to recapture (G,Ω) in some sense, and yet not to select too many blocks as we want analogues of Theorems 3A-3E to hold in the intransitive case (so that we get a nice structure theory).

EXAMPLE 3.2.1. Let Ω be the result of taking $\mathbb{R}$ and replacing each $n \in \mathbb{Z}$ by a three point set $\{n_1,n_2,n_3\}$ with $n_1 < n_2 < n_3$. Let $G = A(\Omega)$. Then (G,Ω) is intransitive since

$$\cdots \quad -1_1 \; -1_2 \; -1_3 \quad (\longrightarrow) \quad 0_1 \; 0_2 \; 0_3 \quad (\longrightarrow) \quad 1_1 \; 1_2 \; 1_3 \quad (\longrightarrow) \quad \cdots$$

$0_2G = \{n_2 \colon n \in \mathbb{Z}\} \subsetneq \Omega$. The following are blocks of (G,Ω) as is easily checked: $\Delta_n = \{r \in \mathbb{R} \colon n < r < n + 1\}$, $\{n_1,n_2,n_3\}$, $\Delta_n \cup \{n_2,n_3,(n + 1)_1\}$, $\{n_1,n_2\}$ $(n \in \mathbb{Z})$. In some sense, the first two sets of blocks are more natural than the second two. Note that Theorem 3A fails here since $0_2 \in \{0_1,0_2\}, \{0_2,0_3\}$ and both $\{0_1,0_2\}$ and $\{0_2,0_3\}$ are blocks of (G,Ω).

EXAMPLE 3.2.2. Let Ω be the result of taking $\mathbb{R}$ and replacing each rational by a copy of $\mathbb{R}$ and each irrational by a

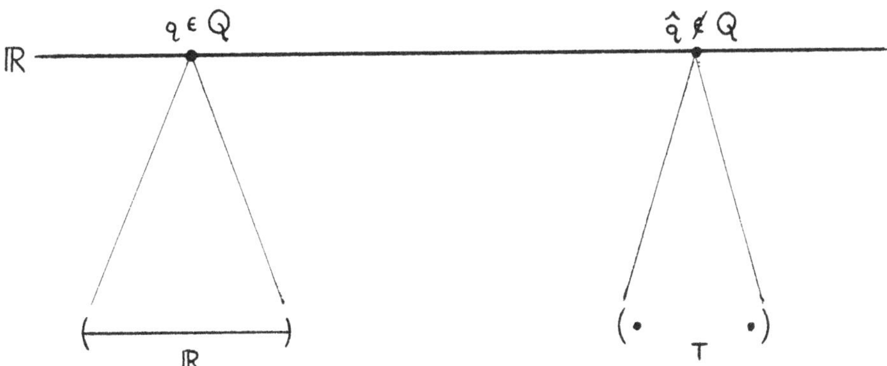

copy of a two point chain T. So $\Omega = \{(r,q): r \in \mathbb{R}, q \in \mathbb{Q}\} \cup \{(\tau,\hat{q}): \tau \in T, \hat{q} \in \mathbb{R}\backslash\mathbb{Q}\}$ ordered by $(\alpha,s) < (\beta,t)$ if $s < t$ (in $\mathbb{R}$), or $s = t$ & $\alpha < \beta$ (in $\mathbb{R}$ if $s \in \mathbb{Q}$, in T if $s \in \mathbb{R}\backslash\mathbb{Q}$). Let $G = A(\Omega)$. Now $\Delta_q = \{(r,q): r \in \mathbb{R}\}$ $(q \in \mathbb{Q})$ and $T_{\hat{q}} = \{(\tau,\hat{q}): \tau \in T\}$ $(\hat{q} \in \mathbb{R}\backslash\mathbb{Q})$ are "natural" blocks of (G,Ω). Note that if Δ is a block of (G,Ω) and $(\alpha,s),(\beta,t) \in \Delta$ with $s \neq t$, then $\Delta = \Omega$.

We wish to examine the property that blocks enjoy in the transitive case for a possible analogue in the intransitive case. If Δ is a block of a transitive ordered permutation group (G,Ω) and $\delta,\delta_1,\delta_2 \in \Delta$, there exist $g_1,g_2 \in G_{(\Delta)}$ such that $\delta g_1 \leq \delta_1$ and $\delta g_2 \geq \delta_2$;

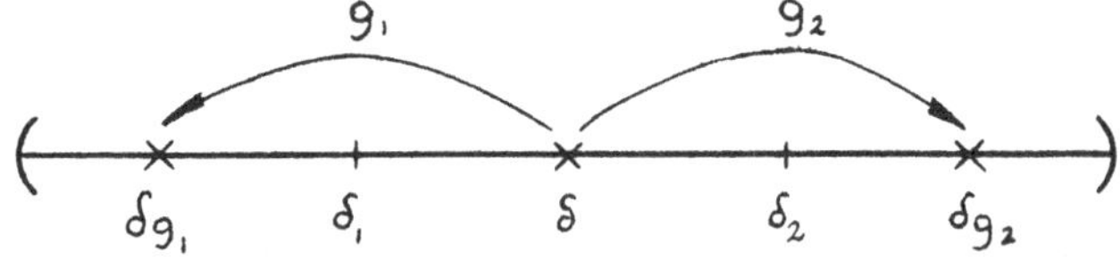

i.e., $\delta G_{(\Delta)}$ is coterminal in Δ. If (G,Ω) is an ordered permutation group and Δ is a block of (G,Ω), we say Δ is <u>*extensive*</u> if for some $\delta \in \Delta$, $\delta G_{(\Delta)}$ is coterminal in Δ. Hence if (G,Ω) is transitive, every block of (G,Ω) is extensive. In Example 3.2.1, the non-singleton extensive blocks are Ω and Δ_n

$(n \in \mathbb{Z})$; in Example 3.2.2 they are Ω and Δ_q $(q \in \mathbb{Q})$. These are obviously blocks we will want to select. Observe that if Δ is an extensive block of an ordered permutation group (G,Ω) and $\bar{\delta} \in \bar{\Delta}$, then $\bar{\delta}G_{(\Delta)}$ is coterminal in $\bar{\Delta}$.

The extensive blocks are not enough to understand (G,Ω) in general. For example, if (G,Ω) is given by Example 3.2.2 and $G' = \{g \in G\colon (\forall q \in \mathbb{Q})(\forall \hat{q} \in \mathbb{R}\setminus\mathbb{Q})(\Delta_q g = \Delta_q \;\&\; T_{\hat{q}}g = T_{\hat{q}})\}$, then (G',Ω) is an ℓ-permutation group whose non-singleton extensive blocks are just Δ_q $(q \in \mathbb{Q})$, and we have lost any evidence of $\{T_{\hat{q}}\colon \hat{q} \in \mathbb{R}\setminus\mathbb{Q}\}$. We must therefore find some other natural blocks.

The definition of "extensive" is internal to a block. We seek an external counterpart.

Let (G,Ω) be an ordered permutation group and Δ be a block of (G,Ω). Let <u>$F(\Delta)$</u> be the maximal convex subset of Ω which contains Δ but is disjoint from all Δg with $\Delta g \neq \Delta$. In Example 3.2.2, $F(\Delta_q) = \Delta_q$ and $F(T_{\hat{q}}) = T_{\hat{q}}$ $(q \in \mathbb{Q},\ \hat{q} \in \mathbb{R}\setminus\mathbb{Q})$; in Example 3.2.1, $F(\{n_1,n_2,n_3\}) = \Delta_{n-1} \cup \{n_1,n_2,n_3\} \cup \Delta_n$ which is not a block.

<u>LEMMA</u> <u>3.2.3</u>. *Let Δ be a block of an ordered permutation group (G,Ω) such that $\{\Delta g\colon \Delta g > \Delta\}$ has no least element. Then $F(\Delta)$ is a block of (G,Ω) and is the largest block $\Lambda \supseteq \Delta$ such that $G_{(\Lambda)} = G_{(\Delta)}$.*

If Δ is a block of an ordered permutation group, $\{\Delta g\colon \Delta g > \Delta\}$ has no least element, and $F(\Delta) = \Delta$, we say that Δ is a <u>*fat*</u> block. Note that Ω is always a fat block (of (G,Ω)). In Example 3.2.2, the fat blocks are Ω, Δ_q and $T_{\hat{q}}$ $(q \in \mathbb{Q},\ \hat{q} \in \mathbb{R}\setminus\mathbb{Q})$; in Example 3.2.1, Ω is the only fat block. Note that in Example 3.2.2, each Δ_q is both extensive and fat whereas each $T_{\hat{q}}$ is merely fat.

Actually, the conclusion of Lemma 3.2.3 can be strenthened to:

<u>LEMMA</u> <u>3.2.3'</u>. *Assume the hypotheses of Lemma 3.2.3. Then $F(\Delta)$ is a fat block of (G,Ω).*

Another way to obtain fat blocks is as follows. If $T \subseteq \Omega$, then T is said to be *dense in itself* if whenever $\tau_1, \tau_2 \in T$ and $\tau_1 < \tau_2$, there exists $\tau \in T$ such that $\tau_1 < \tau < \tau_2$. If $T = \Omega$, we will conform to our previous notation (p. vii) and write Ω is *dense*.

LEMMA 3.2.4. *Let (G,Ω) be an ordered permutation group and $\alpha \in \Omega$. If $T = \alpha G$ is dense in itself and Δ is a convex subset of Ω maximal with $T \cap \Delta = \{\alpha\}$, then Δ is a fat block of (G,Ω).*

Fortunately, the extensive and fat blocks are sufficient for our purposes. A block Δ of an ordered permutation group (G,Ω) is *natural* if it is extensive or fat. Hence, if (G,Ω) is transitive, all blocks are natural. We will prove the following analogues of Theorem 3A and Corollary 3B:

THEOREM 3A*. *Let (G,Ω) be an ordered permutation group and $\alpha \in \Omega$. The natural blocks of (G,Ω) that contain α form a chain under inclusion.*

COROLLARY 3B*. *Let (G,Ω) be an ordered permutation group and $\alpha \in \Omega$. The set of natural blocks of (G,Ω) that contain α is closed under arbitrary union and intersection.*

Let Δ_1 and Δ_2 be natural blocks of (G,Ω). We say Δ_2 *covers* Δ_1 if $\Delta_1 \subsetneq \Delta_2$ and there does not exist a natural block Δ of (G,Ω) such that $\Delta_1 \subsetneq \Delta \subsetneq \Delta_2$. Note that in Example 3.2.1 the fat block Ω covers an extensive block (indeed, Ω covers each Δ_n $(n \in \mathbb{Z})$); in Example 3.2.2. the fat block Ω covers each Δ_q $(q \in \mathbb{Q})$ and the fat block $T_{\hat{q}}$ covers the extensive blocks $\{(\tau,\hat{q})\}$ $(\tau \in T)$. This phenomenon is quite general (and is used in the proof of Corollary 3B*):

LEMMA 3.2.5. *Let Δ be a non-extensive block of an ordered permutation group (G,Ω) and $\delta \in \Delta$. Then $Conv_\Omega(\delta G_{(\Delta)})$ is a* *natural block maximal* $\subsetneq \Delta$. *Indeed, if* $\Delta_1 \subsetneq \Delta_2$ *are cofinal natural blocks of* (G,Ω), *then* Δ_2 *is non-extensive,* Δ_1 *is extensive and* Δ_2 *covers* Δ_1.

We also have analogues of Theorems 3C and 3D:

THEOREM 3C*. *Let* (G,Ω) *be an ordered permutation group and* α,β *be distinct elements of* Ω. *There is a unique covering pair* $\Delta_1 \subsetneq \Delta_2$ *of natural blocks of* (G,Ω) *such that* $\alpha \in \Delta_1$ *and* $\beta \in \Delta_2 \setminus \Delta_1$. *Moreover, if* $\Delta_1 \subsetneq \Delta_2$ *is a covering pair of natural blocks containing* α, *there exists* $\beta \in \Delta_2 \setminus \Delta_1$.

THEOREM 3D*. *Let* (G,Ω) *be an ordered permutation group and* $\alpha \in \Omega$. *Each natural block containing* α *is the intersection of lower halves of covering pairs of natural blocks that contain it and dually.*

Unlike the transitive case, there is not a one-to-one correspondence between congruences whose classes are natural blocks and the natural blocks that contain a given $\alpha \in \Omega$. So Theorems 3A*, 3C*, 3D* and Corollary 3B* cannot simply be lifted to "natural" congruences. (Clearly we will only want to bother with such congruences.) More precisely, a congruence $\mathcal{C}$ of an ordered permutation group (G,Ω) is called *natural* if all of its classes are natural blocks. We will show:

LEMMA 3.2.6. *Let* Δ *be a natural block of an ordered permutation group* (G,Ω). *Then there exists a largest natural congruence* $\mathcal{C}$ *of* (G,Ω) *for which* Δ *is a* $\mathcal{C}$*-class.*

In order to get an analogue of Theorem 3E with the "primitive" component independent of $\alpha \in \Omega$, we will need a homogeneity condition on the $\mathcal{C}^K$ classes. A natural congruence $\mathcal{C}$ of (G,Ω) is *homogeneous* if whenever Δ_1, Δ_2 are $\mathcal{C}$-classes, there is $g \in G$ such that $\Delta_1 g = \Delta_2$. In Example 3.2.2, define $\mathcal{C}$ and $\mathcal{D}$ by:

$(\alpha,s)\mathfrak{C}(\beta,t)$ if $(\alpha,s)=(\beta,t)$ or $s=t\in\mathbb{Q}$ and $(\alpha,s)\mathfrak{D}(\beta,t)$ if $s=t$. Then $\mathfrak{C}$ and $\mathfrak{D}$ are natural congruences of (G,Ω) and $\mathfrak{C}\subsetneq\mathfrak{D}$. Moreover, $\mathfrak{D}$ covers $\mathfrak{C}$; i.e., there is no natural congruence $\mathfrak{E}$ of (G,Ω) such that $\mathfrak{C}\subsetneq\mathfrak{E}\subsetneq\mathfrak{D}$. However, $\mathfrak{D}$ is not homogeneous (Δ_q and $T_{\hat{q}}$ are $\mathfrak{D}$-classes but there is no $g\in G$ with $\Delta_q g = T_{\hat{q}}$ $(q\in\mathbb{Q},\ \hat{q}\in\mathbb{R}\setminus\mathbb{Q})$). To overcome this problem, we define natural partial congruences.

We say that $\mathfrak{C}$ is a *natural partial congruence* of (G,Ω) with *domain* $\Omega'\subseteq\Omega$ if $\mathfrak{C}$ is an equivalence relation on Ω', $\Omega'G=\Omega'$, $\alpha\mathfrak{C}\beta$ implies $\alpha g\mathfrak{C}\beta g$ $(\alpha,\beta\in\Omega',\ g\in G)$, and each class is a natural block of (G,Ω). In particular, we are insisting that the $\mathfrak{C}$-classes are convex subsets of Ω (not just of Ω'). Note that if (G,Ω) is transitive, $\Omega'=\Omega$ so each natural partial congruence is a congruence (in the sense of Section 3.1). In Example 3.2.1, the relation $\mathfrak{C}$ defined by: $r\mathfrak{C}s$ if $(\exists n\in\mathbb{Z})$ $r,s\in(n,n+1)$ is a natural partial congruence on $\bigcup_{n\in\mathbb{Z}}(n,n+1)=\Omega'\subsetneq\Omega$. The $\mathfrak{C}$ classes are just the Δ_n $(n\in\mathbb{Z})$. In Example 3.2.2, $\mathfrak{E}$ and $\mathfrak{F}$ are natural partial congruences on $\cup\{\Delta_q: q\in\mathbb{Q}\}$ and $\cup\{T_{\hat{q}}:\hat{q}\in\mathbb{R}\setminus\mathbb{Q}\}$ respectively, where $(\alpha,s)\mathfrak{E}(\beta,t)$ if $s=t\in\mathbb{Q}$ and $(\alpha,s)\mathfrak{F}(\beta,t)$ if $s=t\in\mathbb{R}\setminus\mathbb{Q}$.

Let (G,Ω) be an ordered permutation group. Define $k(G,\Omega)$ to be the set of pairs of natural partial congruences $(\mathfrak{C}_K,\mathfrak{C}^K)$ for which (i) $\mathfrak{C}^K$ covers $\mathfrak{C}_K$ (i.e., $\mathfrak{C}_K\subsetneq\mathfrak{C}^K$ and there is no natural partial congruence $\mathfrak{E}$ with $\mathfrak{C}_K\subsetneq\mathfrak{E}\subsetneq\mathfrak{C}^K$) and (ii) $\mathfrak{C}^K$ is homogeneous. The members of $k(G,\Omega)$ are called *values*. We will show:

LEMMA 3.2.7. *If (G,Ω) is an ordered permutation group and $(\mathfrak{C}_K,\mathfrak{C}^K)\in k(G,\Omega)$, then $\mathfrak{C}_K$ and $\mathfrak{C}^K$ have the same domain.*

In view of Lemma 3.2.7, we will write *dom(K)* for the domain of $\mathfrak{C}_K$ $(=\mathrm{dom}(\mathfrak{C}^K))$.

We also have the desired analogues of Theorems 3A-3E. They will follow immediately from Theorems 3A*-3D*, Lemma 3.2.7 and the transfer principle:

LEMMA 3.2.8. *Let* (G,Ω) *be an ordered permutation group and* $\alpha \in \Omega$. *The map* $(\mathcal{C}_K, \mathcal{C}^K) \mapsto (\alpha\mathcal{C}_K, \alpha\mathcal{C}^K)$ *is an order-morphism of* $\{K \in k(G,\Omega): \alpha \in dom(K)\}$ *onto the set of covering pairs of natural blocks of* (G,Ω) *that contain* α.

THEOREM 3A†. *Let* (G,Ω) *be an ordered permutation group and* $\mathcal{C}$ *be a natural partial congruence of* (G,Ω). *Then the set of natural partial congruences of* (G,Ω) *that contain* α *form a chain (under inclusion). Hence if* $K,K' \in k(G,\Omega)$, *then* $K \leq K'$ *or* $K' \leq K$ *or* $dom(K) \cap dom(K') = \emptyset$, *where* $K < K'$ *if* $\mathcal{C}^K \subseteq \mathcal{C}_{K'}$.

COROLLARY 3B†. *If* (G,Ω) *is an ordered permutation group, then the set of natural partial congruences of* (G,Ω) *is closed under arbitrary unions and non-empty intersections.*

THEOREM 3C†. *Let* (G,Ω) *be an ordered permutation group. For each* $\alpha,\beta \in \Omega$ *with* $\alpha \neq \beta$, *there is a unique* $K \in k(G,\Omega)$ *such that* $\alpha\mathcal{C}^K\beta$ *but* $\beta \notin \alpha\mathcal{C}_K$. *Moreover, for each* $K \in k(G,\Omega)$, *there are* $\alpha,\beta \in \Omega$ *such that* $\alpha\mathcal{C}^K\beta$ *but* $\beta \notin \alpha\mathcal{C}_K$.

If we write <u>*Val*(α,β)</u> for the unique $K \in k(G,\Omega)$ guaranteed by Theorem 3C†, we obtain Theorem 3C with the word "transitive" deleted (but now $k(G,\Omega)$ has a more general interpretation in the intransitive case).

THEOREM 3D†. *Let* (G,Ω) *be an ordered permutation group and* $\mathcal{C}$ *be a natural partial congruence of* (G,Ω). *Then*
$\cap\{\mathcal{C}_K: K \in k \;\&\; \mathcal{C} \subseteq \mathcal{C}_K\} = \mathcal{C} = \cup\{\mathcal{C}^K: K \in k \;\&\; \mathcal{C}^K \subseteq \mathcal{C}\}$, *where* $k = k(G,\Omega)$.

An ordered permutation group (G,Ω) is said to be *primitive* if it has no non-trivial natural congruences. This coincides with the definition given in Section 1.6 if (G,Ω) is transitive.

<u>THEOREM 3E</u>[†]. *If* (G,Ω) *is an ordered permutation group,* $K \in k(G,\Omega)$ *and* $\alpha \in dom(\mathcal{C}^K)$, *then* $(\hat{G}_{(\alpha\mathcal{C}^K)}, \alpha\mathcal{C}^K/\mathcal{C}_K)$ *is a primitive ordered permutation group independent of the choice of* $\alpha \in dom(\mathcal{C}^K)$. *If* (G,Ω) *is an* ℓ*-permutation group, so is* $(\hat{G}_{(\alpha\mathcal{C}^K)}, \alpha\mathcal{C}^K/\mathcal{C}_K)$.

As before, we write $\underline{(G_K,\Omega_K)}$ for $(\hat{G}_{(\alpha\mathcal{C}^K)}, \alpha\mathcal{C}^K/\mathcal{C}_K)$ and call $\{(G_K,\Omega_K): K \in k(G,\Omega)\}$ the set of *<u>primitive</u> <u>components</u>* of (G,Ω). We will often say that a value K enjoys a property if (G_K,Ω_K) does.

We have two kinds of ordered permutation groups that are obviously primitive (we will find others in the next chapter). Recall that (G,Ω) is static if $G = \{e\}$ (see Section 1.6). These are clearly primitive under our extended definition. (G,Ω) is said to be *<u>integral</u>* if $\Omega = T \overleftarrow{\times} \mathbb{Z}$ for some chain T and G acts like $\mathbb{Z}$ on Ω (i.e., if $g \in G$, $(\exists m \in \mathbb{Z})[(\tau,n)g = (\tau,n+m)]$ $(n \in \mathbb{Z},\ \tau \in T)$). These too are clearly primitive.

Let us return to Examples 3.2.1 and 3.2.2. In Example 3.2.1, $k(G,\Omega)$ has two elements K_0,K_1. The $\mathcal{C}^{K_0}$ classes are Δ_n $(n \in \mathbb{Z})$; The $\mathcal{C}_{K_0}$ classes are $\{r\}$ $(r \in \bigcup_{n\in\mathbb{Z}} \Delta_n)$; the $\mathcal{C}^{K_1}$ class is Ω; $\mathcal{C}_{K_1}$ is defined by $\alpha\,\mathcal{C}_{K_1}\,\beta$ if $\alpha = \beta$ or $\alpha,\beta \in \Delta_n$ for some $n \in \mathbb{Z}$. Then $\mathcal{C}^{K_0} \subsetneq \mathcal{C}^{K_1}$, $(G_{K_0},\Omega_{K_0}) = (A(\mathbb{R}),\mathbb{R})$ and (G_{K_1},Ω_{K_1}) is integral where T is a four point set, with $T \times \{n\} \cong \{n_1,n_2,n_3,(n+\frac{1}{2})\mathcal{C}_{K_1}\}$. So we have recaptured (G,Ω) from $k(G,\Omega)$ and the primitive components. In Example 3.2.2, $k(G,\Omega)$ has three elements K_0,K_1,K_2. The $\mathcal{C}^{K_0}$ classes are Δ_q $(q \in \mathbb{Q})$, the $\mathcal{C}^{K_1}$ classes are $T_{\hat{q}}$ $(\hat{q} \in \mathbb{R}\backslash\mathbb{Q})$ and the $\mathcal{C}_{K_i}$ classes are singletons with $dom(\mathcal{C}_{K_i}) = dom(\mathcal{C}^{K_i})$ $(i = 0,1)$. The $\mathcal{C}^{K_2}$ class is Ω and the $\mathcal{C}_{K_2}$ classes are either $\mathcal{C}^{K_0}$ or $\mathcal{C}^{K_1}$ classes. So $K_0,K_1 < K_2$ but K_0 and K_1 are incomparable. $(G_{K_0},\Omega_{K_0}) = (A(\mathbb{R}),\mathbb{R})$, $(G_{K_1},\Omega_{K_1}) = (\{e\},T)$ and

$(G_{K_2}, \Omega_{K_2}) = (A(\mathbb{Q}), \mathbb{R})$, where T is the original two point set. Again we have recaptured (G,Ω) from $k(G,\Omega)$ and the primitive components.

Finally, in this section, we prove an analogue of Lemma 3.2.5.

LEMMA 3.2.9. *Let (G,Ω) be an ordered permutation group and K_1 be a static or integral value. Then the set of non-static values in $k_1 = \{K \in k(G,\Omega): K < K_1\}$ is cofinal in k_1. Hence no static or integral value covers a static value.*

This entire section is due to McCleary [76]. The proofs of the results follow from the definitions and are easy--the hard part was selecting which blocks and congruences would be useful. For the sake of completeness, we provide the omitted proofs for the dedicated reader.

Proof of Lemmas 3.2.3 and 3.2.3': Let Δ be a block of (G,Ω) such that $\{\Delta g: \Delta g > \Delta\}$ has no least element and let $F(\Delta)$ be the maximal convex subset of Ω such that $\Delta \subseteq F(\Delta)$ and $F(\Delta) \cap \Delta g = \emptyset$ if $\Delta g \neq \Delta$. Let $\beta > F(\Delta)$. By the maximality of $F(\Delta)$ there exist $\delta \in \Delta$ and $f \in G$ such that $\Delta f > F(\Delta)$ and $\beta \geq \delta f$. Let $h_1, h_2 \in G$ be such that $\Delta f > \Delta h_1 > \Delta h_2 > \Delta$. Then $\Delta f > F(\Delta)h_1 > \Delta h_2 > F(\Delta)$ so $\beta > F(\Delta)h_1$. Therefore $\{F(\Delta)h: F(\Delta)h > F(\Delta)\}$ has no least element and has infimum $\sup(F(\Delta))$. If further $\beta g^{-1} \in F(\Delta)$, then $\Delta f \leq \Delta g$ since $\beta \geq \delta f$. So $F(\Delta)g > \Delta h_2 > F(\Delta)$. Hence $F(\Delta)g \cap F(\Delta) = \emptyset$ and $F(\Delta)$ is a fat block.

If Λ is a block with $\Delta \subseteq \Lambda \not\subseteq F(\Delta)$, then $\Delta g \cap \Lambda \neq \emptyset$ for some $g \in G$ with $\Delta g \neq \Delta$. Hence $g \in G_{(\Lambda)} \setminus G_{(\Delta)}$. Thus if $G_{(\Lambda)} = G_{(\Delta)}$ and $\Delta \subseteq \Lambda$, then $\Lambda \subseteq F(\Delta)$. But since $F(\Delta)$ is a block and $\Delta \subseteq F(\Delta)$, $G_{(\Delta)} \subseteq G_{(F(\Delta))}$. By the definition of $F(\Delta)$, $G_{(F(\Delta))} \subseteq G_{(\Delta)}$. So $F(\Delta)$ is the largest block $\Delta_1 \supseteq \Delta$ such that $G_{(\Delta_1)} = G_{(\Delta)}$.

Proof of Lemma 3.2.4: Let $\alpha \in \Omega$, $T = \alpha G$ be dense in itself and Δ be a convex subset of Ω maximal with respect to $T \cap \Delta = \{\alpha\}$. Then for all $g \in G$, $T \cap \Delta g = \{\alpha g\}$. If $\Delta g \neq \Delta$, $\alpha < \alpha g$ without loss of generality. Since T is dense in itself, there is $\tau \in T$ with $\alpha < \tau < \alpha g$. Now $\Delta < \tau < \Delta g$ so $\Delta g \cap \Delta = \emptyset$. Thus Δ is a block of (G,Ω). Let $\beta \in \Omega$ with $\Delta < \beta$. By the maximality of Δ, there is $\tau \in T$ with $\Delta < \tau \leq \beta$. Now $\alpha < \tau$ so there is $\tau' \in T$ with $\alpha < \tau' < \tau$. Let $g \in G$ be such that $\alpha g = \tau'$ $(T = \alpha G)$. But $\tau' \in \Delta g$ so $\Delta < \Delta g < \tau \leq \beta$. Hence Δ is a fat block of (G,Ω).

Proof of Theorem 3A:* Assume $\alpha \in \Delta_1 \cap \Delta_2$ where Δ_1, Δ_2 are natural blocks such that $\Delta_1 \not\subseteq \Delta_2$ and $\Delta_2 \not\subseteq \Delta_1$. We have $\inf(\Delta_1) < \inf(\Delta_2) < \sup(\Delta_1) < \sup(\Delta_2)$ without loss of generality. The proof of Theorem 3A gives the desired contradiction if Δ_2 is extensive. If Δ_2 is not extensive, $\{\Delta_2 g : \Delta_2 g < \Delta_2\}$ has no largest element and supremum $\inf(\Delta_2)$. Thus for some $g \in G$, $\Delta_2 g \subseteq \Delta_1$. But then $\alpha g \in \Delta_1$; so $\Delta_1 g = \Delta_1$. Hence $\sup(\Delta_1 g) = \sup(\Delta_1) \geq \sup(\Delta_2 g)$, contradicting $\sup(\Delta_1) < \sup(\Delta_2)$.

Proof of Lemma 3.2.5: Let $T = \mathrm{Conv}_\Omega(\delta G_{(\Delta)})$ where Δ is a non-extensive block and $\delta \in \Delta$. Then T is extensive. If Λ is a block and $T \subsetneq \Lambda \subsetneq \Delta$, then $G_{(\Lambda)} = G_{(\Delta)}$ since $G_{(T)} = G_{(\Delta)}$. By Lemma 3.2.3, Λ is not fat and it is clearly not extensive ($\delta G_{(\Lambda)} = \delta G_{(\Delta)}$ which is coterminal in $T \subsetneq \Lambda$). Hence Λ is unnatural.

Assume $\Delta_1 \subsetneq \Delta_2$ are cofinal natural blocks of (G,Ω). Then $\inf(\Delta_1)g = \inf(\Delta_1)$ for all $g \in G$ and hence all $g \in G_{(\Delta_2)}$. Thus Δ_2 is not extensive and as $G_{(\Delta_1)} = G_{(\Delta_2)}$, Δ_1 is not fat (by Lemma 3.2.3). Therefore Δ_1 is extensive; indeed, $\Delta_1 = \mathrm{Conv}(\delta_1 G_{(\Delta_2)})$ if $\delta_1 \in \Delta_1$. Hence Δ_2 covers Δ_1.

Proof of Corollary 3B:* Let $\{\Delta_i : i \in I\}$ be a collection of natural blocks such that $\alpha \in \Delta_i$ $(i \in I)$. By Theorem 3A*, $\mathcal{X} = \{\Delta_i : i \in I\}$ is a chain so we may assume that I is totally ordered and $\Delta_i \subsetneq \Delta_j$ if $i < j$. If $\mathcal{X}$ has a greatest element, $\cup\mathcal{X}$ is clearly natural. If it doesn't, there is a chain $\mathcal{Y}$ of extensive blocks of (G,Ω) such that $\cup\mathcal{X} = \cup\mathcal{Y}$ (by Lemma 3.2.5). But $\cup\mathcal{Y}$ is extensive so the set of natural blocks containing α is closed under arbitrary unions. The same argument shows that we may assume that I has no least element and each Δ_i is extensive. We will show $\Lambda = \cap\mathcal{X}$ is fat. Clearly Λ is a block. If $\sup(\Lambda) < \beta$, then for some $j \in I$, $\sup(\Delta_j) < \beta$. Let $i < j$. By Lemma 3.2.5 applied to $(G_{(\Delta_j)}, \Delta_j)$, $\sup(\Delta_i) < \sup(\Delta_j)$. Thus $\sup(\Lambda) < \sup(\Delta_j)$. Let $\delta_j \in \Delta_j$ with $\sup(\Lambda) < \delta_j$. Δ_j is extensive so $\delta_j \leq \alpha g$ for some $g \in G_{(\Delta_j)}$. Now Λ is a block so $\Lambda < \Lambda g < \beta$. Hence Λ is fat and Corollary 3B* is proved.

Theorem 3C* now follows immediately as does Theorem 3D*, and this completes the block part.

Proof of Lemma 3.2.6: Clearly we will want each Δg $(g \in G)$ to be a $\mathfrak{C}$ class. If Λ is a convex subset of Ω maximal with respect to $\Lambda \cap \bigcup_{g \in G} \Delta g = \emptyset$ and $\Lambda \neq \emptyset$, then Λ is a block of (G,Ω). If Λ is a natural block, we will make it a $\mathfrak{C}$ class. If it isn't, let the $\mathfrak{C}$ classes in Λ be $\{\mathrm{Conv}_\Omega(\lambda G_{(\Lambda)}) : \lambda \in \Lambda\}$. These are clearly natural and, by Lemma 3.2.5, $\mathfrak{C}$ is maximal.

We now turn to the partial congruence lemmas.

Proof of Lemma 3.2.7: Since $\mathfrak{C}_K \subseteq \mathfrak{C}^K$, $\mathrm{dom}(\mathfrak{C}_K) \subseteq \mathrm{dom}(\mathfrak{C}^K)$. If $\mathrm{dom}(\mathfrak{C}_K) \neq \mathrm{dom}(\mathfrak{C}^K)$, let $\mathfrak{E}$ be the partial convex congruence with $\mathrm{dom}(\mathfrak{E}) = \mathrm{dom}(\mathfrak{C}^K)$ such that $\alpha \mathfrak{E} \beta$ if $\alpha = \beta$, or $\alpha \mathfrak{C}_K \beta$ if $\alpha, \beta \in \mathrm{dom}(\mathfrak{C}_K)$. Then $\mathfrak{E}$ is natural and $\mathfrak{C}_K \subsetneq \mathfrak{E} \subsetneq \mathfrak{C}^K$, a contradiction.

Proof of Lemma 3.2.8: We first show that if $\alpha \in \mathrm{dom}(K)$, then $\alpha\mathcal{C}^K$ covers $\alpha\mathcal{C}_K$. Clearly $\alpha\mathcal{C}_K \subseteq \alpha\mathcal{C}^K$. If $\alpha\mathcal{C}_K = \alpha\mathcal{C}^K$, then since $\mathcal{C}^K$ is homogeneous and $\mathrm{dom}(\mathcal{C}_K) = \mathrm{dom}(\mathcal{C}^K)$, $\beta\mathcal{C}_K = \beta\mathcal{C}^K$ for all $\beta \in \mathrm{dom}(K)$. Hence $\mathcal{C}_K = \mathcal{C}^K$, a contradiction. Thus $\alpha\mathcal{C}_K \subsetneqq \alpha\mathcal{C}^K$. If Δ is a natural block and $\alpha\mathcal{C}_K \subsetneqq \Delta \subsetneqq \alpha\mathcal{C}^K$, let $\mathcal{C}$ be the natural partial congruence whose classes are Δg $(g \in G)$ and the $\mathcal{C}_K$ classes disjoint from $\bigcup\{\Delta g\colon g \in G\}$. Then $\mathcal{C}_K \subsetneqq \mathcal{C} \subsetneqq \mathcal{C}^K$, a contradiction. Thus $(\alpha\mathcal{C}_K, \alpha\mathcal{C}^K)$ is a covering pair of natural blocks. Therefore the map $(\mathcal{C}_K, \mathcal{C}^K) \mapsto (\alpha\mathcal{C}_K, \alpha\mathcal{C}^K)$ maps $k_\alpha = \{K \in k(G,\Omega)\colon \alpha \in \mathrm{dom}(K)\}$ into the set of covering pairs of natural blocks of (G,Ω) that contain α. If $\alpha \in \mathrm{dom}(K_1) \cap \mathrm{dom}(K_2)$, then $\alpha\mathcal{C}^{K_1} \subseteq \alpha\mathcal{C}^{K_2}$ without loss of generality (by Theorem 3A*). If $\alpha\mathcal{C}^{K_1} = \alpha\mathcal{C}^{K_2}$, then $\mathcal{C}^{K_1} = \mathcal{C}^{K_2}$ by the homogeneity of $\mathcal{C}^{K_1}$ and $\mathcal{C}^{K_2}$. Hence the map is an ordermorphism (into).

Finally, if (Δ_1, Δ_2) is a covering pair of natural blocks containing α, there is a largest natural congruence $\mathcal{C}$ of $(\hat{G}_{(\Delta_2)}, \Delta_2)$ having Δ_1 as a $\mathcal{C}$ class by Lemma 3.2.6. Since Δ_2 covers Δ_1, $\mathcal{C}$ is the largest proper natural congruence of $(\hat{G}_{(\Delta_2)}, \Delta_2)$. Let $\mathcal{C}_2$ have $\{\Delta_2 g\colon g \in G\}$ as its classes and $\mathcal{C}_1$ have $\{\Delta g\colon g \in G \;\&\; \Delta \text{ is a } \mathcal{C} \text{ class}\}$ as its classes. Then $(\mathcal{C}_1, \mathcal{C}_2) \in k_\alpha$ and is mapped to (Δ_1, Δ_2). Thus the map is onto and the proof is complete.

Proof of Lemma 3.2.9: Let $K, K_1 \in k(G,\Omega)$ with $K < K_1$ (i.e., $\mathcal{C}^K \subseteq \mathcal{C}_{K_1}$). Let $\alpha \in \mathrm{dom}(K)$. If K_1 is integral, $\alpha\mathcal{C}_{K_1}$ is not a fat block. Hence it is extensive. If K_1 is static, $\alpha\mathcal{C}^{K_1}$ is not extensive. By Lemmas 3.2.5 and 3.2.8, $\alpha\mathcal{C}_{K_1}$ is extensive. Thus if K_1 is integral or static and K is static, $\alpha\mathcal{C}^K \subsetneqq \alpha\mathcal{C}_{K_1}$ and there exists $g \in G_{(\alpha\mathcal{C}_{K_1})}$ with $\alpha\mathcal{C}^K < \alpha g$.

Then $K < \mathrm{Val}(\alpha,\alpha g) < K_1$. Since $\mathrm{Val}(\alpha,\alpha g)$ is not static, the lemma is proved.

For further results on intransitive ordered permutation groups, see McCleary [76].

CHAPTER 4
PRIMITIVE ORDERED PERMUTATION GROUPS

Our goal in this chapter is to give a complete classification of primitive ℓ-permutation groups. The classification will prove extremely useful in later chapters and is one of the cornerstones in the study of ordered permutation groups.

Recall that an ordered permutation group is primitive if it has no non-trivial natural congruences. In the transitive case this means no non-trivial congruences (since all are extensive and hence natural). We will first deal with the transitive case and then show how to derive the intransitive case from it.

Let (G,Ω) be a transitive ordered permutation group, $\alpha \in \Omega$ and $e < z \in A(\bar{\Omega})$. (G,Ω) is said to be *periodic* (*with period* z) if $\{\alpha z^n : n \in \mathbb{Z}\}$ is coterminal in $\bar{\Omega}$ and $C_{A(\bar{\Omega})}(G) = \langle z \rangle$. A canonical example of a periodic ℓ-permutation group is Example 1.9.2: $(G,\mathbb{R})$ where $G = \{g \in A(\mathbb{R}):$ $(\alpha + 1)g = \alpha g + 1$ for all $\alpha \in \mathbb{R}\}$. In this example z, the period, is translation by $+1$; i.e., $z: \alpha \mapsto \alpha + 1$. $(G,\mathbb{R})$ is primitive (If $\alpha < \beta$, there exists $g \in G$ such that $\alpha < \alpha g < \beta < \beta g$; so (α,β) is never a block of $(G,\mathbb{R})$.).

We will prove:

THEOREM 4A. *Let* (G,Ω) *be a transitive primitive* ℓ*-permutation group. Then either:*

(i) (G,Ω) *is doubly transitive,*

(ii) (G,Ω) *is the right regular representation of a subgroup of* $\mathbb{R}$*,*

or (iii) (G,Ω) *is periodic.*

Moreover, (i) and (ii) are always primitive.

As an immediate consequence, we will have:

THEOREM 4B. *If* Ω *is homogeneous and primitive, then* $A(\Omega)$ *is either doubly transitive or the right regular representation of a subgroup of* $\mathbb{R}$.

We will then show that the intransitive primitive ordered permutation groups are derived in a natural way from transitive ones. Recall that static and integral ordered permutation groups are primitive (see p. 87). Let (G,T) be a transitive primitive ordered permutation group with T not ordermorphic to $\mathbb{Z}$. (G,Ω) is *transitively derived* from (G,T) if Ω is obtained by adjoining to T a subset of the set of orbits of G in $\bar{T}$. For example, $(A(\mathbb{Q}),\mathbb{R})$ is primitive, intransitive and transitively derived from $(A(\mathbb{Q}),\mathbb{Q})$. This is, in some sense, canonical:

THEOREM 4C. *Transitively derived, integral and static ordered permutation groups are all primitive, and every primitive ordered permutation group falls into precisely one of these classes.*

Finally, as a hint of the power of Theorem 4B, we deduce:

THEOREM 4D. *Let* Ω *be homogeneous. Then the following are equivalent:*

(a) $A(\Omega)$ *is Abelian.*

(b) $A(\Omega)$ *is totally ordered.*

(c) $A(\Omega)$ *is the right regular representation of a subgroup of* $\mathbb{R}$.

Moreover, there is a sentence ϕ *of the language of group theory such that (for* Ω *homogeneous)* $A(\Omega) \models \phi$ *if and only if* Ω *is doubly homogeneous. Similarly there is a sentence* χ *of the lattice language with* e *such that (for* Ω *homogeneous)* $A(\Omega) \models \chi$ *if and only if* Ω *is doubly homogeneous.*

Theorem 4D shows that it is possible, using group-theoretic sentences about $A(\Omega)$, to distinguish between the classes of imprimitive, doubly transitive and "regular" homogeneous chains. Similarly for the lattice language. This was the strengthening promised on page 51.

4.1. TRANSITIVE PRIMITIVE ORDERED PERMUTATION GROUPS.

In this section we obtain conditions equivalent to primitivity. They will be used in the proof of Theorem 4A. They extend Theorem 1.6.5 and are of interest in their own right.

A subset Δ of Ω is dense in Ω if whenever $\alpha,\beta \in \Omega$ and $\alpha < \beta$, there is $\delta \in \Delta$ with $\alpha < \delta < \beta$ (cf., dense in itself, page 83)--see page vi.

The following theorem is due to Holland [65a].

THEOREM 4.1.1. *Let (G,Ω) be a transitive ℓ-permutation group. Then the following are equivalent:*

(1) (G,Ω) is primitive.

(2) (G,Ω) satisfies the separation property.

(3) For each $\alpha \in \Omega$, G_α is a maximal convex subgroup of G.

(4) For each $\bar\alpha \in \bar\Omega$, $G_{\bar\alpha}$ is a maximal convex subgroup of G.

(5) If $\bar\alpha \in \bar\Omega$, $\bar\alpha G$ is dense in $\bar\Omega$ or $\bar\Omega = \Omega \cong \mathbb{Z}$.

Before proving the theorem we first note (cf. Theorem 2.3.1):

LEMMA 4.1.2. *Let (G,Ω) be an ℓ-permutation group, $\bar\alpha \in \bar\Omega$ and C be a subgroup of G with $C \supseteq G_{\bar\alpha}$. Then C is an ℓ-subgroup of G.*

Proof: Let $f \in C$. If $\bar\alpha f \le \bar\alpha$, then $f \vee e \in G_{\bar\alpha} \subseteq C$. Similarly, if $\bar\alpha f > \bar\alpha$, $f^{-1} \vee e \in C$. Hence $f \vee e = f(f^{-1} \vee e) \in C$. So $f \in C$ implies $f \vee e \in C$. If $f,g \in C$, then $fg^{-1} \vee e \in C$; so $f \vee g = (fg^{-1} \vee e)g \in C$. Since $f \wedge g = (f^{-1} \vee g^{-1})^{-1}$, $f \wedge g \in C$. Thus C is an ℓ-subgroup of G.

We now prove Theorem 4.1.1:

<u>(1) ↔ (2)</u>. This is Theorem 1.6.5.

<u>(1) ↔ (3)</u>. By Lemma 4.1.2 and Theorems 1.6.1 and 1.6.2.

<u>(4) → (3)</u>. A fortiori.

<u>(3) → (4)</u>. For reductio ad absurdum, suppose $G_{\bar\alpha} \subsetneqq C \subsetneqq G$ for some convex subgroup C of G. Let $g_0 \in C \setminus G_{\bar\alpha}$. Without loss of generality, $\bar\alpha < \bar\alpha g_0$. Let $\beta \in \Omega$ with $\bar\alpha \le \beta \le \bar\alpha g_0$. Let $e \le f \in G_\beta$. Then $\bar\alpha f \le \beta f = \beta \le \bar\alpha g_0$; so $fg_0^{-1} \vee e \in G_{\bar\alpha} \subseteq C$. Now $e \le f \le f \vee g_0 = (fg_0^{-1} \vee e)g_0 \in C$. Therefore $f \in C$. Hence $G_\beta \subseteq C \subsetneqq G$. But $g_0 \in C \setminus G_\beta$; so $G_\beta \subsetneqq C \subsetneqq G$. This contradicts (3).

<u>(1) → (5)</u>. If an element $\beta \in \Omega$ has an immediate successor, then so does every element of Ω (by transitivity). Moreover, some element of Ω (and hence all) would have an immediate predecessor. Define $\mathcal{C}$ on Ω by: $\alpha \mathcal{C} \beta$ if there are only finitely many points of Ω between α and β. Then $\mathcal{C}$ is a congruence of (G,Ω) whose classes have more than one element. Hence $\alpha \mathcal{C} \beta$ for all $\alpha, \beta \in \Omega$ if (G,Ω) is primitive. Thus $\Omega \cong \mathbb{Z}$.

If no element of Ω has an immediate predecessor, it follows (by the above) that Ω is dense. Hence (5) holds if $\bar\alpha \in \Omega$. If $\bar\alpha \in \bar\Omega \setminus \Omega$, define $\mathcal{C}$ by: $\alpha \mathcal{C} \beta$ if there is no $g \in G$ such that $\bar\alpha g$ lies between α and β. $\mathcal{C}$ is a congruence of (G,Ω) and clearly there exist $\alpha, \beta \in \Omega$ such that $\beta \notin \alpha\mathcal{C}$. Since (G,Ω) is primitive, the $\mathcal{C}$ classes are singletons; i.e., $\bar\alpha G$ is dense in $\bar\Omega$.

<u>(5) → (2)</u>. If $\Omega \cong \mathbb{Z}$, each bounded non-empty interval of Ω is finite and $G \cong \mathbb{Z}$. (2) follows immediately.

If $\Omega \ncong \mathbb{Z}$, then it is dense (take $\bar\alpha \in \Omega$). If Δ is any bounded non-empty interval of Ω, let $\bar\alpha = \inf(\Delta) \in \bar\Omega$. Let $\alpha, \beta \in \Omega$ with $\alpha < \beta$. By (5), there is $g \in G$ such that $\alpha < \bar\alpha g < \beta$. Hence $\alpha g^{-1} < \bar\alpha < \beta g^{-1}$. For some $e \le f \in G$, $\beta g^{-1} f^{-1} \in \Delta$. Then $\alpha g^{-1} f^{-1} \le \alpha g^{-1} < \bar\alpha$ so $\alpha g^{-1} f^{-1} \notin \Delta$. Hence (G,Ω) satisfies the separation property. Q.E.D.

As a consequence of our proof:

THEOREM 4.1.3. *Let (G,Ω) be a transitive ordered permutation group. Then (4) $\rightarrow$ (3) $\rightarrow$ (1) $\leftrightarrow$ (2) $\rightarrow$ (5), where (1)-(5) are given in the statement of Theorem 4.1.1.*

In the proof of (5) $\rightarrow$ (2) of Theorem 4.1.1, all we needed was: If $\alpha < \beta$, there is $g \in G^+$ such that $\alpha g = \beta$, where $G^+ = \{g \in G\colon g \geq e\}$. Any transitive ordered permutation group that satisfies this condition is called *coherent*. Thus we have:

THEOREM 4.1.4. *If (G,Ω) is a transitive coherent ordered permutation group, then (1), (2) and (5) of Theorem 4.1.1 are equivalent.*

We next strengthen Theorem 4.1.1 in the special case that (G,Ω) is a doubly transitive ordered permutation group (and so is primitive--Corollary 1.6.6). The result is due to Lloyd [64].

THEOREM 4.1.5. *If (G,Ω) is a doubly transitive ordered permutation group, then G_α is a maximal subgroup of G for each $\alpha \in \Omega$.*

Proof: Let $\alpha \in \Omega$ and C be a subgroup of G with $C \supsetneqq G_\alpha$. Let $g \in C\backslash G_\alpha$. Without loss of generality, $\alpha g > \alpha$. Let $f \in G$. If $\alpha f = \alpha$, then $f \in G_\alpha \subseteq C$. If $\alpha f > \alpha$, then there is $h \in G$ such that $\alpha fh = \alpha g$ and $\alpha h = \alpha$ (by double transitivity). Hence $fhg^{-1} \in G_\alpha \subseteq C$. Since $h \in G_\alpha \subseteq C$ and $g \in C$, $f \in C$. If $\alpha f < \alpha$, then $\alpha f^{-1} > \alpha$; so $f^{-1} \in C$. Thus $G = C$. Consequently, G_α is a maximal subgroup of G.

Finally we note:

LEMMA 4.1.6. *If (G,Ω) is a transitive primitive ordered permutation group and $\bar{\alpha} \in \bar{\Omega}$, then so is $(G,\bar{\alpha}G)$.*

Proof: By Theorem 4.1.3, $\bar{\alpha}G$ is dense in $\bar{\Omega}$. If $\Delta \subseteq \bar{\alpha}G$ is a block, then so is $\bar{\Delta}$ (for (G,Ω)). Hence $\bar{\Delta} \cap \Omega$ is a block for (G,Ω). Since (G,Ω) is primitive, $\bar{\Delta} \cap \Omega = \Omega$ or $|\bar{\Delta} \cap \Omega| \leq 1$.

Thus $\Delta = \bar{\alpha}G$ or $|\Delta| = 1$ (by the density of $\bar{\alpha}G$ in $\bar{\Omega}$). Therefore $(G,\bar{\alpha}G)$ is primitive.

4.2. TOTALLY ORDERED TRANSITIVE ORDERED PERMUTATION GROUPS.

We now wish to consider those primitive ordered permutation groups (G,Ω) for which G is totally ordered (i.e., an o-group) under the pointwise ordering.

(G,Ω) is said to be *uniquely transitive* if given $\alpha,\beta \in \Omega$, there is a unique $g \in G$ such that $\alpha g = \beta$; e.g., the right regular representation of an o-group. If $(A(\Omega),\Omega)$ is uniquely transitive, then we will say that Ω is *rigidly homogeneous* (cf. Example 1.1.1).

We wish to prove the following theorems:

THEOREM 4.2.1. *If (G,Ω) is a transitive primitive ordered permutation group such that G is an o-group, then (G,Ω) is the right regular representation of a subgroup of $\mathbb{R}$. Conversely, the right regular representation of any subgroup of $\mathbb{R}$ is primitive.*

THEOREM 4.2.2. *If Ω is rigidly homogeneous, then it is ordermorphic to a subgroup of $\mathbb{R}$ and $(A(\Omega),\Omega)$ is just the right regular representation. Moreover, there exist $2^{2^{\aleph_0}}$ pairwise non-ordermorphic rigidly homogeneous chains.*

Theorem 4.2.2 shows that, in Theorem 4B, there do exist homogeneous primitive chains Ω such that $A(\Omega)$ is the right regular representation of a subgroup of $\mathbb{R}$--in fact, they exist in abundance.

An o-group G is said to be *Archimedean* if $g^n \leq h$ for all $n \in \mathbb{Z}^+$ implies $g \leq e$ $(g,h \in G)$. G is Archimedean if and only if it has no proper convex subgroups. This is also equivalent to: G is o-isomorphic to a subgroup of $\mathbb{R}$ (Fuchs [63, page 45]).

LEMMA 4.2.3. *Let (G,Ω) be an ordered permutation group with G an o-group, and let $\alpha \in \Omega$. If $\alpha f < \alpha g$ $(f,g \in G)$, then $f < g$.*

Proof: If G is an o-group and $f \not< g$, then $f \geq g$. Hence $\beta f \geq \beta g$ for all $\beta \in \Omega$. So if $\alpha f < \alpha g$, $f < g$.

LEMMA 4.2.4. *If (G,Ω) is a uniquely transitive ℓ-permutation group, then G is an o-group, and (G,Ω) is o-isomorphic to (G,G), the right regular representation.*

Proof: If $f,g \in G$ and $f \not\leq g$, then for some $\alpha \in \Omega$, $\alpha f < \alpha g$. Now $fg^{-1} \vee e \in G_\alpha$, and $G_\alpha = \{e\}$ by the unique transitivity. Therefore $fg^{-1} \leq e$; i.e., $f \leq g$. Hence G is an o-group. Fix $\alpha_0 \in \Omega$ and let $\alpha \in \Omega$. There is a unique $g \in G$ such that $\alpha_0 g = \alpha$. The map ϕ: $G \to \Omega$ given by: $g\phi = \alpha_0 g$ is an ordermorphism. Moreover, $(g\phi)f = (\alpha_0 g)f = \alpha_0(gf) = (gf)\phi$ $(f,g \in G)$. Thus ϕ: $(G,G) \to (G,\Omega)$ is an o-isomorphism.

We now prove Theorem 4.2.1.

In any o-group, if C_1 and C_2 are convex subgroups of G, then $C_1 \subseteq C_2$ or $C_2 \subseteq C_1$ (If $C_1 \not\subseteq C_2$, let $e < c_1 \in C_1 \setminus C_2$. If $e < c_2 \in C_2$, then $c_1 \not\leq c_2$ so $c_2 < c_1$. Hence $C_2 \subseteq C_1$). Let $\alpha \in \Omega$. So if $\beta \in \Omega$, then $G_\alpha \subseteq G_\beta$ or $G_\beta \subseteq G_\alpha$. By Theorem 4.1.1, $G_\alpha = G_\beta$ and is the unique maximal convex subgroup of G. If $f \neq e$, let $\sigma \in \Omega$ with $\sigma f \neq \sigma$. Thus $f \not\in G_\sigma = G_\alpha$. Consequently, $G_\alpha = \{e\}$; so (G,Ω) is uniquely transitive and G has no proper convex subgroups. Therefore, G is o-isomorphic to a subgroup of $\mathbb{R}$ and (G,Ω) is the right regular representation (by Lemma 4.2.4).

Conversely, if G is a subgroup of $\mathbb{R}$ and Δ is a block, then $G_{(\Delta)} = G$ or $\{e\}$. If $G_{(\Delta)} = G$, then $\Delta = \Omega$; and if $G_{(\Delta)} = \{e\}$, then $|\Delta| = 1$. Hence (G,G) is primitive and Theorem 4.2.1 is proved.

We now give examples to show that Theorem 4.2.1 is sharp.

<u>EXAMPLE</u> <u>4.2.5</u>. Let G be a simple non-Abelian o-group (see Theorem 6F) and let H be a proper subgroup of G. Then $(G,R(H))$ is an ℓ-permutation group (Example 1.4.4) that is not uniquely transitive (and hence not primitive).

<u>EXAMPLE</u> <u>4.2.6</u>. Let $\Omega = \mathbb{Z} \overleftarrow{\times} \mathbb{Z}$ and (G,Ω) be the right regular representation. (G,Ω) is uniquely transitive but $\mathbb{Z} \overleftarrow{\times} \mathbb{Z}$ is not Archimedean and so not o-isomorphic to a subgroup of $\mathbb{R}$. Indeed $\{(m,0): m \in \mathbb{Z}\}$ is a block so (G,Ω) is not primitive. $A(\mathbb{Z} \overleftarrow{\times} \mathbb{Z})$ is not an o-group since if $(m,n)f = (m - n,n)$, $f \in A(\mathbb{Z} \overleftarrow{\times} \mathbb{Z})$ and $(0,1)f = (-1,1) < (0,1)$ but $(0,-1)f = (1,-1) > (0,-1)$.

In contrast to Example 4.2.6 we prove the first part of Theorem 4.2.2.

So assume $(A(\Omega),\Omega)$ is uniquely transitive. By Lemma 4.2.4, $A(\Omega)$ is an o-group. Let $\alpha_0 \in \Omega$. If $e < f^n < g$ for all $n \in \mathbb{Z}^+$, then $\alpha_0 < \alpha_0 f^n < \alpha_0 g$ for all $n \in \mathbb{Z}^+$ by the unique transitivity. Let $\bar{\alpha} = \sup\{\alpha_0 f^n: n \in \mathbb{Z}^+\}$ and define $h \in A(\Omega)$ by:

$$\alpha h = \begin{cases} \alpha f & \text{if } \alpha < \bar{\alpha} \\ \alpha & \text{otherwise} \end{cases}$$

. Then h fixes $\alpha_0 g$ but not α_0, a contradiction $(G_{\alpha_0 g} = \{e\})$. Therefore $A(\Omega)$ is Archimedean and the first part of the proof of Theorem 4.2.2 follows.

For the rest of this section we will assume that Ω is a rigidly homogeneous subgroup of $\mathbb{R}$. $\mathbb{Z}$ is an example but we now wish to construct dense rigidly homogeneous subgroups of $\mathbb{R}$.

If p is a prime number and G is a torsion-free Abelian group, let <u>$\gamma_p(G)$</u> be the dimension of G/pG as a vector space over $\mathbb{Z}_p$, the Galois field of p elements. If G is a subgroup of $\mathbb{R}$, then $\gamma_p(G) \leq 2^{\aleph_0}$ and if G is rigidly homogeneous $\gamma_p(G) \geq 1$ (Otherwise the map $g \mapsto pg$ would be an element of G fixing 0).

Let $\{\gamma_p\colon p \text{ prime}\}$ be a set of cardinal numbers with $1 \le \gamma_p \le 2^{\aleph_0}$ and $\{I_p\colon p \text{ prime}\}$ be a set of pairwise disjoint sets with $|I_p| = \gamma_p$. Let $I = \cup\{I_p\colon p \text{ prime}\}$ and $\{\xi_i\colon i \in I\}$ be a linearly independent set (over $\mathbb{Q}$) of real numbers. For each prime p, let $\mathbb{Q}_p$ be the set of rational numbers which, in simplest form, have denominators relatively prime to p. Note that $\gamma_q(\mathbb{Q}_p) = \begin{cases} 0 & \text{if } q \ne p \\ 1 & \text{if } q = p \end{cases}$. Let $G_p = \Sigma\{\xi_i\mathbb{Q}_p\colon i \in I_p\}$ and $G = \Sigma\{G_p\colon p \text{ prime}\}$. Then $\gamma_p(G) = \gamma_p$ for each prime p and G is a dense subgroup of $\mathbb{R}$.

<u>THEOREM</u> <u>4.2.7</u>. *Let* $\{\gamma_p\colon p \text{ prime}\}$ *be a set of cardinal numbers with* $1 \le \gamma_p \le 2^{\aleph_0}$ *for all primes* p. *Then there exists a dense rigidly homogeneous subgroup* Ω *of* $\mathbb{R}$ *such that* $\gamma_p(\Omega) = \gamma_p$ *for all primes* p.

<u>*Proof:*</u> First assume $\gamma_p < 2^{\aleph_0}$ for all primes p. By the above, there exists Ω_0, a dense subgroup of $\mathbb{R}$, such that $\gamma_p(\Omega_0) = \gamma_p$ for all primes p with $|\Omega_0| < 2^{\aleph_0}$. We wish to enlarge Ω_0, preserving the invariants $\{\gamma_p\colon p \text{ prime}\}$, so that the only order-preserving permutations of the enlarged group will be translations. To do this we must kill all elements of $A(\mathbb{R})$ which are not translations. Note that since each $f \in A(\mathbb{R})$ is completely determined by its action on $\mathbb{Q}$, $|A(\mathbb{R})| \le |\mathbb{R}^{\mathbb{Q}}| = (2^{\aleph_0})^{\aleph_0} = 2^{\aleph_0}$.

Let $\{f_\lambda\colon 1 \le \lambda < 2^{\aleph_0}\}$ be an enumeration of the elements of $A(\mathbb{R})$ which are not translations. Let $E_0 = \{0\}$ and D_0 be the divisible closure (in $\mathbb{R}$) of Ω_0. We define Ω_λ, D_λ and E_λ inductively so that:

(1) D_λ is a divisible subgroup of $\mathbb{R}$ containing Ω_λ.

(2) $|D_\lambda| \le \max\{|\lambda|, |\Omega_0|\}$.

(3) Ω_λ is a direct sum of Ω_0 and the divisible subgroup E_λ of $\mathbb{R}$.

(4) If $\lambda' < \lambda$, then $E_{\lambda'} \subseteq E_\lambda$ and $D_{\lambda'} \setminus \Omega_{\lambda'} \subseteq D_\lambda \setminus \Omega_\lambda$.

(5) $(\Omega_\lambda f_\lambda \cup \Omega_\lambda f_\lambda^{-1}) \cap (D_\lambda \setminus \Omega_\lambda) \neq \emptyset$.

Then $\Omega = \cup\{\Omega_\lambda : \lambda < 2^{\aleph_0}\}$ will be the required dense rigidly homogeneous subgroup of $\mathbb{R}$. (By (3), $\gamma_p(\Omega_\lambda) = \gamma_p(\Omega_0) = \gamma_p$ for all $\lambda < 2^{\aleph_0}$, so $\gamma_p(\Omega) = \gamma_p$ for all primes p. If $f \in A(\mathbb{R})$ is not a translation, $f = f_\lambda$ for some λ. By (5) and (4), there exists $\alpha \in \Omega_\lambda \subseteq \Omega$ such that αf or $\alpha f^{-1} \in D_\lambda \setminus \Omega_\lambda \subseteq D_\mu \setminus \Omega_\mu$ for all $\mu \geq \lambda$. Now αf (or $\alpha f^{-1}) \notin \cup\{\Omega_\mu : \mu \geq \lambda\} = \Omega$, so Ω is rigidly homogeneous.)

Suppose D_λ, E_λ and Ω_λ have been defined for all $\lambda < \mu$ so as to satisfy (1)-(5). Let $X_\mu^* = \cup\{X_\lambda : \lambda < \mu\}$ where $X = D$, E or Ω. Then D_μ^*, E_μ^* and Ω_μ^* are subgroups of $\mathbb{R}$ with D_μ^* and E_μ^* divisible, and $\Omega_\mu^* = E_\mu^* \oplus \Omega_0$. Let $f = f_\mu$. We wish to find $\zeta \notin D_\mu^*$ such that ζf (or $\zeta f^{-1}) \notin \Omega_\mu^* \oplus \mathbb{Q}\zeta$, in order to continue the induction. This is rather tricky.

Let $A = \{\alpha \in \mathbb{R} : \alpha f = q\alpha + \sigma$ for some $q \in \mathbb{Q}$ and $\sigma \in \Omega_\mu^*\}$.

(i) If $\mathbb{R} \neq A \cup \Omega_\mu^* f^{-1} \cup D_\mu^*$, let $\alpha \in \mathbb{R} \setminus (A \cup \Omega_\mu^* f^{-1} \cup D_\mu^*)$. Since $E_\mu^* \subseteq \Omega_\mu^* \subseteq D_\mu^*$, $E_\mu = E_\mu^* \oplus \alpha\mathbb{Q}$ is indeed a direct sum as is $\Omega_\mu = E_\mu \oplus \Omega_0$. If $\alpha f \in \Omega_\mu$, then $\alpha f = q\alpha + \sigma$ for some $q \in \mathbb{Q}$, $\sigma \in \Omega_\mu^*$. Hence $\alpha \in A$, a contradiction. Thus (1)-(5) hold if we let D_μ be the divisible subgroup of $\mathbb{R}$ generated by αf and $D_\mu^* \oplus \alpha\mathbb{Q}$.

(ii) If $\mathbb{R} = A \cup \Omega_\mu^* f^{-1} \cup D_\mu^*$, we wish to find $\delta \in A \setminus D_\mu^*$ such that $\delta f = q\delta + \sigma$ some $\sigma \in \Omega_\mu^*$ and $q \in \mathbb{Q}$ with q or $\frac{1}{q} \notin \mathbb{Z}$. To do this, we partition $A = A_1 \cup A_{-1} \cup \Omega^* f^{-1} \cup B$ where

$A_1 = \{\alpha \in \mathbb{R} : \alpha f = \alpha + \sigma$ for some $\sigma \in \Omega_\mu^*\}$,

$A_{-1} = \{\alpha \in \mathbb{R} : \alpha f = -\alpha + \sigma$ some $\sigma \in \Omega_\mu^*\}$ and

$B = \{\alpha \in \mathbb{R} : \alpha f = q\alpha + \sigma$ for some $\sigma \in \Omega_\mu^*$ and $q \in \mathbb{Q}$, $q \neq 0, \pm 1\}$.

We now show $B \setminus D_\mu^* \neq \emptyset$. Let $k: \alpha \mapsto \alpha f + \alpha$. Then k is a continuous strictly increasing real valued function. Hence there is at most one $\alpha \in \mathbb{R}$ such that $\alpha k = \sigma$. Therefore $|A_{-1}| \leq |\Omega_\mu^* k^{-1}| \leq |D_\mu^*| < 2^{\aleph_0}$. Let $h: \alpha \mapsto \alpha f - \alpha$. Then h is continuous and since f is not a translation, h is not constant.

Hence $\mathbb{R}h$ contains an interval. But $(\mathbb{R}\setminus A_1)h \supseteq \mathbb{R}h\setminus A_1h \supseteq \mathbb{R}h\setminus\Omega^*_\mu$ since $A_1h \subseteq \Omega^*_\mu$. Since $|\mathbb{R}h| = 2^{\aleph_0}$, $|\mathbb{R}\setminus A_1| = 2^{\aleph_0}$. But $|\Omega^*_\mu f^{-1}| \leq |D^*_\mu| < 2^{\aleph_0}$ and $|A_{-1}| < 2^{\aleph_0}$, so $|B| = 2^{\aleph_0}$. Thus $B\setminus D^*_\mu \neq \emptyset$.

If there exists $\delta \in B\setminus D^*_\mu$ with $\delta f = q\delta + \sigma$ for some $q \notin \mathbb{Z}$, $\sigma \in \Omega^*_\mu$, let $\alpha = \delta$. Otherwise let $\alpha = \delta f$ where δ is an arbitrary element of $B\setminus D^*_\mu$. In the latter case, $\alpha = \delta f = m\delta + \sigma$ for some $\sigma \in \Omega^*_\mu$ and $m \in \mathbb{Z}$, $m \neq 0,\pm 1$. Therefore $\alpha f^{-1} = \delta = (\alpha - \sigma)/m$. But $\sigma \in D^*_\mu$ and $\delta \notin D^*_\mu$, so $\alpha \notin D^*_\mu$. In either case we have $\alpha \notin D^*_\mu$ with αf (or αf^{-1}) having the form $\frac{k}{n}\alpha + \sigma'$ with $\sigma' \in D^*_\mu$ and k,n relatively prime integers with $n \neq 1$.

If $\sigma' \notin \Omega^*_\mu$, let $E_\mu = E^*_\mu \oplus \alpha\mathbb{Q}$, $\Omega_\mu = \Omega_0 \oplus E_\mu$, and D_μ be the divisible subgroup of $\mathbb{R}$ generated by $D^*_\mu \oplus \alpha\mathbb{Q}$ and β ($= \alpha f$ or αf^{-1})--the indicated sums being indeed direct. It is straightforward to show that (1)-(5) now hold for all $\lambda \leq \mu$ (If $\beta \in \Omega_\mu$, then $\frac{k}{n}\alpha + \sigma' = r\alpha + \tau$ for some $r \in \mathbb{Q}$ and $\tau \in \Omega^*_\mu$. Since $\alpha \notin D^*_\mu$, $\frac{k}{n} = r$ and $\sigma' = \tau \in \Omega^*_\mu$, a contradiction).

If $\sigma' \in \Omega^*_\mu$, let p be a prime such that $p|n$. Let $\xi \in \Omega_0$ be such that $\frac{\xi}{p} \in D_0\setminus\Omega_0$ ($\gamma_p > 0$ for all primes p, so such a ξ must exist). Let $\zeta = \alpha - \xi$. Then $\zeta \notin D^*_\mu$. Let $E_\mu = E^*_\mu \oplus \zeta\mathbb{Q}$, $\Omega_\mu = \Omega_0 \oplus E_\mu$ and D_μ be the divisible subgroup of $\mathbb{R}$ generated by D^*_μ, ζ and β ($= \alpha f$ or αf^{-1})--the indicated sums being indeed direct as is easily shown. Again a routine verification shows that (1)-(5) now hold for all $\lambda \leq \mu$. (Note that $\alpha \in \Omega_\mu$ and $\beta \notin \Omega_\mu$--if $\beta \in \Omega_\mu$, $\frac{k}{n}\alpha + \sigma' = \beta = \sigma + r\zeta$ for some $r \in \mathbb{Q}$ and $\sigma \in \Omega^*_\mu$. Since $\alpha \notin D^*_\mu$, $\frac{k}{n} = r$. So $\frac{k}{n}\xi \in \Omega^*_\mu$. But $\frac{k}{n}\xi \in D_0\setminus\Omega_0 \subseteq D^*_\mu\setminus\Omega^*_\mu$, a contradiction.) This completes the proof of the theorem if all $\gamma_p < 2^{\aleph_0}$.

In the general case, let $J = \{p\colon p \text{ prime and } \gamma_p = 2^{\aleph_0}\}$. Let $\gamma'_p = \gamma_p$ if $p \notin J$ and $\gamma'_p = 1$ if $p \in J$. Let Ω_0 be a dense subgroup of $\mathbb{R}$ with $\gamma_p(\Omega_0) = \gamma'_p$ and $|\Omega_0| < 2^{\aleph_0}$. Let D_0

be the divisible closure of Ω_0 and $E_0 = \{0\}$. If $\mu = \nu + p$ for ν a limit ordinal and $p \in J$, let $\tau \in \mathbb{R} \setminus D_{\nu+p-1}$. Let $E^*_\mu = E_{\nu+p-1} \oplus \tau\mathbb{Q}_p$, $\Omega^*_\mu = E^*_\mu \oplus \Omega_0$ and D^*_μ be the divisible closure of $\Omega^*_\mu \cup D_{\nu+p-1}$. For all μ not of this form, let Ω^*_μ, D^*_μ and E^*_μ be defined as in the special case above ($\gamma_p < 2^{\aleph_0}$ for all primes p). Now proceed as before using the new Ω^*_μ, D^*_μ, and E^*_μ in place of the old ones.

By varying $\{\gamma_p: p \text{ prime}\}$ in Theorem 4.2.7, we get $2^{\aleph_0}$ dense rigidly homogeneous chains. We now sketch a proof for $2^{2^{\aleph_0}}$-- actually, $2^{2^{\aleph_0}}$ pairwise non-isomorphic such chains for each prescribed $\{\gamma_p: p \text{ prime}\}$.

In the proof of Theorem 4.2.7 we also wish to construct inductively Δ_λ, divisible subgroups of $\mathbb{R}$ such that $\Delta_\lambda \cap D_\lambda = \{0\}$, $|\Delta_\lambda| = \max\{\aleph_0, |\lambda|\}$ and $\Delta_{\lambda'}$ is a direct summand of Δ_λ if $\lambda' < \lambda$. To do this, let $\{\eta_\nu: \nu < 2^{\aleph_0}\}$ be a basis for $\mathbb{R}$ as a rational vector space and ν_0 the least ν such that $\eta_\nu \notin D_0$. Let $\Delta_0 = \mathbb{Q}\eta_{\nu_0}$. At stage μ of the induction, since $|B| = 2^{\aleph_0} > |\Delta^*_\lambda|$, we can select $\delta, \alpha \in B \setminus (\Delta^*_\mu \cup \Delta^*_\mu f^{-1} \cup D^*_\mu)$ with the desired properties (as before), where $\Delta^*_\mu = \bigcup\{\Delta_\lambda: \lambda < \mu\}$ (Case (i) now becomes $\mathbb{R} \neq A \cup \Omega^*_\mu f^{-1} \cup D^*_\mu \cup \Delta^*_\mu \cup \Delta^*_\mu f^{-1}$). If Ω_μ, E_μ and D_μ are then obtained as before, we let ν_μ be the least ν such that $\eta_\nu \notin D_\mu \oplus \Delta^*_\mu$. Now let $\Delta_\mu = \Delta^*_\mu \oplus \mathbb{Q}\eta_{\nu_\mu}$. If $\Delta = \bigcup\{\Delta_\lambda: \lambda < 2^{\aleph_0}\}$, then Δ is a divisible subgroup of $\mathbb{R}$, $|\Delta| = 2^{\aleph_0}$ and $\Delta \cap \Omega = \{0\}$. Let $\{\zeta_\nu: \nu < 2^{\aleph_0}\}$ be a basis for Δ as a rational vector space and $X \subseteq 2^{\aleph_0}$. Then $\Omega \oplus \Sigma\{\mathbb{Q}\zeta_\nu: \nu \in X\}$ is a dense rigidly homogeneous chain (as is easily seen) having the prescribed γ_p's. Thus the number of such dense rigidly homogeneous chains is greater than or equal to $2^{2^{\aleph_0}}$ (the number of subsets of $2^{\aleph_0}$). Since there can be no more ($|\Omega| \leq 2^{\aleph_0}$), and as any order-preserving isomorphism between two totally ordered subgroups of $\mathbb{R}$ is obtained by multiplication by a positive real number (so the number in any isomorphism class is $2^{\aleph_0}$), there are $2^{2^{\aleph_0}}$ pairwise non-isomorphic dense rigidly homogeneous chains of the prescribed γ_p's.

Other modifications of the method are possible to prove, for example, that there exist rigidly homogeneous chains whose divisible closure is $\mathbb{R}$. If $\mathbb{R}$ cannot be written as the union of less than $2^{\aleph_0}$ nowhere dense subsets (which is certainly true if we assume the Continuum Hypothesis--by Baire's Theorem), then all dense rigidly homogeneous chains have size $2^{\aleph_0}$. However, it is consistent with ZFC (set theory) that the Continuum Hypothesis fails and for each uncountable cardinal $\kappa \leq 2^{\aleph_0}$, there exist 2^{κ} dense rigidly homogeneous chains of size κ. For these and other results about dense rigidly homogeneous chains, see Glass, Gurevich, Holland and Shelah [81], and Roitman [81].

4.3. PERIODIC PRIMITIVE ℓ-PERMUTATION GROUPS.

In this section we classify the remaining transitive primitive ℓ-permutation groups. Recall the example given in the introduction to this chapter. We showed it is primitive. However, it is clearly neither uniquely nor doubly transitive ($0g = 0$ implies $1g = 1$). We wish to show this example is, in some sense, canonical.

THEOREM 4.3.1. *If (G,Ω) is a transitive primitive ℓ-permutation that is neither uniquely nor doubly transitive, then (G,Ω) is periodic; i.e., there exists $e < z \in A(\bar{\Omega})$ such that $\{\alpha z^n : n \in \mathbb{Z}\}$ is coterminal in $\bar{\Omega}$ (for any $\alpha \in \Omega$) and $C_{A(\bar{\Omega})}(G) = \langle z \rangle$. Moreover, if $\alpha,\beta \in \Omega$ and $\Delta = \beta G_\alpha \neq \{\beta\}$, then $(G_\alpha|\Delta,\Delta)$ is non-pathologically doubly transitive.*

This will complete the proof of Theorem 4A (and hence Theorem 4B). It also yields the following extension of Lemma 4.1.6:

COROLLARY 4.3.2. *Let* (G,Ω) *be a transitive primitive* ℓ*-permutation group and* $\bar{\alpha} \in \bar{\Omega}$, *then so is* $(G,\bar{\alpha}G)$, *and* $(G,\bar{\alpha}G)$ *is doubly transitive (regular, periodic) if* (G,Ω) *is.*

Throughout this section (due to McCleary [72a] and [73c]), we will assume that (G,Ω) is a transitive ℓ-permutation group. We will build on the ideas of Section 1.6 to obtain the theorem. Recall Theorem 1.6.12: If (G,Ω) is primitive, then it is balanced; i.e., the paired orbit of a long orbit is long.

For $\alpha \in \Omega$, let $\overline{Fx(G_\alpha)} = \{\bar{\alpha} \in \bar{\Omega}: G_\alpha \subseteq G_{\bar{\alpha}}\}$. If (G,Ω) is a transitive primitive ℓ-permutation group, $\overline{Fx}(G_\alpha) = \{\bar{\alpha} \in \bar{\Omega}: G_\alpha = G_{\bar{\alpha}}\}$ by Theorem 4.1.1.

LEMMA 4.3.3. *Let* (G,Ω) *be a transitive primitive* ℓ*-permutation group, and* $Z = C_{A(\bar{\Omega})}(G)$. *Let* $z \in Z$ *and* $\bar{\alpha} = \alpha z$. *Then* $\bar{\alpha} \in \overline{Fx(G_\alpha)}$. *Conversely, if* $\bar{\alpha} \in \overline{Fx(G_\alpha)}$ *and* z *is defined as the unique member of* $A(\bar{\Omega})$ *such that* $(\alpha g)z = \bar{\alpha}g$ $(g \in G)$, *then* $z \in Z$.

Proof: Let $g \in G_\alpha$, $z \in Z$ and $\bar{\alpha} = \alpha z$. Then $\bar{\alpha}g = \alpha zg = \alpha gz = \alpha z = \bar{\alpha}$, so $g \in G_{\bar{\alpha}}$. Hence $\bar{\alpha} \in \overline{Fx}(G_\alpha)$.

Conversely, since $\bar{\alpha} \in \overline{Fx}(G_\alpha)$, $G_\alpha = G_{\bar{\alpha}}$. Hence the map $z_0: \Omega \to \bar{\Omega}$ given by $(\alpha g)z_0 = \bar{\alpha}g$ $(g \in G)$ is well-defined. Let $\beta \in \Omega$ and $f \in G$. Let $g \in G$ be such that $\alpha g = \beta$. Then $\beta f z_0 = (\alpha g f)z_0 = \bar{\alpha}gf = \beta z_0 f$. Thus z_0 commutes with every element of G. If $\beta_1, \beta_2 \in \Omega$ with $\beta_1 < \beta_2$, let $g \in G^+$ be such that $\beta_1 g = \beta_2$. Let $f_1 \in G$ with $\alpha f_1 = \beta_1$. Then $\alpha f_1 g = \beta_2 > \beta_1 = \alpha f_1$ so $f_1 g f_1^{-1} \notin G_\alpha = G_{\bar{\alpha}}$. Hence $\bar{\alpha}f_1 g \neq \bar{\alpha}f_1$. Now $\beta_1 z_0 = \bar{\alpha}f_1 < \bar{\alpha}f_1 g = \beta_2 z_0$ so z_0 is one-to-one and preserves order. Therefore it extends to a unique element z of $A(\bar{\Omega})$ which must still commute with G; i.e., $z \in Z$.

COROLLARY 4.3.4. *Let* (G,Ω) *be a transitive primitive* ℓ*-permutation group,* $\alpha \in \Omega$ *and* $Z = C_{A(\bar{\Omega})}(G)$. *Then the map* $z \mapsto \alpha z$ *is an ordermorphism of* Z *onto* $\overline{Fx(G_\alpha)}$. *Hence* Z *is an o-group.*

Part (i) of the next corollary strengthens Lemma 1.6.8.

COROLLARY 4.3.5.

(i) If (G,Ω) is doubly transitive, $Z = \{e\}$.

(ii) If (G,Ω) is the right regular representation of a subgroup of $\mathbb{R}$, $Z \cong \mathbb{Z}$ or $Z = \mathbb{R}$.

Proof: If (G,Ω) is doubly transitive $\overline{Fx}(G_\alpha) = \{\alpha\}$, and in case (ii), $\overline{Fx}(G_\alpha) = \bar{\Omega}$. If Ω is a non-discrete subgroup of $\mathbb{R}$, $\bar{\Omega} = \mathbb{R}$ so the corollary follows.

We now come to the key lemma. Recall that in our canonical example of a periodic primitive ℓ-permutation group, G_α has a first positive orbit, namely $(\alpha,\alpha + 1) = (\alpha,\alpha z)$, the G_α orbit of $\alpha + \frac{1}{2}$ $(\alpha \in \mathbb{R})$. The paired orbit is $(\alpha - 1,\alpha) = (\alpha z^{-1},\alpha)$. Both are long and the only point of $\bar{\Omega}$ between them is α. This phenomenon is quite general as we now show.

LEMMA 4.3.6. *If (G,Ω) is a transitive primitive ℓ-permutation group that is neither uniquely nor doubly transitive, then G_α has a first positive long orbit Δ_1 and α is the only point of $\bar{\Omega}$ between Δ_1 and its paired orbit, Δ_1'.*

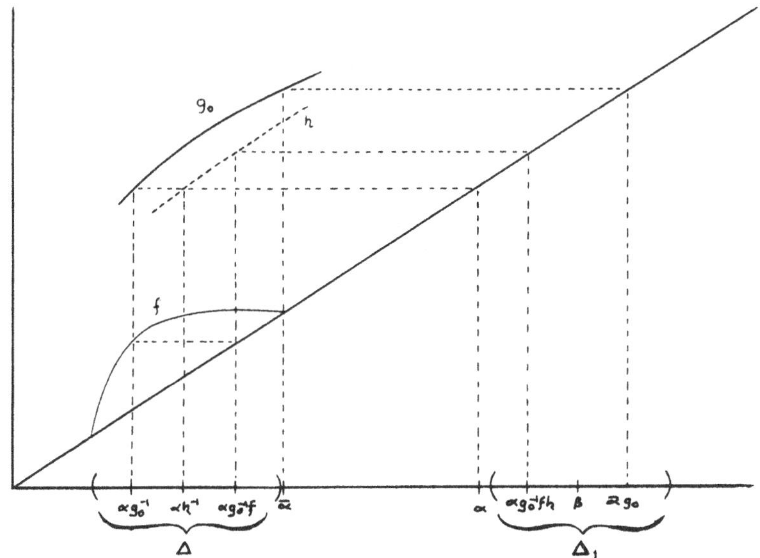

Proof: Since (G,Ω) is not uniquely transitive, $G_\alpha \neq \{e\}$. Hence G_α has a long orbit Δ. Since (G,Ω) is balanced (Theorem 1.6.12), we may assume that Δ is a negative orbit and so not cofinal in Ω. Let $\bar{\alpha} = \sup(\Delta) \in \bar{\Omega}$. Since Δ is a G_α orbit, $G_\alpha \subseteq G_{\bar{\alpha}}$. By Theorem 4.1.1, $G_\alpha = G_{\bar{\alpha}}$. Choose $g_0 \in G$ so that $\alpha g_0^{-1} \in \Delta$ and let $\Delta_1 = \mathrm{Conv}_\Omega(\bar{\alpha} g_0 G_\alpha)$. We now show that Δ_1 is the required long orbit. Since Ω is not discrete and $\alpha g_0^{-1} < \bar{\alpha}$, $(\alpha,\bar{\alpha}g_0) \neq \emptyset$. Let $\beta \in \Omega \cap (\alpha,\bar{\alpha}g_0)$. By Theorem 4.1.1, $\bar{\alpha}G$ is dense in $\bar{\Omega}$ so we may pick $h \in G^+$ such that $\alpha < \bar{\alpha}h < \beta$. Then $\alpha h^{-1} < \bar{\alpha}$ so there is $f \in G_\alpha^+ = G_{\bar{\alpha}}^+$ such that $\alpha h^{-1} \leq \alpha g_0^{-1} f$. Thus $g_0^{-1} fh \wedge e \in G_\alpha$. Now $(\bar{\alpha}g_0)(g_0^{-1} fh \wedge e) \leq \bar{\alpha} fh = \bar{\alpha} h < \beta < \bar{\alpha} g_0$, so $\beta \in \Delta_1$. Therefore $(\alpha,\bar{\alpha}g_0) \subseteq \bar{\Delta}_1$. Since $\alpha \notin \Delta_1$, $\inf(\Delta_1) = \alpha$ as required.

We are now ready to complete the proof of Theorem 4.3.1. So assume that (G,Ω) is a transitive primitive ℓ-permutation group that is neither uniquely nor doubly transitive. By Lemma 4.3.6, G_α has a first positive long orbit Δ_1 which is not cofinal in $\bar{\Omega}$ since (G,Ω) is not doubly transitive (Lemma 1.10.2). Let $\bar{\alpha} = \sup(\Delta_1) \in \overline{\mathrm{Fx}}(G_\alpha)$. By Lemma 4.3.3 and Corollary 4.3.4, we obtain $e < z \in C_{A(\bar{\Omega})}(G)$ such that $\alpha z = \bar{\alpha}$. For each $n \in \mathbb{Z}$, let $\Delta_n = \Omega \cap (\alpha z^{n-1}, \alpha z^n)$. Note $\bar{\Delta}_1 = (\alpha,\alpha z)$ and $\bar{\Delta}_n = \bar{\Delta}_1 z^{n-1}$. Since $\langle z\rangle \subseteq C_{A(\bar{\Omega})}(G)$, each Δ_n is a G_α orbit. Moreover, Δ_n is paired with Δ_{-n+1}. But if $\Lambda = \bigcup\{\Delta_n : n \in \mathbb{Z}\}$, Λ is a block of (G,Ω). Since (G,Ω) is primitive, $\Lambda = \Omega$; i.e., $\{\alpha z^n : n \in \mathbb{Z}\}$ is coterminal in $\bar{\Omega}$ for some (and hence all) $\alpha \in \Omega$. Therefore $\overline{\mathrm{Fx}}(G_\alpha) = \{\alpha z^n : n \in \mathbb{Z}\}$, so $\langle z\rangle = C_{A(\bar{\Omega})}(G)$ by Corollary 4.3.4. Consequently, (G,Ω) is periodic.

Finally, we prove $(G_\alpha|\Delta,\Delta)$ is non-pathologically doubly transitive where $\Delta = \beta G_\alpha$ ($\neq \{\beta\}$). Firstly by periodicity, G_α is faithful on Δ. To prove double transitivity, it is enough to show that if $\beta_1 < \beta_2 < \beta_3$ in Δ, there exists $e < g \in G_\alpha \cap G_{\beta_1}$ such that $\beta_2 g = \beta_3$ (by Lemma 1.10.2). There is $e \leq h \in G_\alpha$ such that $\beta_2 h = \beta_3$. $\bar{\Delta} = (\alpha z^m, \alpha z^{m+1})$ for some $m \in \mathbb{Z}$. Now

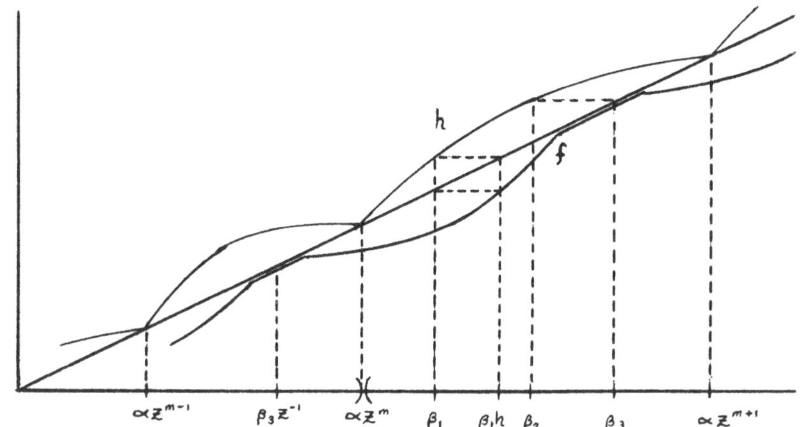

$\alpha z^{m-1} < \beta_3 z^{-1} < \alpha z^m < \beta_1 \leq \beta_1 h < \beta_2 h = \beta_3$. Thus there is $e \geq f \in G_{\beta_3}$ such that $\beta_1 hf = \beta_1$. Let $g = hf \vee e \in G_{\beta_1}$. Then $\beta_2 g = \beta_3$ and $\alpha g = \max\{\alpha hf, \alpha\} = \max\{\alpha f, \alpha\} = \alpha$. Consequently, $(G_\alpha|\Delta,\Delta)$ is doubly transitive. Next we can choose $g_1 \in G$ such that $\alpha z^m g_1 < \alpha z^m$ and $\beta_1 g_1 > \beta_1$. Then $g_1 \vee e \in G_\alpha$ and fixes all points in $\Delta \cap (\alpha z^m, \alpha z^m g_1^{-1}) \neq \emptyset$. Therefore $e \neq g_1 \vee e|\Delta$ is bounded below (in Δ). But $g_1 \vee e|\Delta z^{-1}$ is bounded above (in Δz^{-1}) so, by periodicity, $g_1 \vee e|\Delta$ is too (in Δ). Hence $g_1 \vee e|\Delta$ has bounded support. Q.E.D.

We note that the proof given of Theorem 4A works equally well for coherent ordered permutation groups provided we change doubly transitive to *weakly doubly transitive*: Given $\alpha, \beta_1, \beta_2 \in \Omega$ with $\alpha < \beta_1 < \beta_2$, there exists $g \in G_\alpha$ such that $\beta_1 g \geq \beta_2$, and $f \in G_{\beta_2}$ such that $\beta_1 f \leq \alpha$. So

THEOREM 4.3.7. *Let* (G,Ω) *be a transitive primitive coherent ordered permutation group. Then:*
either

(i) (G,Ω) *is weakly doubly transitive,*

or (ii) (G,Ω) *is the right regular representation of a subgroup of* $\mathbb{R}$,

or (iii) (G,Ω) *is periodic.*

Moreover, (i) and (ii) are always primitive.

This classification may hold for all transitive primitive ordered permutation groups. We know of no counterexample. Howevery, we do know of examples of incoherent doubly transitive ordered permutation groups.

Theorem 4A makes no claim that periodic ℓ-permutation groups are primitive. We now give an example to show that periodic does not imply primitivity.

EXAMPLE 4.3.8. Let $\Omega = \mathbb{R}\setminus\mathbb{Z}$ and $G = \{g \in A(\Omega): (\alpha + 1)g = \alpha g + 1$ for all $\alpha \in \Omega\}$. Then (G,Ω) is a transitive periodic ℓ-permutation with period z, translation by $+1$. However, $(n,n+1)$ is a block of (G,Ω) for any $n \in \mathbb{Z}$, so (G,Ω) is not primitive. Note that $G_{\frac{1}{2}}$ has a first long orbit $(\frac{1}{2},1)$ but this is not of the form $(\frac{1}{2}, \frac{1}{2}z)$.

We now obtain further information about periodic primitive ℓ-permutation groups.

If (G,Ω) is a periodic primitive ℓ-permutation group with period z and $\alpha \in \Omega$, either $\alpha z^m \notin \Omega$ for all $m \in \mathbb{Z}^+$ or there is a least $n \in \mathbb{Z}^+$ such that $\alpha z^n \in \Omega$. In this latter case, $\alpha z^m \in \Omega$ if and only if $n|m$. By Theorem 1.6.11, the existence and value of such an n is independent of the choice of $\alpha \in \Omega$. In the first case we say that (G,Ω) has *Config(∞)* and in the second, *Config(n)*.

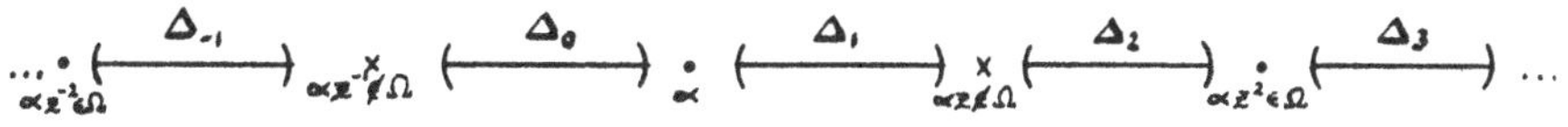

$\ldots \underset{\alpha z^{-2}\in\Omega}{\bullet} \;(\overset{\Delta_{-1}}{\longrightarrow})\; \underset{\alpha z^{-1}\notin\Omega}{\times} \;(\overset{\Delta_{0}}{\longrightarrow})\; \underset{\alpha}{\bullet} \;(\overset{\Delta_{1}}{\longrightarrow})\; \underset{\alpha z\notin\Omega}{\times} \;(\overset{\Delta_{2}}{\longrightarrow})\; \underset{\alpha z^{2}\in\Omega}{\bullet} \;(\overset{\Delta_{3}}{\longrightarrow})\; \ldots$

Config(2)

Our canonical example, $G = \{g \in A(\mathbb{R}): (\alpha + 1)g = \alpha g + 1$ for all $\alpha \in \mathbb{R}\}$ has Config(1).

It is straightforward to show that if (G,Ω) is a periodic primitive ℓ-permutation group, then any two long orbits have the same cardinality; and if (G,Ω) has Config(n), $\Delta_j \cong \Delta_k$ if $j \equiv k \pmod n$, and $C_{A(\Omega)}(G) = \langle z^n \rangle$.

If (G,Ω) is a periodic primitive ℓ-permutation group with period z, then $G \subseteq C_{A(\Omega)}(z)$. If $G = C_{A(\Omega)}(z)$, we say that G is *full*.

LEMMA 4.3.9. *Let (G,Ω) be a full periodic primitive ℓ-permutation group with period z. If there exist $\alpha \in \Omega$, $m \in \mathbb{Z}$ and $g \in G$ such that $\alpha g = \alpha z^m$, then $z^m \in G$.*

Proof: If $\alpha g = \alpha z^m$ for some $m \in \mathbb{Z}$, $\alpha \in \Omega$ and $g \in G$, let $\beta \in \Omega$. Let $f \in G$ be such that $\beta = \alpha f$. Then $\beta z^m = \alpha f z^m = \alpha z^m f = \alpha g f \in \Omega$. Hence $z^m \in A(\Omega)$. So $z^m \in C_{A(\Omega)}(z) = G$.

We conclude this section by proving:

THEOREM 4.3.10. *If $n \in \mathbb{Z}^+ \cup \{\infty\}$, there exists a full periodic primitive ℓ-permutation group of Config(n).*

Proof: Let $\{\mathbb{Q}_m : m \in \mathbb{Z}\}$ be distinct cosets of $\mathbb{Q}$ in $\mathbb{R}$ and T be the Dedekind completion of $\mathbb{R} \overleftarrow{\times} \mathbb{Z}$. So $T = (\{\zeta\} \overleftarrow{\cup} \mathbb{R}) \overleftarrow{\times} \mathbb{Z}$.

If $n = \infty$, let $\Omega_n = \{(\zeta,0)\} \cup \{(\alpha,m) \in T: \alpha \in \mathbb{Q}_m\}$. If $n \in \mathbb{Z}^+$, let $\Omega_n = \{(\zeta,mn): m \in \mathbb{Z}\} \cup \{(\alpha,m) \in T: (\exists k \in \mathbb{Z})(1 \leq k \leq n$ & $k \equiv m \pmod n$ & $\alpha \in \mathbb{Q}_k)\}$. In either case, $\bar{\Omega}_n = T$.

Let $e < z \in A(T)$ be given by: $(\alpha,m)z = (\alpha,m+1)$ $(\alpha \in \{\zeta\} \cup \mathbb{R};\ m \in \mathbb{Z})$. Let $G = C_{A(\Omega_n)}(z)$. In either case, it is quite easy to show that (G,Ω_n) is a full periodic ℓ-permutation group (with period z) of Config(n). //

4.4. INTRANSITIVE PRIMITIVE ORDERED PERMUTATION GROUPS.

Recall that (G,Ω) is transitively derived if there exists T such that (G,T) is a transitive primitive ordered permutation group and Ω is obtained by adding to T some subset of orbits of G in $\bar{T}$. We now wish to establish that every primitive ordered permutation group is integral, static or transitively derived, and that any integral, static or transitively derived ordered permutation group is indeed primitive (Theorem 4C). As an immediate consequence we will have:

COROLLARY 4.4.1. *Let (G,Ω) be a primitive ℓ-permutation group. Then (G,Ω) is static, integral or transitively derived from an ℓ-permutation group that is doubly transitive, the right regular representation of a subgroup of $\mathbb{R}$, or periodic primitive with Config(∞) or Config(n) $(n \in \mathbb{Z}^+)$.*

Theorem 4C (due to McCleary [76]) shows that once we have classified all transitive primitive ordered permutation groups, it is straightforward to classify all primitive ordered permutation groups.

Observe that if (G,Ω) is transitively derived from (G,T), then, by Theorem 4.1.3, $\bar{\alpha}G$ is dense in $\bar{T}$ for all $\bar{\alpha} \in \bar{T}$. Moreover, $(G,\bar{\alpha}G)$ is also transitive and primitive (Lemma 4.1.6). Thus (G,Ω) is transitively derived from the ordered permutation group on any one of its orbits.

We now prove Theorem 4C.

We already saw that static and integral ordered permutation groups are primitive (see page 87); so assume (G,Ω) is transitively derived from (G,T). Let Δ be a natural block of (G,Ω). Since (G,T) is transitive and primitive, $\Delta \cap T = T$ or has cardinality at most 1. Hence $\Delta = \Omega$ or is a singleton (T is dense in $\bar{T}$ by Theorem 4.1.3). Thus (G,Ω) is primitive.

Conversely, let (G,Ω) be a primitive ordered permutation group. If $G = \{e\}$, then (G,Ω) is static; so assume $G \neq \{e\}$. Therefore there is $\alpha \in \Omega$ such that $\alpha G \neq \{\alpha\}$. Then $\mathrm{Conv}_\Omega(\alpha G)$ is an extensive block. Thus $\mathrm{Conv}_\Omega(\alpha G) = \Omega$.

<u>Case 1</u>. Assume $\alpha G \cong \mathbb{Z}$, say $\alpha G = \{\alpha_n : n \in \mathbb{Z}\}$ with $\alpha_0 = \alpha$. We claim that $G_\alpha = \{e\}$.

For reductio ad absurdum, assume $\Delta = \mathrm{Conv}_\Omega(\delta G_\alpha)$ *with* $|\Delta| > 1$. Then Δ is an extensive convex set and $G_{\alpha_n} = G_\alpha$ for all $n \in \mathbb{Z}$. Hence $\Delta \cap \alpha G = \emptyset$ so $\Delta \subseteq (\alpha_m, \alpha_{m+1})$ for some $m \in \mathbb{Z}$. If $h \in G \setminus G_{(\Delta)}$, then $h \notin G_{\alpha_m}$. Since $\alpha_m h \in \alpha G$, $\alpha_m h = \alpha_n$ for some $n \neq m$. Therefore $\Delta \cap \Delta h = \emptyset$; so Δ is an extensive block, a contradiction ($\alpha \notin \Delta$ and $|\Delta| > 1$). Thus $G_\alpha = \{e\}$. Let $T = [\alpha_0, \alpha_1)$. Then $\Omega \cong T \overleftarrow{\times} \mathbb{Z}$ and (G,Ω) is integral.

<u>Case 2</u>. Assume $\alpha G \not\cong \mathbb{Z}$. Let $T = \alpha G$. If Δ were a block of (G,T), then Δ would be extensive. Hence $\mathrm{Conv}_\Omega(\Delta)$ would be an extensive block of (G,Ω). Thus (G,T) is transitive and primitive.

It remains to show $\Omega \subseteq \bar{T}$. By Lemma 3.2.4, if $\beta \in T$ and Δ is a convex subset of Ω maximal with respect to $\Delta \cap T = \{\beta\}$, then Δ is a fat block of (G,Ω). Hence $\Delta = \{\beta\}$.

Let Λ be a convex subset of Ω maximal with respect to containing no point of T. Suppose $\Lambda \neq \emptyset$. We claim that Λ is a fat block of (G,Ω) whence $|\Lambda| = 1$. Since T is an orbit of G, Λ is a block. Let $\Delta = \{\tau \in T: \Lambda < \tau < \Lambda f$ for all $\Lambda f > \Lambda\}$. Then $\Delta = \emptyset$ or is a block of (G,T). Since (G,T) is primitive, $|\Delta| \leq 1$. If $\Delta = \{\delta\}$, then $\{\delta\} \cup \Lambda$ is a convex subset of Ω. By the previous paragraph, $\Lambda \cup \{\delta\} = \{\delta\}$, a contradiction; so $\Delta = \emptyset$. If for some $\sigma \in \Omega$, $\Lambda < \sigma < \Lambda f$ for all $\Lambda f > \Lambda$, then by the maximality of Λ, there exists $\tau_1 \in T$ such that $\Lambda < \tau_1 \leq \sigma$. Hence $\tau_1 \in \Delta$, a contradiction. Consequently, $\sup(\Lambda) = \inf\{\Lambda f: \Lambda f > \Lambda\}$; i.e., Λ is a fat block of (G,Ω).

Now let (A,B) be a cut in T and $\Lambda = \{\sigma \in \Omega: A < \sigma < B\}$. Then Λ is a convex subset of Ω maximal with respect to containing no point of T. Thus $|\Lambda| = 1$ or $\Lambda = \emptyset$. Hence $\bar{T} = \bar{\Omega}$ and (G,Ω) is transitively derived from (G,T).

We can now locate the non-extensive natural blocks of an ordered permutation group by combining Theorem 4C and Lemmas 3.2.5 and 3.2.8.

COROLLARY 4.4.2. *The non-extensive natural blocks of an ordered permutation group are precisely the* $\mathcal{C}^K$*-classes for which* K *is static.*

For more technical consequences, see McCleary [76].

4.5. THE PROOF OF THEOREM 4D.

Throughout this section, let Ω be a homogeneous chain. We now exploit Theorem 4B to deduce Theorem 4D. This proof will be the first of very many we will give heavily using Theorems 4A and 4B.

By Theorem 4B and Lemma 1.6.8, the only primitive homogeneous chains that are Abelian are rigidly homogeneous chains. Moreover, if Ω is imprimitive, let Δ be a non-trivial block.

Let $e \neq g \in A(\Omega)_{(\Delta)}$ and $h \notin A(\Omega)_{(\Delta)}$. Then $\text{supp}(h^{-1}gh) \subseteq \Delta h$ and $\Delta h \cap \Delta = \emptyset$, so $h^{-1}gh \wedge g = e$. Hence $A(\Omega)$ is Abelian if and only if Ω is rigidly homogeneous if and only if $A(\Omega)$ is totally ordered.

If Ω is imprimitive, let Δ_i $(i = 1,2,3)$ be distinct classes of a non-trivial congruence of $(A(\Omega),\Omega)$. Let $e \neq g_i \in A(\Omega)$ with $\text{supp}(g_i) \subseteq \Delta_i$ $(i = 1,2,3)$. Then for all $h \in A(\Omega)$, $h^{-1}g_1h$ is disjoint from at least one of g_2 and g_3. Thus $A(\Omega) \models (\exists g_1 \neq e)(\exists g_2 \neq e)(\exists g_3 \neq e)(\forall h)([h^{-1}g_1h,g_2] = e$ or $[h^{-1}g_1h,g_3] = e)$. If we let θ denote this sentence, then clearly $A(\Omega) \models \neg\theta$ if (and only if) Ω is doubly homogeneous, and $A(\Omega) \models \theta$ if $A(\Omega)$ is Abelian. If $\psi \doteq \theta \;\&\; (\exists f)(\exists g)([f,g] \neq e)$, then $A(\Omega) \models \psi$ if and only if Ω is imprimitive.

If Ω is imprimitive and $g_1, g_2 > e$ are defined as above, then there does not exist $e < h \in A(\Omega)$ having one bump such that

$$A(\Omega) \models h \wedge g_1 \neq e \;\&\; h \wedge g_2 \neq e \;\&\; (\exists f_1)(\exists f_2)$$
$$(f_1 \wedge g_1 \neq e \;\&\; f_2 \wedge g_2 \neq e \;\&\; h \wedge f_1 = e \;\&\; h \wedge f_2 = e).$$

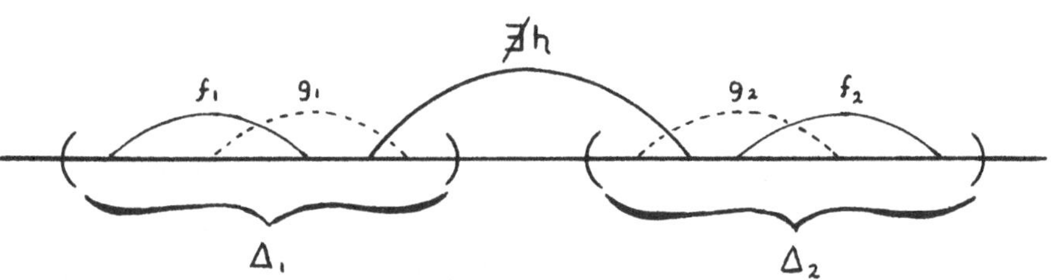

(If h has one bump and $g_1 \wedge h \neq e \neq g_1 \wedge f_1$ but $f_1 \wedge h = e$, then $\Delta_1 h \cap \Delta_1 \neq \emptyset$; hence $\Delta_1 h = \Delta_1$ and so $h \wedge g_2 = e$.) Since "h has one bump" is expressible by a sentence in the lattice language (Lemma 2.1.1), we have that if Ω is imprimitive, then $A(\Omega) \models \bar{\chi}$ where $\bar{\chi}$ is the lattice sentence

$(\exists g_1 > e)(\exists g_2 > e)(g_1 \wedge g_2 = e \;\&\; (\forall h > e)[\mathrm{bump}(h) \;\&\; h \wedge g_1 \neq e \;\&$

$h \wedge g_2 \neq e \to \neg(\exists f_1)(\exists f_2)(f_1 \wedge g_1 \neq e \;\&\; f_2 \wedge g_2 \neq e \;\&$

$h \wedge f_1 = e \;\&\; h \wedge f_2 = e)])$.

Clearly, $A(\Omega) \models \bar{\chi}$ if and only if Ω is imprimitive; and $A(\Omega) \models \neg\bar{\chi} \;\&\; (\exists a > e)(\exists b > e)(a \wedge b = e)$ if and only if Ω is doubly homogeneous. This completes the proof of Theorem 4D.

A group is said to be an *O-group* if there is a total order which can be put on G making it an o-group. The following theorem due to P. M. Cohn [57] is similar to Theorem 4D.

THEOREM 4.5.1. *Let Ω be a (not necessarily homogeneous) chain. Then $A(\Omega)$ is an O-group if and only if it is Abelian.*

Proof: By Appendix I Corollary 10, $A(\Omega)$ can be right totally ordered. Hence if $A(\Omega)$ is Abelian, this order makes $A(\Omega)$ an O-group.

Conversely, $A(\Omega) \cong \prod_{i \in I} A(\Lambda_i)$ (as a group), where $\{\Lambda_i : i \in I\}$ is the set of distinct orbitals of $A(\Omega)$; i.e., $\{\Lambda_i : i \in I\} = \{\mathrm{Conv}_\Omega(\alpha A(\Omega)) : \alpha \in \Omega\}$. Since $A(\Omega)$ is an O-group if and only if each $A(\Lambda_i)$ is, we may assume $\Omega = \Lambda_{i_0}$ some $i_0 \in I$. That is, we assume that Ω is an extensive block of $A(\Omega)$. If $e < g \in A(\Omega)$ has bounded support, say $\mathrm{supp}(g) \subseteq (\alpha,\beta)$, let $e < f \in A(\Omega)$ with $\alpha f \geq \beta$ (Ω is extensive so such an f exists). Let $h \in A(\Omega)$ with $\mathrm{supp}(h) \subseteq \bigcup\{\mathrm{supp}(f^{-n}gf^n) : n \in \mathbb{Z}\}$ and $h|(\alpha f^{2n},\beta f^{2n}) = f^{-2n}gf^{2n}$ and $h|(\alpha f^{2n+1},\beta f^{2n+1}) = f^{-(2n+1)}g^{-1}f^{2n+1}$ $(n \in \mathbb{Z})$. Then $f^{-1}hf = h^{-1}$ so $A(\Omega)$ is not an O-group ($h \succ e$ implies $f^{-1}hf \succ e$). Consequently, if $A(\Omega)$ is an O-group, it has no elements of bounded support. Hence Ω is primitive or has exactly one natural non-trivial congruence $\mathcal{C}$ that is static (Lemma 3.2.9). By Corollary 4.4.1, the unique non-static value is integral or transitively derived from a rigidly homogeneous chain ($A(\Omega)$ has no elements of bounded support). In either case, $A(\Omega)$ is clearly Abelian.

CHAPTER 5
THE WREATH PRODUCT

Let (G,Ω) be a transitive ordered permutation group. By Theorem 3A, the set of congruences of (G,Ω) form a chain. As in Section 3.1, let $k = k(G,\Omega)$ be the set of covering pairs of congruences of (G,Ω). If $K = (\mathcal{C}_K, \mathcal{C}^K) \in k$, then $\{(G_K,\Omega_K)\colon K \in k\}$ is the set of primitive components of (G,Ω). We will define an ordered permutation group called the <u>*Wreath product*</u> of $\{(G_K,\Omega_K)\colon K \in k\}$ (written <u>$Wr\{(G_K,\Omega_K)\colon K \in k(G,\Omega)\}$</u>) so that:

<u>THEOREM 5A</u>. *Let (G,Ω) be a transitive ordered (ℓ-)permutation group, $k = k(G,\Omega)$, and $\{(G_K,\Omega_K)\colon K \in k\}$ be the set of primitive components of (G,Ω). Let $(W,\hat{\Omega}) = Wr\{(G_K,\Omega_K)\colon K \in k\}$. Then $(W,\hat{\Omega})$ is a transitive ordered (ℓ-)permutation group whose set of primitive components is $\{(G_K,\Omega_K)\colon K \in k\}$. Moreover, if (H,Λ) is any transitive ordered (ℓ-)permutation group having $\{(G_K,\Omega_K)\colon K \in k\}$ as its set of primitive components, then (H,Λ) can be o-(ℓ-)embedded in $(W,\hat{\Omega})$.*

We first define the Wreath product of two (not necessarily primitive) ordered permutation groups.

The following lemma is easily verified (where we write $\{g_\tau\}$ as a shorthand for $\{g_\tau\colon \tau \in T\}$):

<u>LEMMA 5.1</u>. *Let (G,Ω) and (H,T) be ordered permutation groups. Let $\Lambda = \Omega \overleftarrow{\times} T$ and $W = \{(\{g_\tau\},h)\colon h \in H \;\&\; g_\tau \in G$ for all $\tau \in T\}$. Define a binary operation on W by:*
$(\{f_\tau\},h_1)(\{g_\tau\},h_2) = (\{a_\tau\},h_1h_2)$, where $a_\tau = f_\tau g_{\tau h_1}$ $(\tau \in T)$.
Then W is a group with identity $(\{e_\tau\},e)$--where $e_\tau = e$ $(\tau \in T)$--and $(\{g_\tau\},h)^{-1} = (\{f_\tau\},h^{-1})$ where $f_\tau = (g_{\tau h^{-1}})^{-1}$ $(\tau \in T)$. (W,Λ)

is an ordered permutation group if $(\alpha,\sigma)(\{g_\tau\},h) = (\alpha g_\sigma, \sigma h)$ $(\alpha \in \Omega,\ \sigma \in T)$. *Moreover,*

(i) (W,Λ) *is transitive if* (G,Ω) *and* (H,T) *are, and*

(ii) (W,Λ) *is an* ℓ*-permutation group if* (G,Ω) *and* (H,T) *are;* $(\{g_\tau\},h) \vee (\{e_\tau\},e) = (\{f_\tau\},h \vee e)$ *where*

$$\alpha f_\sigma = \begin{cases} \alpha g_\sigma & \text{if } \sigma h > \sigma \\ \alpha & \text{if } \sigma h < \sigma. \\ \alpha(g_\sigma \vee e) & \text{if } \sigma h = \sigma \end{cases}$$

(W,Λ) is called the *Wreath product* of (G,Ω) and (H,T) and is written *(G,Ω) Wr (H,T)*. We will also write *$W = G$ Wr H* if we wish to emphasise the group part. If $w = (\{g_\tau\},h) \in (G,\Omega)\ \mathrm{Wr}\ (H,T)$, then h is called the *global component* of w, and $\{g_\tau : \tau \in T\}$ the set of *local components* of w.

EXAMPLE 5.2. Let $(G,\Omega) = (H,T) = (\mathbb{Z},\mathbb{Z})$, the right regular representation of $\mathbb{Z}$. Let $f_n = \begin{cases} 1 & \text{if } n = 0 \\ 0 & \text{if } n \neq 0 \end{cases}$ and

$g_n = \begin{cases} 1 & \text{if } n = 1 \\ -1 & \text{if } n = 0 \\ 0 & \text{if } n \neq 0,1 \end{cases}$. Let $w_1 = (\{f_n\},1)$ and $w_2 = (\{g_n\},0)$.

$e = (\{0_n\},0) < w_1$; $w_2 \vee (\{0_n\},0) = (\{m_n\},0)$ where

$m_n = \begin{cases} 1 & \text{if } n = 1 \\ 0 & \text{if } n \neq 1 \end{cases}$; $w_1 w_2 = (\{k_n\},1) \neq (\{k_n'\},1) = w_2 w_1$ where

$k_n = \begin{cases} 2 & \text{if } n = 0 \\ -1 & \text{if } n = -1 \\ 0 & \text{if } n \neq -1,0 \end{cases}$ and $k_n' = \begin{cases} 1 & \text{if } n = 1 \\ 0 & \text{if } n \neq 1 \end{cases}$.

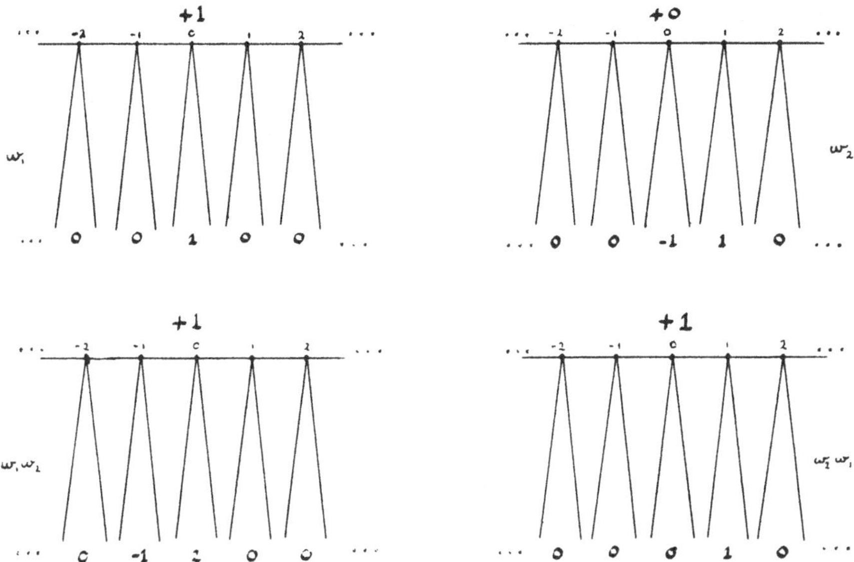

(To obtain w_1w_2, superimpose the w_2 picture moved left by +1 (the global component of w_1) on the w_1 picture. To obtain w_2w_1, superimpose the w_1 picture moved left by +0 (the global component of w_2) on the w_2 picture.)

A useful (ℓ-)subgroup of (G,Ω) Wr (H,T) is <u>(G,Ω) *wr* (H,T)</u>, the (<u>*restricted*</u>) <u>*wreath*</u> <u>*product*</u> of (G,Ω) and (H,T), and comprises those elements $(\{g_\tau\},h) \in W = G \text{ Wr } H$ such that $g_\tau = e$ for all but a finite number of $\tau \in T$.

Note that G Wr H and G wr H are splitting extensions of $\prod_{\tau\in T} G$ and $\sum_{\tau\in T} G$ respectively, by H, given by $(g(h\theta))_\tau = g_{\tau h^{-1}}$ $(\tau \in T,\ h \in H)$.

Let $\mathcal{C}$ be a congruence of a transitive ordered (ℓ-)permutation group (G,Ω). Let $\underline{L(\mathcal{C})} = \{g \in G: (\forall\alpha \in \Omega)\alpha g \mathcal{C} \alpha\}$, a convex normal ($\ell$-)subgroup of G. So $L(\mathcal{C})$ is the lazy subgroup of the action $(G,\Omega/\mathcal{C})$--recall that $(\beta\mathcal{C})g = \beta g\mathcal{C}$ $(\beta \in \Omega)$ gives an action. We call $(G/L(\mathcal{C}),\Omega/\mathcal{C})$ the global ordered permutation group of (G,Ω) relative to $\mathcal{C}$, or the $\underline{\mathcal{C}}$ *global part* of (G,Ω) for short. Let Δ be a $\mathcal{C}$ class. Recall that $(\hat{G}_{(\Delta)},\Delta)$, the $(\mathcal{S},\mathcal{C})$ component of (G,Ω), is an ordered permutation group (independent of the particular $\mathcal{C}$ class chosen--to within isomorphism), where $\hat{G}_{(\Delta)} = G_{(\Delta)}/G_\Delta$. It is called the $\underline{\mathcal{C}}$ *local part* of (G,Ω).

THEOREM 5.3. *Let $\mathcal{C}$ be a congruence of a transitive ordered (ℓ-)permutation group (G,Ω), and Δ be a $\mathcal{C}$ class. Then (G,Ω) can be o-(ℓ-)embedded in $(W,\Lambda) = (\hat{G}_{(\Delta)},\Delta)$ Wr $(G/L(\mathcal{C}),\Omega/\mathcal{C})$.*

Proof: For each $\mathcal{C}$ class A, choose $c_A \in G$ such that $\Delta c_A = A$, with $c_\Delta = e$. These are the *coordinate maps* and will remain fixed throughout the proof. Define $\phi\colon \Omega \to \Lambda = \Delta \overleftarrow{\times} (\Omega/\mathcal{C})$ by: $\alpha\phi = (\alpha c^{-1}_{\alpha\mathcal{C}},\alpha\mathcal{C})$. We show that ϕ preserves order and is one-to-one (it is clearly onto).

If $\alpha < \beta$, either $\alpha\mathcal{C} \neq \beta\mathcal{C}$ in which case $\alpha\phi < \beta\phi$ trivially, or $\alpha\mathcal{C} = \beta\mathcal{C}$. In this latter case, $\alpha\phi = (\alpha c^{-1}_{\alpha\mathcal{C}},\alpha\mathcal{C}) = (\alpha c^{-1}_{\beta\mathcal{C}},\beta\mathcal{C}) < (\beta c^{-1}_{\beta\mathcal{C}},\beta\mathcal{C}) = \beta\phi$, as desired

Let $g \mapsto \bar{g}$ be the natural map of G onto $G/L(\mathcal{C})$, and for each $\mathcal{C}$ class A, let $\hat{g}_A = c_A g c^{-1}_{Ag}|\Delta \in \hat{G}_{(\Delta)}$. Let $\psi\colon g \mapsto (\{\hat{g}_A\},\bar{g})$. We now show that ψ is an o-(ℓ-)embedding.

If $f < g$, then $\alpha\mathcal{C}f \leq \alpha\mathcal{C}g$ for all $\alpha \in \Omega$. Therefore $\bar{f} \leq \bar{g}$. If $A\bar{f} = A\bar{g}$, then $\alpha\hat{f}_A = \alpha c_A f c^{-1}_{Af} = \alpha c_A f c^{-1}_{Ag} \leq \alpha\hat{g}_A$ for all $\alpha \in A$ (so $f\psi \leq g\psi$). Moreover, if $\beta f < \beta g$, either $\beta\mathcal{C}\bar{f} < \beta\mathcal{C}\bar{g}$ or $\beta\hat{f}_{\beta\mathcal{C}} < \beta\hat{g}_{\beta\mathcal{C}}$ by the previous sentence. Hence $f\psi < g\psi$, so ψ is one-to-one and preserves order. If $h \in G$, then $\widehat{fh}_A = c_A fh c^{-1}_{Afh} = c_A f c^{-1}_{Af} c_{Af} h c^{-1}_{Afh} = \hat{f}_A \hat{h}_{A\bar{f}}$. Thus $(fh)\psi = (\{\widehat{fh}_A\},\overline{fh}) = (\{\hat{f}_A\},\bar{f})(\{\hat{h}_A\},\bar{h})$ so ψ is a group homomorphism. Clearly $(f \vee e)\psi = f\psi \vee e$ if $f \vee e$ exists. Consequently, ψ is an o-(ℓ-)embedding.

Now if $\alpha \in \Omega$ and $g \in G$, $(\alpha g)\phi = (\alpha g c^{-1}_{\alpha g \mathcal{C}}, \alpha g \mathcal{C}) = (\alpha c^{-1}_{\alpha \mathcal{C}}, \alpha \mathcal{C})(\{\hat{g}_A\}, \bar{g}) = (\alpha\phi)(g\psi)$. Thus we have an o-($\ell$-)embedding of (G,Ω) in $(\hat{G}_{(\Delta)}, \Delta)$ Wr $(G/L(\mathcal{C}), \Omega/\mathcal{C})$.

In the above proof, observe that ϕ is onto. Also $\hat{g}_A$ corresponds to a local shuffle inside the $\mathcal{C}$ class A, whereas $\bar{g}$ represents a shuffle of the $\mathcal{C}$ classes. Hence we are thinking of $g \in G$ as a set of local pieces, and one global piece.

We now obtain some information about the ℓ-ideals (i.e., convex normal ℓ-subgroups) of the wreath product.

LEMMA 5.4 (The Hourglass Lemma). *Let (G,Ω) and (H,T) be ℓ-permutation groups with (H,T) transitive. Then every ℓ-ideal N of G wr H is comparable (by inclusion) to $\sum_{\tau \in T} G$, and every ℓ-ideal of G Wr H not contained in $\prod_{\tau \in T} G$ contains $\sum_{\tau \in T} G$.*

Proof: Write G^σ for the set of $(\{g_\tau\}, e) \in G$ wr H with $g_\tau = e$ if $\tau \neq \sigma$ $(\sigma \in T)$. If N is an ℓ-ideal of G wr H with $N \not\subseteq \sum_{\tau \in T} G$, there is $(\{g_\tau\}, h) \in N$ with $h > e$. Let $\sigma \in T$ with $\sigma h > \sigma$. By convexity, $G^\sigma \subseteq N$. But if $\tau \in T$, $G^\tau = (\{e\}, f)^{-1} G^\sigma (\{e\}, f) \subseteq N$, where $f \in H$ and $\sigma f = \tau$. Hence $\sum_{\tau \in T} G \subseteq N$. The rest of the lemma is proved similarly.

The following lemma (the associative law) is straightforward verification:

LEMMA 5.5. *Let (G,Ω), (H,T) and (L,B) be ordered (ℓ-)permutation groups. Then $[(G,\Omega)$ Wr $(H,T)]$ Wr (L,B) is o-(ℓ-)isomorphic to (G,Ω) Wr $[(H,T)$ Wr $(L,B)]$.*

We will write *(G,Ω) Wr (H,T) Wr (L,B)* for either of the ordered (ℓ-)permutation groups obtained by Lemma 5.5.

We now wish to extend our construction to the Wreath product of (possibly infinitely many) ordered permutation groups. In order to motivate it, we give another way of viewing the Wreath product of (G,Ω) and (H,T).

If $(W,\Lambda) = (G,\Omega) \text{ Wr } (H,T)$, then define $\equiv$ on $\Lambda\ (= \Omega \overleftarrow{\times} T)$ by: $(\alpha,\sigma) \equiv (\beta,\tau)$ if $\sigma = \tau$. Clearly $\equiv$ is a congruence of (W,Λ). So $W \subseteq W_1 = \{f \in A(\Lambda)\colon (\forall\lambda_1)(\forall\lambda_2)[\lambda_1 f \equiv \lambda_2 f$ if and only if $\lambda_1 \equiv \lambda_2]\}$, and (W_1,Λ) is an ordered permutation group having $\equiv$ as a congruence. Each $f \in W_1$ induces an element $\bar{f} \in A(T)$ as follows: let $\tau \in T$ and choose $\alpha \in \Omega$. Then $\tau\bar{f} = \sigma$ where $(\alpha,\sigma) \equiv (\alpha,\tau)f$. Since W_1 respects $\equiv$, σ is independent of the choice of $\alpha \in \Omega$. Also each $f \in W_1$ induces $\hat{f}\colon T \to A(\Omega)$ via: $\alpha\hat{f}_\tau = \beta$ if $(\alpha,\tau)f = (\beta,\tau\bar{f})$--where we write $\hat{f}_\tau$ for the image of τ under $\hat{f}$. Thus we obtain an o-(ℓ-)isomorphism between W_1 and $A(\Omega) \text{ Wr } A(T)$ given by $f \mapsto (\hat{f},\bar{f})$. Hence
$(W,\Lambda) = (G,\Omega) \text{ Wr } (H,T) = \{f \in W_1\colon \bar{f} \in H \ \&\ \hat{f}_\tau \in G$ for all $\tau \in T\}$.

Let k be any chain and (G_K,Ω_K) be an ordered permutation group for each $K \in k$. Let $\underline{\Omega^+} = \prod\{\Omega_K\colon K \in k\}$ and choose an arbitrary fixed reference point in Ω^+, call it $\underline{0}$. For each $\alpha \in \Omega^+$, let $\underline{\mathit{supp}(\alpha)} = \{K \in k\colon \alpha_K \neq \underline{0}_K\}$. Let
$\underline{\hat{\Omega}} = \{\alpha \in \Omega^+\colon \text{supp}(\alpha)$ is an inversely well-ordered subset of $k\}$. Note that if $\alpha,\beta \in \hat{\Omega}$ are distinct, $\emptyset \neq \underline{k(\alpha,\beta)} = \{K \in k\colon \alpha_K \neq \beta_K\}$ $\subseteq \text{supp}(\alpha) \cup \text{supp}(\beta)$, and so $k(\alpha,\beta)$ is also inversely well-ordered. It therefore has a greatest element say K_0. We make $\hat{\Omega}$ a chain via: $\alpha < \beta$ if $\alpha_{K_0} < \beta_{K_0}$.

We next define some "natural" equivalence relations on $\hat{\Omega}$. For each $K \in k$, define $\equiv^K$ and $\equiv_K$ on $\hat{\Omega}$ by:

$$\alpha \equiv^K \beta \text{ if } \alpha_{K'} = \beta_{K'} \text{ for all } K' > K$$
$$\alpha \equiv_K \beta \text{ if } \alpha_{K'} = \beta_{K'} \text{ for all } K' \geq K.$$

Hence if $\alpha \neq \beta$ and K_0 is the largest element of $k(\alpha,\beta)$, then $\alpha \equiv^K \beta$ if $K \geq K_0$ and $\alpha \equiv_K \beta$ if $K > K_0$. Clearly $\equiv^K$ and $\equiv_K$ have convex classes (all $K \in k$). We wish them to be congruences, so let $\underline{W_1} = \{g \in A(\hat{\Omega})\colon (\forall K \in k)(\forall\alpha,\beta \in \hat{\Omega})[(\alpha \equiv^K \beta \iff \alpha g \equiv^K \beta g)\ \&$ $(\alpha \equiv_K \beta \iff \alpha g \equiv_K \beta g)]\}$. Then $\equiv^K$ and $\equiv_K$ are congruences of $(W_1,\hat{\Omega})$. Observe that $\alpha^{\equiv^K}/{\equiv_K}$ is just Ω_K for each $\alpha \in \hat{\Omega}$ and $K \in k$.

For each $K \in k$ and $\alpha \in \hat{\Omega}$, let $\underline{\alpha^K} \in \prod_{K'>K} \Omega_{K'}$ with $(\alpha^K)_{K'} = \alpha_{K'}$ $(K' > K)$; i.e., α^K is α above K. Note that $\alpha^K = \beta^K$ precisely when $\alpha \equiv^K \beta$.

For each $g \in W_1$, $\alpha \in \hat{\Omega}$ and $K \in k$, g induces an element of $A(\Omega_K)$ as we now show. Let $\sigma \in \Omega_K$ and define $\underline{g_{K,\alpha^K}}$ by: $\sigma g_{K,\alpha^K} = (\alpha' g)_K \in \Omega_K$, where $\alpha' \equiv^K \alpha$ and $\alpha'_K = \sigma$.

<u>LEMMA</u> <u>5.6</u>. *With the above notation,* $g_{K,\alpha^K} \in A(\Omega_K)$ *for each* $\alpha \in \hat{\Omega}$, $g \in W_1$ *and* $K \in k$.

<u>Proof</u>: If $\alpha \equiv^K \beta$ and $\beta_K = \sigma$, then $\alpha' \equiv_K \beta$. Since $g \in W_1$, $\beta g \equiv_K \alpha' g$; so $(\beta g)_K = (\alpha' g)_K$ and g_{K,α^K} is well-defined. If $\tau \in \Omega_K$ with $\sigma < \tau$, and $\gamma \equiv^K \alpha$ with $\gamma_K = \tau$, then $\gamma \equiv^K \alpha'$ but $\alpha'_K < \gamma_K$. Hence $\alpha' < \gamma$ and $(\alpha' g)_K < (\gamma g)_K$. Therefore g_{K,α^K} is one-to-one and preserves order. It clearly maps Ω_K onto Ω_K proving the lemma.

Let $\underline{W} = \{g \in W_1 : g_{K,\alpha^K} \in G_K$ for all $K \in k$ & $\alpha \in \hat{\Omega}\}$. Then $(W,\hat{\Omega})$ is called the *<u>Wreath</u>* *<u>product</u>* of $\{(G_K,\Omega_K): K \in k\}$ and is written $\underline{Wr\{(G_K,\Omega_K): K \in k\}}$. We may think of the elements of W as $k \times \hat{\Omega}$ matrices $(g_{K,\alpha})$ with $g_{K,\alpha} = g_{K,\beta}$ if $\alpha^K = \beta^K$. Note that the group operation on these matrices is given by:

$$(*)\quad (f_{K,\alpha})(g_{K,\alpha}) = (h_{K,\alpha}) \text{ where } h_{K,\alpha} = f_{K,\alpha} g_{K,\alpha f}$$

i.e., $h_{K,\alpha^K} = f_{K,\alpha^K} g_{K,(\alpha f)^K}$ $(K \in k,\ \alpha \in \hat{\Omega})$--not the usual matrix multiplication (cf. the definition of multiplication for the Wreath product of two ordered permutation groups).

<u>LEMMA</u> <u>5.7</u>. *If each* (G_K,Ω_K) *is an* ℓ*-permutation group, so is* $(W,\hat{\Omega}) = Wr\{(G_K,\Omega_K): K \in k\}$.

<u>Proof</u>: Let $g \in W$. We must show that $g \vee e$ exists in W. A routine verification shows that the pointwise supremum of g and

$$e \text{ is } h \text{ where } h_{K,\alpha} = \begin{cases} g_{K,\alpha} & \text{if } \alpha^K g > \alpha^K \\ e & \text{if } \alpha^K g < \alpha^K \\ g_{K,\alpha} \vee e & \alpha^K g = \alpha^K \end{cases}. \text{ Clearly}$$

$h \in W$, as desired. //

Our definition of the generalised Wreath product depended on the choice of the reference point $\underline{0}$. We wish to show that if each (G_K, Ω_K) is transitive, then the dependence is illusory. Specifically:

LEMMA 5.8. *Assume that each* (G_K, Ω_K) *is transitive. Then so is* $(W, \hat{\Omega}) = Wr\{(G_K, \Omega_K): K \in k\}$. *Moreover, if* $\underline{0}'$ $\Omega^\dagger = \prod\{\Omega_K: K \in k\}$ *and* $(W', \hat{\Omega}')$ *is the Wreath product of* $\{(G_K, \Omega_K): K \in k\}$ *with reference point* $\underline{0}'$, *then* $(W, \hat{\Omega})$ *and* $(W', \hat{\Omega}')$ *are o-(ℓ-)isomorphic.*

Proof: Let $\alpha, \beta \in \hat{\Omega}$. For each $K \in k$, there is $g_K \in G_K$ such that $\alpha_K g_K = \beta_K$ where we take $g_K = e$ if $K \notin k(\alpha,\beta)$. Now if g is defined by: $(\gamma g)_K = \gamma_K g_K$ $(\gamma \in \hat{\Omega};\ K \in k)$, then $\alpha g = \beta$. Moreover, if $\gamma \in \hat{\Omega}$, $\text{supp}(\gamma g) \subseteq \text{supp}(\gamma) \cup k(\alpha,\beta)$ which is inversely well-ordered. Hence $g|\hat{\Omega}$ maps $\hat{\Omega}$ into $\hat{\Omega}$. It is straightforward now to show that $g|\hat{\Omega} \in A(\hat{\Omega})$. Since $(g|\hat{\Omega})_{K,\gamma} = (g|\hat{\Omega})_K = g_K \in G_K$, $g|\hat{\Omega} \in W$ and $(W, \hat{\Omega})$ is transitive.

For each $K \in k$, there is $f_K \in G_K$ such that $0_K f_K = 0'_K$. Define $\phi: \Omega^\dagger \to \Omega^\dagger$ by: $(\alpha\phi)_K = \alpha_K f_K$ $(\alpha \in \Omega)$. Now ϕ is an ordermorphism and if $\alpha \in \hat{\Omega}$, $\alpha\phi \in \hat{\Omega}'$. Indeed, $\phi|\hat{\Omega}$ maps $\hat{\Omega}$ onto $\hat{\Omega}'$. Define $\psi: W_1 \to W'_1$ by: $g\psi = \phi^{-1} g \phi$. Clearly $(\alpha\phi)(g\psi) = (\alpha g)\phi$, and $\psi|W$ is an o-(ℓ-)isomorphism of W onto W' (since $f_K \in G_K$ for all $K \in k$). Hence (ψ, ϕ) is the desired o-(ℓ-)isomorphism.

We will prove the following strong form of Theorem 5A:

THEOREM 5.9. *Let* (G, Ω) *be a transitive ordered (ℓ-)permutation group,* $k = k(G, \Omega)$, *and* $\{(G_K, \Omega_K): K \in k\}$ *be the set of primitive components of* (G, Ω). *Let* $(W, \hat{\Omega}) = Wr\{(G_K, \Omega_K): K \in k\}$.

There exists an o-(ℓ-)embedding (ψ,ϕ) *of* (G,Ω) *into* $(W,\hat{\Omega})$ *such that (i) each* $\equiv_K$ *class contained in* $(\sigma\phi)\equiv^K$ *meets* $\Omega\phi$ $(K \in k, \sigma \in \Omega)$, *and (ii)* $(\sigma\phi)w \equiv^K \sigma\phi$ *implies there is* $g \in G$ *with* $\hat{\beta}(g\psi) \equiv_K \hat{\beta}w$ *for all* $\hat{\beta} \in (\sigma\phi)\equiv^K$ $(\sigma \in \Omega, w \in W)$.

In Theorem 5.3, the ϕ part was actually onto; in Theorem 5.9 ϕ need not be onto. However, what is claimed in (i) is that $\Omega\phi$ is dense in $\hat{\Omega}$ except possibly whole segments may be added to $\hat{\Omega}$ in those places where $k' \subseteq k$ is not coinitial in k, has no least element, and there are $\sigma_K \in \Omega$ $(K \in k')$ with $\sigma_{K_1}\mathcal{C}^{K_1} \subseteq \sigma_{K_2}\mathcal{C}^{K_2}$ if $K_1 \leq K_2$ in k' but $\bigcap_{K \in k'} \sigma_K\mathcal{C}^K = \emptyset$. (ii) shows that $(W,\hat{\Omega})$ has the same primitive components as (G,Ω).

Before we can prove Theorem 5.9, we need a technical lemma.

Let L be a subgroup of a group H. T(L) is a *transversal* of L in H if T(L) contains exactly one element of each right coset of L in H; i.e., $|T(L) \cap Lh| = 1$ for all $h \in H$.

LEMMA 5.10. *If* H *is any group, there is a set of transversals* $\{T(L)\colon L$ *a subgroup of* $H\}$ *with* $T(L_2) \subseteq T(L_1)$ *if* $L_1 \subseteq L_2$, *and* $T(L) \cap L = \{e\}$ *for all subgroups* L.

Proof: Well-order the set of elements of H with e first. Let L be a subgroup of H and $h \in H$. Choose the least element of Lh for the unique member of T(L) in Lh. Clearly, $\{T(L)\colon L$ subgroup of $H\}$ has the required properties.

Proof of Theorem 5.9: Fix $\sigma_0 \in \Omega$ and take $\Omega_K = \sigma_0\mathcal{C}^K/\mathcal{C}_K$ $(K \in k)$. Let T be a transversal function for the subgroups of G guaranteed by Lemma 5.10. For each $\sigma \in \Omega$, let $c(\sigma,K)$ be the unique element of $T(G_{(\sigma_0\mathcal{C}^K)})$ such that $\sigma_0 c(\sigma,K)\mathcal{C}^K\sigma$. These $c(\sigma,K)$ $(\sigma \in \Omega, K \in k)$ are the *coordinate maps* (cf. proof of Theorem 5.3). Note that $c(\sigma_0,K) = e$ by Lemma 5.10. Define $\phi\colon \Omega \to \Omega^\dagger = \prod_{K \in k} \Omega_K$ by $(\sigma\phi)_K = \sigma c(\sigma,K)^{-1}\mathcal{C}_K \in \Omega_K$ $(\sigma \in \Omega, K \in k)$. Let $\underline{0} = \sigma_0\phi$ (so $\underline{0}_K = \sigma_0\mathcal{C}_K$) and let $\hat{\Omega}$ be the subset of $\Omega^\dagger$

comprising those elements whose support (with respect to $\underline{0}$) is inversely well-ordered. We now show that $\Omega\phi \subseteq \hat{\Omega}$ and that ϕ is one-to-one and order-preserving.

Let $\sigma \in \Omega$ and $\emptyset \neq k' \subseteq \mathrm{supp}(\sigma\phi)$. Choose $f \in G$ with $\sigma_0 f = \sigma$, and let $H = \bigcup_{K \in k'} G_{(\sigma_0 \mathcal{C}^K)}$. Since $\{G_{(\sigma_0\mathcal{C}^K)} : K \in k'\}$ is a tower of subgroups, H is a subgroup of G. Let $T(H) \cap Hf = \{g\}$, say. Then $fg^{-1} \in H$ so $fg^{-1} \in G_{(\sigma_0\mathcal{C}^{K_1})}$ for some $K_1 \in k'$. We now show by reductio ad absurdum that K_1 is the largest element of k' whence $\mathrm{supp}(\sigma\phi)$ is inversely well-ordered (and so $\sigma\phi \in \hat{\Omega}$). If $K_1 < K \in k'$, then $\sigma_0\mathcal{C}^{K_1} \subsetneqq \sigma_0\mathcal{C}^K \subseteq H$ and, by the choice of T, $g \in T(H) \subseteq T(G_{(\sigma_0\mathcal{C}^K)})$. But $fg^{-1} \in G_{(\sigma_0\mathcal{C}^{K_1})} \subseteq G_{(\sigma_0\mathcal{C}_K)}$, so $\sigma_0 fg^{-1}\mathcal{C}_K\sigma_0$. Hence $\sigma g^{-1}\mathcal{C}_K\sigma_0$, so $\sigma\mathcal{C}^K\sigma_0 g$. Thus $g = c(\sigma,K)$. Therefore, $(\sigma\phi)_K = \sigma c(\sigma,K)^{-1}\mathcal{C}_K = \sigma g^{-1}\mathcal{C}_K = \sigma_0\mathcal{C}_K = \underline{0}_K$. This contradicts the fact that $K \in \mathrm{supp}(\sigma\phi)$. Consequently, $\Omega\phi \subseteq \hat{\Omega}$.

Now let $\sigma,\tau \in \Omega$ with $\sigma < \tau$. Let $K = \mathrm{Val}(\sigma,\tau)$. Since $\sigma\mathcal{C}^K\tau$, $c(\sigma,K) = c(\tau,K)$. So $(\sigma\phi)_K = \sigma c(\sigma,K)^{-1}\mathcal{C}_K < \tau c(\sigma,K)^{-1}\mathcal{C}_K = \tau c(\tau,K)^{-1}\mathcal{C}_K = (\tau\phi)_K$. If $K' > K$, then $\mathcal{C}_{K'} \supseteq \mathcal{C}^K$. Thus $(\sigma\phi)_{K'} = \sigma c(\sigma,K')^{-1}\mathcal{C}_{K'} = \tau c(\sigma,K')^{-1}\mathcal{C}_{K'} = \tau c(\tau,K')^{-1}\mathcal{C}_{K'} = (\tau\phi)_{K'}$. Hence $\sigma\phi < \tau\phi$ completing the claim.

We next show that $\sigma\mathcal{C}^K\tau$ if and only if $\sigma\phi \equiv^K \tau\phi$, and $\sigma\mathcal{C}_K\tau$ if and only if $\sigma\phi \equiv_K \tau\phi$ $(K \in k)$.

Suppose $\sigma\mathcal{C}^K\tau$. Then $\sigma\mathcal{C}_{K'}\tau$ for all $K' > K$; so $c(\sigma,K') = c(\tau,K')$ if $K' > K$. Thus $(\sigma\phi)_{K'} = \sigma c(\sigma,K')^{-1}\mathcal{C}_{K'} = (\tau\phi)_{K'}$. Hence $\sigma\phi \equiv^K \tau\phi$. Conversely, if $\tau \notin \sigma\mathcal{C}^K$, let $K' = \mathrm{Val}(\sigma,\tau)$. Then $K' > K$ and $c(\sigma,K') = c(\tau,K')$. Thus $(\sigma\phi)_{K'} = \sigma c(\sigma,K')^{-1}\mathcal{C}_{K'} = \sigma c(\tau,K')^{-1}\mathcal{C}_{K'} \neq (\tau\phi)_{K'}$. Hence $\sigma\phi \not\equiv^K \tau\phi$. The proof for $\sigma\mathcal{C}_K\tau$ is similar.

We now show that ϕ satisfies (i).

So let $\sigma \in \Omega$ and $\hat{\alpha} \in \hat{\Omega}$ with $\sigma\phi \equiv^K \hat{\alpha}$. $\hat{\alpha}_K \in \Omega_K = \sigma_0\mathcal{C}^K/\mathcal{C}_K$ so $\hat{\alpha}_K = \tau\mathcal{C}_K$ for some $\tau \in \sigma_0\mathcal{C}^K$. Let

$\tau' = \tau c(\sigma,K) \in \Omega$. Then $\tau' \in \sigma_0 c(\sigma,K)\mathfrak{C}^K = \sigma\mathfrak{C}^K$. Thus $\tau'\phi \equiv^K \sigma\phi \equiv^K \hat{\alpha}$ by the preceding paragraph. But $(\tau'\phi)_K = \tau' c(\tau',K)^{-1}\mathfrak{C}_K = \tau' c(\sigma,K)^{-1}\mathfrak{C}_K = \tau\mathfrak{C}_K = \hat{\alpha}_K$. Hence $\tau'\phi \equiv_K \hat{\alpha}$, as required.

We must now define $\psi: G \to W$. Let $\hat{\alpha} \in \hat{\Omega}$, $K \in k$ and $g \in G$. Let $$(\hat{\alpha}(g\psi))_K = \begin{cases} ((\sigma g)\phi)_K & \text{if } \sigma\phi \equiv_K \hat{\alpha} \text{ for some } \sigma \in \Omega \\ \hat{\alpha}_K & \text{otherwise} \end{cases}.$$ This is well-defined (since $\sigma\phi \equiv_K \tau\phi$ implies $\sigma\mathfrak{C}_K\tau$ and so $\sigma g\mathfrak{C}_K\tau g$; whence $(\sigma g)\phi \equiv_K (\tau g)\phi$). Moreover, by the definition, $(\sigma\phi)(g\psi) = (\sigma g)\phi$. It therefore remains to show that $g\psi \in W$ and ψ is an o-(ℓ-)embedding.

Let $\hat{\alpha} \in \hat{\Omega}$ and let $\emptyset \neq k' \subseteq \text{supp}(\hat{\alpha}(g\psi))$. If $k' \subseteq \text{supp}(\hat{\alpha})$, k' has a greatest element. If $k' \not\subseteq \text{supp}(\hat{\alpha})$, let $K \in k' \setminus \text{supp}(\hat{\alpha})$. There is $\sigma \in \Omega$ with $\sigma\phi \equiv_K \hat{\alpha}$. For all $K' > K$, $(\hat{\alpha}(g\psi))_{K'} = ((\sigma g)\phi)_{K'}$. Hence $\emptyset \neq \{K' \in k': K' \geq K\} \subseteq \{K' \in k: K' \geq K \ \&\ K' \in \text{supp}((\sigma g)\phi)\} \subseteq \text{supp}((\sigma g)\phi)$. Since $\text{supp}((\sigma g)\phi)$ is inversely well-ordered, it follows that k' has a greatest element. Therefore $\text{supp}(\hat{\alpha}(g\psi))$ is inversely well-ordered, so $\hat{\alpha}(g\psi) \in \hat{\Omega}$. It is now straightforward to check that $g\psi \in A(\hat{\Omega})$ and respects each $\equiv^K$ and $\equiv_K$ $(K \in k)$. But $(g\psi)_{K,\hat{\alpha}} \in G_K$ for each $K \in k$, $\hat{\alpha} \in \hat{\Omega}$. Consequently, $g\psi \in W$. Now $$(\hat{\alpha}(f\psi)(g\psi))_K = \left\{\begin{matrix} (((\sigma f)g)\phi)_K & \text{if } \sigma\phi \equiv_K \hat{\alpha} \text{ for some } \sigma \in \Omega \\ \hat{\alpha}_K & \text{otherwise} \end{matrix}\right\} =$$ $(\hat{\alpha}((fg)\psi))_K$, so ψ is a homomorphism. Moreover, it is easily verified that $f\psi < g\psi$ precisely when $f < g$ and that $(g \vee e)\psi = g\psi \vee e$. Hence (ψ,ϕ) is an embedding of (G,Ω) into $(W,\hat{\Omega})$.

Finally, we must prove (ii). Let $\sigma \in \Omega$, $w \in W$ and $(\sigma\phi)w \equiv^K \sigma\phi$. Then $w_{K,\sigma\phi} \in G_K$. So $w_{K,\sigma\phi} = f_K$ for some $f \in G_{(\sigma_0\mathfrak{C}^K)}$, where f_K denotes the image of f in G_K. Let $g = c(\sigma,K)^{-1}fc(\sigma,K) \in G_{(\sigma\mathfrak{C}^K)}$. If $\hat{\beta} \equiv^K \sigma\phi$, let $\tau \in \Omega$ with

$\tau\phi \equiv_K \hat{\beta}$ (by (i)). Hence $c(\sigma,K) = c(\tau,K)$. Thus $(\hat{\beta}w)_K = ((\tau\phi)w)_K = (\tau c(\sigma,K)^{-1}\mathfrak{C}_K)w_{K,\sigma\phi} = \tau c(\sigma,K)^{-1}\mathfrak{C}_K f_K = \tau g c(\sigma,K)^{-1}\mathfrak{C}_K = ((\tau g)\phi)_K = ((\tau\phi)(g\psi))_K = (\hat{\beta}(g\psi))_K$. Consequently, $\hat{\beta}w \equiv_K \hat{\beta}(g\psi)$ for all $\hat{\beta} \equiv^K \sigma\phi$. Q.E.D.

Note that the ψ given in the proof is by no means unique. However, the one given in the proof is called a *standard embedding* of G in W.

This chapter covers some of the material to be found in Holland [69], Holland and McCleary [69], and McCleary [70].

PART III

APPLICATIONS TO ORDERED PERMUTATION GROUPS

CHAPTER 6
SIMPLE ℓ-PERMUTATION GROUPS

An ℓ-group G having no proper ℓ-ideals is called <u>*ℓ-simple*</u> (Recall that an ℓ-ideal is a normal convex ℓ-subgroup). Such ℓ-groups have a faithful transitive representation (Appendix I Corollary 8). In this chapter we wish to determine which transitive primitive ℓ-permutations are ℓ-simple and which are simple (no non-trivial normal subgroups; i.e., simple as groups). By Theorem 4A, there are four cases to consider if we split the doubly transitive case into two, pathological and non-pathological. The regular case is trivial. They are subgroups of $\mathbb{R}$ and hence are ℓ-simple but not simple. The non-pathological doubly transitive case is also quite straightforward. Recall that if Ω is doubly homogeneous, every element of $B(\Omega)$ is a commutator (Corollary 2.2.7) and $B(\Omega)$ is simple (Theorem 2G). For a group H, write <u>$[H,H]$</u> for the subgroup of H generated by $\{[h_1,h_2]: h_1,h_2 \in H\}$ (Recall $[h_1,h_2] = h_1^{-1}h_2^{-1}h_1h_2$)--so $[B(\Omega),B(\Omega)] = B(\Omega)$ by Corollary 2.2.7. We generalise Theorem 2G. Note first that $G \cap B(\Omega)$ is an ℓ-ideal of G for any ℓ-permutation group (G,Ω).

<u>THEOREM</u> <u>6A</u>. *Let (G,Ω) be a doubly transitive ℓ-permutation group with $G \cap B(\Omega) \neq \{e\}$. Then G is ℓ-simple if and only if $G \subseteq B(\Omega)$.*

<u>THEOREM</u> <u>6B</u>. *Let (G,Ω) be a doubly transitive ℓ-permutation group with $G \cap B(\Omega) \neq \{e\}$. Then $[G \cap B(\Omega), G \cap B(\Omega)]$ is simple.*

The periodic primitive case is also quite straightforward. Recall that if (G,Ω) is periodic primitive with period z and

$G = C_{A(\Omega)}(z)$, we say that (G,Ω) is full. In this case, if (G,Ω) has Config(m) for some $m < \infty$, $z^m \in G$ by Lemma 4.3.9 and $\langle z^m \rangle$ is a non-trivial normal subgroup of G. Hence if (G,Ω) is a full periodic primitive ℓ-permutation group, G is simple only if (G,Ω) has Config(∞).

<u>THEOREM</u> <u>6C</u>. *If (G,Ω) is periodic primitive, then G is ℓ-simple.*

<u>THEOREM</u> <u>6D</u>. *Let (G,Ω) be full periodic primitive. Then G is simple if and only if (G,Ω) has Config(∞).*

In the regular and periodic cases, $G \cap B(\Omega) = \{e\}$. Hence, in view of Theorems 6A and 6C, we might hope that if (G,Ω) is a transitive primitive ℓ-permutation group, then G is ℓ-simple if and only if $G \cap B(\Omega) = \{e\}$ or G. Were this true, G would have to be ℓ-simple whenever (G,Ω) is pathological. However:

<u>THEOREM</u> <u>6E</u>. *There is a pathological ℓ-permutation group (G,Ω) such that G is not ℓ-simple.*

Finally, we prove:

<u>THEOREM</u> <u>6F</u>. *There is a simple o-group that has no faithful primitive representation.*

The proofs of Theorems 6A, 6B and 6D exploit similar ideas to those we used to prove Theorem 2G. In Corollary 2.1.4 we expressed "bounded support" in $A(\Omega)$ for Ω homogeneous. Recall that if (G,Ω) is a transitive ℓ-permutation group and $e \neq g_1, g_2 \in G$, then $g_1 \mathcal{L} g_2$ if and only if g_1 is to the left of g_2 (Lemma 2.1.3). We now generalise Corollary 2.1.4.

LEMMA 6.1. *If (G,Ω) is a transitive ℓ-permutation group and $e \neq g \in G$, then g has bounded support if and only if $(\exists h)(\forall f \geq e)(g\mathcal{L}(f^{-1}h^{-1}ghf))$.*

Proof: If $\text{supp}(g) \subseteq (\alpha,\beta)$, let $h \in G$ be such that $\alpha h = \beta$. Hence if $f \geq e$, $\beta \leq \alpha hf < \text{supp}(f^{-1}h^{-1}ghf)$. By Lemma 2.1.3, $g\mathcal{L}(f^{-1}h^{-1}ghf)$.

Conversely, if $g\mathcal{L}(f^{-1}h^{-1}ghf)$ for all $f \geq e$, then g is to the left of $h^{-1}gh$ by Lemma 2.1.3. Thus $\text{supp}(h^{-1}gh)$ is bounded below. By the fundamental triviality, so is $\text{supp}(g)$. Hence g has bounded support. //

Let G be an ℓ-group and $e \neq g \in G$. We call g *insular* if $(\exists h)(\forall f \geq e)(g\mathcal{L}(f^{-1}h^{-1}ghf))$. Insularity is an ℓ-group concept not an ℓ-permutation one. We now prove the following strong version of Theorem 6A due to Holland [65b].

THEOREM 6.2. *Let $G \neq \{e\}$ be an ℓ-group. Then G is ℓ-simple and contains an insular element if and only if for some chain Ω, (G,Ω) is a doubly transitive ℓ-permutation group with $G \subseteq B(\Omega)$.*

We first need a lemma:

LEMMA 6.3. *Let (G,Ω) be a transitive primitive ℓ-permutation group and L an ℓ-ideal of G. If $L \neq \{e\}$, (L,Ω) is a transitive ℓ-permutation group. If (G,Ω) is doubly transitive, so is (L,Ω) and $L \supseteq B(\Omega) \cap G$.*

Proof: Since $L \neq \{e\}$, $\Delta = \alpha L \neq \{\alpha\}$ for some $\alpha \in \Omega$. By Lemma 1.6.9, Δ is convex since L is a convex ℓ-subgroup of G. If $g \in G$ and $\Delta g \cap \Delta \neq \emptyset$, there exist $f,h \in L$ such that $\alpha fg = \alpha h$. Let $\beta \in \Delta$. So $\beta = \alpha k$ for some $k \in L$. $\beta g = \alpha kg = \alpha hg^{-1}(f^{-1}k)g$. $f^{-1}k \in L$ so $g^{-1}(f^{-1}k)g \in L$ $(L \triangleleft G)$. Hence $\beta g \in \Delta$. So $\Delta g = \Delta$. Therefore Δ is a block of (G,Ω). Since (G,Ω) is primitive $\Delta = \Omega$; i.e., $\alpha L = \Omega$. Consequently, (L,Ω) is transitive. If (G,Ω) is doubly transitive, let $\alpha < \beta < \gamma$ in Ω. Let $e < g \in G_\alpha$ be such that $\beta g = \gamma$. Choose

$e < f \in L$ such that $\beta f = \gamma$. Then $g \wedge f \in L_\alpha$ and $\beta(g \wedge f) = \gamma$. By Lemma 1.10.2, (L,Ω) is doubly transitive.

Let $e < g \in B(\Omega) \cap G$, say $\mathrm{supp}(g) \subseteq (\alpha,\beta)$. Since (L,Ω) is transitive, there exists $e < f \in L$ such that $\alpha f = \beta$. By Lemma 1.9.4, $e < g < f$ so $g \in L$. But in any ℓ-group $h = (h \vee e)(h^{-1} \vee e)^{-1}$. Hence $B(\Omega) \cap G \subseteq L$.

We now prove Theorem 6.2. If (G,Ω) is doubly transitive and $G \subseteq B(\Omega)$, then as each ℓ-ideal $\neq \{e\}$ contains $G \cap B(\Omega) = G$ (by Lemma 6.3), G is ℓ-simple.

Conversely, suppose G is ℓ-simple and contains an insular element. By Appendix I Corollary 8, there is a chain T such that (G,T) is a transitive ℓ-permutation group. By Lemma 6.1, $G \cap B(T) \neq \{e\}$. Since $G \cap B(T)$ is an ℓ-ideal of G, $G \subseteq B(T)$. However, (G,T) may not be primitive. Let $e < f \in G$, say $\mathrm{supp}(f) \subseteq (\alpha,\beta)$. Let $\mathcal{C}$ be any non-trivial congruence of (G,T). Now $L(\mathcal{C}) = \{g \in G\colon (\forall\gamma)\gamma g \mathcal{C} \gamma\}$ is an ℓ-ideal of G and hence $L(\mathcal{C}) = \{e\}$. But $\beta \notin \alpha\mathcal{C}$ (Otherwise $e \neq f \in L(\mathcal{C})$). Since the union of a family of congruences is a congruence (Corollary 3B), (G,T) has a maximal proper congruence $\mathcal{C}_0$, namely the union of all non-trivial congruences. Let $\Omega = T/\mathcal{C}_0$. $L(\mathcal{C}_0) = \{e\}$ so (G,Ω) is a transitive primitive ℓ-permutation group. By Lemma 6.1, $G \cap B(\Omega) \neq \{e\}$ so $G \subseteq B(\Omega)$. By Theorem 4A, (G,Ω) is doubly transitive.

To prove Theorem 6B, it is enough to show (by Lemma 1.10.1):

<u>LEMMA</u> <u>6.4</u>. *If (G,Ω) is a triply transitive ordered permutation group and $G \cap B(\Omega) \neq \{e\}$, then $([G \cap B(\Omega), G \cap B(\Omega)],\Omega)$ is a triply transitive ordered permutation group. Moreover, if $h_1,h_2 \in G \cap B(\Omega)$ and $e \neq g \in G$, there exist $f_1,f_2,f_3,f_4 \in G \cap B(\Omega)$ such that $[h_1,h_2] = f_1^{-1}g^{-1}f_1 \cdot f_2^{-1}gf_2 \cdot f_3^{-1}g^{-1}f_3 \cdot f_4^{-1}gf_4$. Thus $[G \cap B(\Omega), G \cap B(\Omega)]$ is contained in every proper normal subgroup of G and is simple.*

Proof: Let $H = G \cap B(\Omega)$. Since $[H,H] \triangleleft G$ and $\{e\} \neq [H,H] \subseteq B(\Omega)$, $([H,H],\Omega)$ is triply transitive by Lemma 1.10.5.

Let $h_1,h_2 \in H$ and $e \neq g \in G$. Let $\alpha \in \text{supp}(g)$ and $\text{supp}(h_i) \subseteq (\sigma,\tau)$ $(i = 1,2)$. Without loss of generality, $\alpha < \alpha g$. There exists $f \in [H,H]$ such that $\sigma f = \alpha$ and $\tau f = \alpha g$. Now $\text{supp}(f^{-1}h_i f) \subseteq (\alpha,\alpha g)$ so $\text{supp}(g^{-1}f^{-1}h_i fg) \subseteq (\alpha g,\alpha g^2)$ $(i = 1,2)$. Since $f^{-1}h_2 f$ and $g^{-1}f^{-1}h_1 fg$ have disjoint supports, they commute. Thus $[h_1,h_2] = h_1^{-1}f\cdot f^{-1}h_2^{-1}f\cdot f^{-1}h_1h_2 =$ $h_1^{-1}f\cdot(g^{-1}f^{-1}h_1fgf^{-1}\cdot f\cdot g^{-1}f^{-1}h_1^{-1}fg)\cdot f^{-1}h_2^{-1}f\cdot f^{-1}h_1h_2 =$ $h_1^{-1}fg^{-1}f^{-1}h_1\cdot fgf^{-1}\cdot f\cdot f^{-1}h_2^{-1}f\cdot g^{-1}f^{-1}h_1^{-1}fg\cdot f^{-1}h_1h_2 =$ $f_1^{-1}g^{-1}f_1\cdot f_2^{-1}gf_2\cdot f_3^{-1}g^{-1}f_3\cdot f_4^{-1}gf_4$ where $f_1 = f^{-1}h_1$, $f_2 = f^{-1}$, $f_3 = f^{-1}h_2$ and $f_4 = f^{-1}h_1h_2$. Since $f_j \in H$ $(j = 1,2,3,4)$, every normal subgroup of H contains $[H,H]$.

It remains only to show that $[H,H]$ is simple. Since $([H,H],\Omega)$ is triply transitive, it follows (with $H_1 = [H,H]$ in place of H in the previous paragraph) that any non-trivial normal subgroup of H_1 contains $[H_1,H_1]$. But $[H_1,H_1] \triangleleft H$ and so contains H_1. Hence H_1 is simple.

Lemma 6.4 is due to Graham Higman [54].

We now prove Theorem 6C by a standard compactness argument. Let z be the period, $\alpha \in \Omega$ and $e < f,g \in G$. Let $\bar\beta \in [\alpha,\alpha z]$. By Theorem 4.1.1, $\bar\beta G$ is dense in $\bar\Omega$. So if $\gamma \in \text{supp}(g)$, there exists $h \in G$ such that $\gamma < \bar\beta h < \gamma g$. Hence $\bar\beta \in \text{supp}(hgh^{-1})$. Let $g_{\bar\beta} = hgh^{-1}$. Then $\bar\beta \in \text{supp}(g_{\bar\beta})$ so $\{\text{supp}(g_{\bar\beta})\colon \bar\beta \in [\alpha,\alpha z]\}$ is an open cover of the compact subset $[\alpha,\alpha z]$ of $\bar\Omega$. Hence there are $\bar\beta_1,\ldots,\bar\beta_n \in [\alpha,\alpha z]$ such that $\bar g = g_{\bar\beta_1}\cdots g_{\bar\beta_n}$ fixes no element of $\bar\Omega$ in $[\alpha,\alpha z]$. By the periodicity, $\bar g$ fixes no point of $\bar\Omega$. Thus for some $m \in \mathbb{Z}^+$, $\alpha\bar g^m \geq \alpha z f$. If $\beta \in [\alpha,\alpha z]$, then $\beta f \leq \beta\bar g^m$; so $f \leq \bar g^m$ on $[\alpha,\alpha z]$ and hence on Ω. Consequently, f belongs to the ℓ-ideal of G generated by g. Therefore G is ℓ-simple.

We next prove Theorem 6D. So assume that (G,Ω) is full periodic primitive with period z and Config(∞). For each $\alpha \in \Omega$, let $\mathcal{N}(\alpha)$ be the filter of neighbourhoods of α (with respect to the open interval topology). So $G_{\mathcal{N}(\alpha)} = \{g \in G: (\exists \Delta \in \mathcal{N}(\alpha)) g \in G_{\Delta}\}$ is an ℓ-subgroup of G. Clearly, $\cup\{G_{\mathcal{N}(\alpha)}: \alpha \in \Omega\}$ generates G (as a group). Hence to prove that G is simple, it is enough to show that if $e \neq g \in G$, then $G_{\mathcal{N}(\alpha)} \subseteq N$ for each $\alpha \in \Omega$, where N is the normal subgroup of G generated by g. Fix $\alpha \in \Omega$ and $g \in G$.

We first show there is $e \neq f \in N$ such that $\alpha f = \alpha$. If $\alpha g = \alpha$ we are home; so assume $\alpha < \alpha g$ without loss of generality. Since (G,Ω) has Config(∞), $\alpha g^{-1}, \alpha g \notin \{\alpha z^m: m \in \mathbb{Z}\}$. So for some $m,n \in \mathbb{Z}$, $\alpha g^{-1}z^m, \alpha g z^n \in (\alpha,\alpha z) = \bar{\Delta}$. But $\alpha g^{-1}z^m \neq \alpha g z^n$; hence there exist $\sigma,\tau \in \Delta$ such that $\alpha g z^n \in (\sigma,\tau)$ but $\alpha g^{-1}z^m \notin (\sigma,\tau)$. Since $(G_\alpha|\Delta,\Delta)$ is non-pathologically doubly transitive (Theorem 4.3.1), there is $e < h \in G_\alpha$ such that $\alpha g z^n \in \text{supp}(h) \cap \bar{\Delta} \subseteq (\sigma,\tau)$. Let $f = [g,h] = g^{-1}\cdot h^{-1}gh \in N$. Then $\alpha z^m f = \alpha z^m g^{-1}h^{-1}gh = \alpha g^{-1}z^m h^{-1}gh = \alpha h z^m = \alpha z^m$; so $\alpha f = \alpha$. However, $\alpha g z^n f = \alpha z^n gh \neq \alpha g z^n$, so $f \neq e$ and the claim is proved.

Now let $h \in G_{\mathcal{N}(\alpha)}$ and $e \neq f \in N$ be such that $\alpha f = \alpha$. But $h|(\alpha,\alpha z)$ has bounded support (If $(\sigma,\tau) \in \mathcal{N}(\alpha)$ and $\delta h = \delta$ for all $\delta \in (\sigma,\tau)$, $\tau < \text{supp}(h|(\alpha,\alpha z)) < \sigma z < \alpha z$). By the fullness, if $\Lambda = (\alpha,\alpha z)$ we may identify $A(\Lambda)$ with $G_{(\Lambda)}$. Since $f \in G_{(\Lambda)}$, $h|\Lambda$ belongs to the normal subgroup of $A(\Lambda)$ generated by $f|\Lambda$ by Theorem 2G. Hence $h \in M$, the normal subgroup of G generated by f. Since $f \in N$, $M \subseteq N$. Thus $h \in N$ and the proof is complete.

Recall Example 1.10.3. $(G,\mathbb{R})$ is pathological. A minor modification of the proof of Theorem 6D gives:

<u>THEOREM</u> <u>6.5</u>. *Let* $G = \{g \in A(\mathbb{R}): (\exists m \in \mathbb{Z}^+)(\forall \alpha)(\alpha + m)g = \alpha g + m\}$. *Then* G *is simple (but* $(G,\mathbb{R})$ *is pathological).*

Theorems 6C, 6D and 6.5 are due to McCleary ([72a] and [78a]) as is the next example [73b].

EXAMPLE 6.6. Let $H = \{h \in A(\mathbb{R}): (\forall\alpha \in \mathbb{R})(\exists m \in \mathbb{Z}^+)(\forall n \in \mathbb{Z})$ $(\alpha + mn)g = \alpha g + mn\}$. Then $(H,\mathbb{R})$ is an ℓ-simple pathological ℓ-permutation group.

We leave the details for the reader (or see McCleary [73b]). For other examples of ℓ-simple pathological ℓ-permutation groups, see Glass and McCleary [76]. It is not known if Example 6.6 and the examples in Glass and McCleary [76] are indeed simple (as groups).

We now prove Theorem 6E. Let $\Omega = \mathbb{R}^+$ and $\bar{G} = \{g \in A(\mathbb{R}^+)$: $(\exists m \in \mathbb{Z}^+)(\forall\alpha \in \mathbb{R}^+)(\alpha + m)g = \alpha g + m\}$. It is straightforward to see that $(\bar{G},\mathbb{R}^+)$ is a pathological ℓ-permutation group. However, $R(\Omega) \cap \bar{G}$ is a proper ℓ-ideal of $\bar{G}$, so $\bar{G}$ is not ℓ-simple.

We give another informative example (Glass [74] and McCleary [81c]):

THEOREM 6.7. *Let F_κ be the free ℓ-group on κ generators (κ a cardinal greater than 1, possibly infinite). Then F_κ has a faithful pathological representation.*

Proof: We will only prove this theorem when $\kappa = \aleph_0$. The proof can easily be adapted to the case that κ is any infinite cardinal. The proof for κ finite is more delicate--see McCleary [81c].

We now prove that for some ψ, $(F_{\aleph_0},\mathbb{Q},\psi)$ is a pathological ℓ-permutation group.

Let $A = \{\langle\alpha,\beta\rangle: \alpha,\beta \in \mathbb{Q}\ \&\ \alpha < \beta\}$ and $B = A \times A$. Since $|B| = \aleph_0$, there is $\phi: \omega \to B$ one-to-one and onto. Let $\langle\alpha_m,\beta_m,\gamma_m,\delta_m\rangle = \phi(m)$. Define a sequence $\{J_m: m \in \omega\}$ of bounded open intervals of $\mathbb{Q}$ such that $J_0 \subseteq J_1 \subseteq J_2 \subseteq \ldots$ and $[\alpha_m,\beta_m] \cup [\gamma_m,\delta_m] \subseteq J_m$ $(m \in \omega)$. Let $\{I_n: n \in \mathbb{Z}\}$ be any set of bounded non-empty open intervals of $\mathbb{Q}$ such that $I_n < I_{n'}$ if $n < n'$ and $I_{-m} < J_m < I_m$ $(m \in \mathbb{Z}^+)$. Now J_m is ordermorphic to $\mathbb{Q}$

so there is $f_m \in A(J_m)$ such that $\alpha_m f_m = \gamma_m$ and $\beta_m f_m = \delta_m$. By Corollary 2L, there is an ℓ-embedding of $F_{\aleph_0}$ in $A(\mathbb{Q})$ and so in $A(I_n)$ (since I_n is ordermorphic to $\mathbb{Q}$) $(n \in \mathbb{Z})$. Let $\{x_m : m \in \omega\}$ be the generators of $F_{\aleph_0}$ and $f_{m,n}$ the image of x_m in $A(I_n)$ $(n \in \mathbb{Z})$. Now define $g_m \in A(\mathbb{Q})$ by

$$\alpha g_m = \begin{cases} \alpha f_m & \text{if } \alpha \in J_m \\ \alpha f_{m,n} & \text{if } \alpha \in I_n \ \& \ |n| > m. \\ \alpha & \text{otherwise} \end{cases}$$

Let W be the ℓ-subgroup of $A(\mathbb{Q})$ generated by $\{g_m : m \in \omega\}$. By the coding of B, $(W,\mathbb{Q})$ is a pathological ℓ-permutation group. The map $\psi: x_m \mapsto g_m$ gives $\psi: F_{\aleph_0} \cong W$; so $(F_{\aleph_0},\mathbb{Q},\psi)$ is the required ℓ-permutation group.

We conclude this chapter by proving Theorem 6F.

Let $(G,\mathbb{R})$ be the ℓ-permutation group of all bounded piecewise linear order-preserving permutations of $\mathbb{R}$ (cf. Example 1.2.3). Then $(G,\mathbb{R})$ is doubly transitive. Since $G \subseteq B(\mathbb{R})$, $[G,G]$ is simple and $([G,G],\mathbb{R})$ is a doubly transitive ordered permutation group. However, we may make G an o-group via: $g \in G$ will be declared greater than e if its left most slope exceeds 1. Denote this o-group by G^*. So G and G^* have the

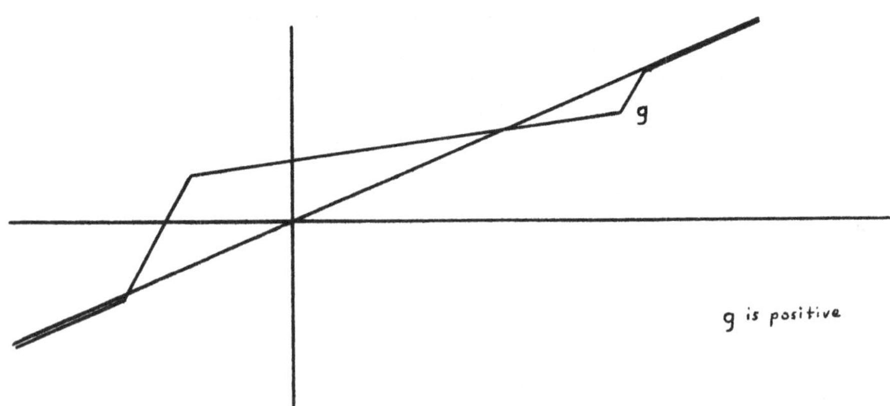

same underlying set and group operations, but different orders. Now the *group* $[G,G]$ is simple; i.e., $[G^*,G^*]$ is simple. Hence $[G^*,G^*]$ is a simple o-group and so has a transitive representation as an ℓ-permutation group. But it cannot have a primitive representation (Since it is an o-group, this would have to be the regular representation; hence $[G^*,G^*]$ would have to be Abelian which it clearly is not--Lemma 1.6.8).

Actually, $G = [G,G]$ as we now show (Consequently, the group of all bounded piecewise linear order-preserving permutations of $\mathbb{R}$ is simple).

We will write each $e \neq g \in G$ as $\begin{Bmatrix} \xi_0 & \xi_1 \cdots & \xi_{n+1} \\ & \eta_0 \cdots & \eta_n \end{Bmatrix}$ where $\xi_0, \ldots, \xi_{n+1} \in \mathbb{R}$ are such that g is linear on each interval $[\xi_i, \xi_{i+1}]$ with slope η_i $(0 \leq i \leq n)$, and $\text{supp}(g) \subseteq (\xi_0, \xi_{n+1})$. Let X be the subset of G comprising those elements of the form $\begin{Bmatrix} \xi_0 & \xi_1 & \xi_2 \\ & \eta_0 & \eta_1 \end{Bmatrix}$. Note that $g^{-1} \in X$ if $g \in X$. To prove that $G = [G,G]$ we establish

(I) each element of G is a product of elements of X,

and (II) each element of X is a product of two commutators.

Proof of (I): By induction on n. Suppose

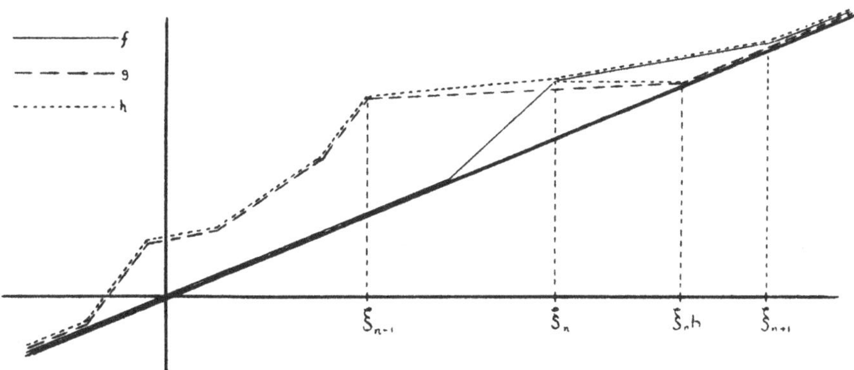

$h = \begin{Bmatrix} \xi_0 & \xi_1 \cdots & \xi_n & \xi_{n+1} \\ & \eta_0 \cdots & & \eta_n \end{Bmatrix}$ with $\eta_n \neq 1$. Assume $\xi_n h > \xi_n$. Let

$f = \begin{Bmatrix} \xi_{n-1} & \xi_n & \xi_{n+1} \\ & \beta & \eta_n \end{Bmatrix}$ and $g = \begin{Bmatrix} \xi_0 & \xi_1 \cdots & \xi_{n-1} & \xi_n h \\ & \eta_0 \cdots & \eta_{n-2} & \gamma \end{Bmatrix}$. Now

$\xi_n f = \xi_n h$ and $\beta = \dfrac{\xi_n f - \xi_{n-1}}{\xi_n - \xi_{n-1}}$. Since $\gamma = \dfrac{\xi_n h - \xi_{n-1} h}{\xi_n h - \xi_{n-1}}$,

$\beta\gamma = \dfrac{\xi_n h - \xi_{n-1} h}{\xi_n - \xi_{n-1}} = \eta_{n-1}$. Thus $fg = \begin{Bmatrix} \xi_0 & \xi_1 \cdots & \xi_{n-2} & \xi_{n-1} & \xi_n & \xi_{n+1} \\ & \eta_0 \cdots & & \eta_{n-2} & \beta\gamma & \eta_n \end{Bmatrix}$

$= h$. Since $f \in X$ and g has no more than n linear pieces in the convexification of its support, induction shows that h is a product of elements of X. If $\xi_n h < \xi_n$, the above argument shows that h^{-1} is a product of elements of X. Since X is closed under inverses, (I) follows.

Proof of (II): Consider the element $h = \begin{Bmatrix} \xi_0 & \xi_1 & \xi_2 \\ & \eta_0 & \eta_1 \end{Bmatrix} \in X$.

Since the inverse of a commutator is a commutator, we may assume $\eta_0 > 1$. We first suppose that $\eta_0\eta_1 > 1$. If

$f = \begin{Bmatrix} \alpha_0 & \alpha_1 & \alpha_2 \\ & \beta_0 & \beta_1 \end{Bmatrix}$ and $g = \begin{Bmatrix} \alpha_0 & \delta & \alpha_1 f \\ & \gamma_0 & \gamma_1 \end{Bmatrix}$ with $\beta_0, \gamma_0 > 1$ and

$\delta g = \alpha_1$, then a routine calculation shows that

$[g^{-1}, f^{-1}] = \begin{Bmatrix} \delta f^{-1} & \delta & \alpha_1 f \\ & \dfrac{\gamma_0}{\gamma_1} & \gamma_1 \end{Bmatrix}$. Hence $h = [g^{-1}, f^{-1}]$ if $\gamma_1 = \eta_1$,

$\gamma_0 = \eta_0\eta_1$, $\delta f^{-1} = \xi_0$, $\delta = \xi_1$ and $\alpha_1 f = \xi_2$. This is consistent:

let $\beta_0 = \dfrac{1 - \frac{1}{\eta_0}}{1 - \eta_1}$, $\alpha_0 = \dfrac{\beta_0\xi_0 - \xi_1}{\beta_0 - 1}$ and $\alpha_1 = \dfrac{\beta_0\gamma_0}{\beta_0 - 1}(\xi_1 - \xi_0) + \alpha_0$.

So any such h is a commutator. If $\eta_0\eta_1 \leq 1$, let

$f = \begin{Bmatrix} \alpha_0 & \alpha_1 & \alpha_2 \\ & \beta_0 & \beta_0\eta_1 \end{Bmatrix}$ with $\beta_0, \beta_0^2\eta_1 > 1 > \beta_0\eta_1$. So $f \in X$ is a commutator. Let $g = \begin{Bmatrix} \alpha_0 & \alpha_1 g^{-1} & \alpha_2 g^{-1} & \alpha_3 \\ \gamma_0 & \frac{\gamma_0}{\eta_0} & \gamma_2 & \end{Bmatrix}$. Then

$$gfg^{-1} = \begin{Bmatrix} \alpha_0 & \alpha_1 f^{-1} g^{-1} & \alpha_1 g^{-1} & \alpha_2 g^{-1} \\ \beta_0 & \eta_0 \beta_0 & \beta_0 \eta_1 & \end{Bmatrix}, \text{ a commutator too.}$$

Let $f' = \begin{Bmatrix} \alpha_0 & \delta & \alpha_2 g^{-1} \\ \frac{1}{\beta_0} & \frac{1}{\beta_0 \eta_1} & \end{Bmatrix}$ a commutator. Note that $\delta g > \alpha_1 f$ as can be verified. Thus $h' = gfg^{-1} \cdot f' = \begin{Bmatrix} \alpha_1 f^{-1} g^{-1} & \alpha_1 g^{-1} & \delta g f^{-1} g^{-1} \\ \eta_0 & \eta_1 & \end{Bmatrix}$.

If we put $\alpha_1 f^{-1} g^{-1} = \xi_0$, $\alpha_1 g^{-1} = \xi_1$ and $\delta g f^{-1} g^{-1} = \xi_2$, we get $h = h'$, a product of two commutators. This is consistent: let $\alpha_0 = \frac{\beta_0 \xi_0 - \xi_1}{\beta_0 - 1}$, $\alpha_1 = \gamma_0 \xi_1 - \alpha_0(\gamma_0 - 1)$ and

$$\delta = \alpha_0 + \frac{(\alpha_1 - \alpha_0)[\eta_0(\beta_0 - 1) + (1 - \beta_0 \eta_1)]}{(1 - \eta_1)};$$ so (II) is proved. //

This example is due to Chehata [52].

The only examples of ℓ-simple ℓ-groups that we know of that have no faithful primitive representation are o-groups! Other examples than the one above can be found in Dlab [68] and Feil [80]. Indeed, Dlab produces uncountably many.

CHAPTER 7
UNIQUENESS OF REPRESENTATION

We saw in Chapter 2 that if Ω and T are doubly homogeneous chains and $A(\Omega)$ is ℓ-isomorphic to $A(T)$, then T is an orbit of $A(\Omega)$ in $\bar{\Omega}$ (Theorem 2D). We now wish to consider the more general case. We know that every ℓ-group G can be represented as an ℓ-permutation group (G,Ω,ψ)--Holland's Theorem, Appendix I. In this chapter we wish to consider whether Ω is unique. More precisely, if (G,Ω) and (G,T,ψ) are ℓ-permutation groups, then are they ℓ-isomorphic? In order to have any hope of an affirmative answer, we must insist that the ℓ-permutation groups are transitive (If Λ is any well-ordered set, then $(G,\Lambda \overleftarrow{\cup} \Omega)$ is an ℓ-permutation group if (G,Ω) is, and $\lambda g = \lambda$ for all $\lambda \in \Lambda$, $g \in G$). Transitivity is not enough, however:

Consider $(A(\mathbb{R}),\mathbb{R})$ a doubly transitive ℓ-permutation group. Let $\mathcal{F}$ be any non-principal ultrafilter on $\mathbb{Z}^+$ and let $P = \{g \in A(\mathbb{R})\colon (\exists X \in \mathcal{F})(\forall n \in X)\frac{1}{n}g = \frac{1}{n}\}$, a convex ℓ-subgroup of $A(\mathbb{R})$. Since $\mathcal{F}$ is non-principal, each $X \in \mathcal{F}$ is cofinal in $\mathbb{Z}^+$. Hence $P \subsetneq A(\mathbb{R})_0$. Therefore $\bigcap_{g \in A(\mathbb{R})} g^{-1}Pg \subseteq \bigcap_{g \in A(\mathbb{R})} g^{-1}A(\mathbb{R})_0 g = \{e\}$. As in Example 7.1.8 below, $R(P)$ is a chain. Thus $(A(\mathbb{R}),R(P))$ is an ℓ-permutation group. The orbits of $A(\mathbb{R})_0$ give rise to a non-trivial congruence of $(A(\mathbb{R}),R(P))$, so $(A(\mathbb{R}),R(P))$ is not ℓ-isomorphic to $(A(\mathbb{R}),\mathbb{R})$.

In general, if $\mathcal{C}$ is a non-trivial congruence of an ℓ-permutation group (G,Ω) and $L(\mathcal{C}) = \{g \in G\colon \alpha g \mathcal{C} \alpha \text{ for all } \alpha \in \Omega\}$, then $L(\mathcal{C})$ is an ℓ-ideal of G and $(G,\Omega/\mathcal{C})$ is an ℓ-permutation group if and only if $L(\mathcal{C}) = \{e\}$ (In the above example, if $\mathcal{C}$ is the congruence defined, then $L(\mathcal{C}) = \{e\}$). So, in order to obtain any uniqueness results, we must further insist that (G,Ω) is *weakly primitive*; i.e., $L(\mathcal{C}) = \{e\}$ only if the $\mathcal{C}$-classes are singletons. This is a sort of minimal condition on Ω.

A transitive ordered permutation group is said to be *locally primitive* if it has a minimal non-singleton congruence. The classes of such a congruence are called *primitive segments*. We will say that a transitive ordered permutation group locally enjoys a property if it is locally primitive and the minimal primitive component enjoys the property.

A transitive ordered permutation group (G,Ω) is said to have the *support property* if for each congruence $\mathcal{C}$ of (G,Ω) whose classes are non-singletons and for each $\alpha \in \Omega$, there exists $e \neq g \in G$ such that $\text{supp}(g) \subseteq \alpha\mathcal{C}$. Note that the interval support property implies the support property but not conversely (Example 1.9.2). Also if (G,Ω) is a transitive locally primitive ℓ-permutation group having the support property, then it is weakly primitive.

If (G,Ω) is a transitive locally primitive ℓ-permutation group, let $\mathcal{C}_0$ be the minimal non-trivial congruence of (G,Ω). The *local completion* Ω^0 of Ω is defined to be $\{\bar{\alpha} \in \bar{\Omega}: \bar{\alpha} \in \bar{\Delta}$ for some $\mathcal{C}_0$-class $\Delta\}$. Observe that $\Omega^0 \supseteq \Omega$, and if (G,Ω) is primitive, the local completion is just the Dedekind completion.

The main results we wish to prove are:

THEOREM 7A. *Let (G,Ω) be a transitive locally primitive ℓ-permutation group having the support property. If (G,Ω) is not locally pathological and (G,T,ψ) is a weakly primitive transitive ℓ-permutation group, then for some $\bar{\alpha} \in \Omega^0$, there is an ordermorphism $\phi: \bar{\alpha}G \cong T$ such that $g\psi = \phi^{-1}g\phi$ for all $g \in G$ and (ψ,ϕ) is an ℓ-permutation group isomorphism between $(G,\bar{\alpha}G)$ and (G,T,ψ).*

In the above example, $(A(\mathbb{R}),R(P)/\mathcal{C}) \cong (A(\mathbb{R}),\mathbb{R})$. More generally:

THEOREM 7B. *If* (G,Ω) *satisfies the hypotheses of Theorem 7A and* (G,T,ψ) *is a transitive* ℓ*-permutation group, then there exists a congruence* $\mathfrak{C}$ *of* (G,T,ψ) *such that* $(G,T/\mathfrak{C},\psi)$ *is weakly primitive and Theorem 7A applies.*

Theorems 7A and 7B are unpublished improvements (due to McCleary) of results of Holland [65a]. They will be proved by associating points of Ω and T with certain algebraically identifiable convex ℓ-subgroups of G. This is quite different from the proof of Theorem 2D, but is in the same spirit.

As immediate consequences we obtain:

COROLLARY 7C. *Let* (G,Ω) *and* (G,T,ψ) *be transitive primitive* ℓ*-permutation groups. Then* (G,Ω) *and* (G,T,ψ) *are both regular, both periodic, both pathological, or both (non-pathological) doubly transitive.*

COROLLARY 7D. *If* (G,Ω) *and* (G,T,ψ) *satisfy the hypotheses of Theorem 7A and one of the primitive segments of* (G,Ω) *is Dedekind complete, then* $\Omega \cong T$.

COROLLARY 7E. *If* (G,Ω) *and* (G,Ω,ψ) *satisfy the hypotheses of Theorem 7A, then* ψ *is the restriction of an inner automorphism* $A(\Omega^0)$.

We will also prove an analogue for the ordered permutation group case due to McCleary [78b]. It necessarily has a more restrictive hypothesis.

THEOREM 7F. *Let* (G,Ω) *and* (H,T) *be triply transitive ordered permutation groups containing strictly positive elements of bounded support. Then each group isomorphism* ψ *between* G *and* H *is induced by a unique ordermorphism or anti-ordermorphism* $\phi\colon \bar{\Omega} \cong \bar{T}$ *via* $g\psi = \phi^{-1}g\phi$ $(g \in G)$. *Thus* ψ *either preserves or reverses the pointwise order.*

7.1. ℓ-PERMUTATION GROUPS.

Let C be a convex ℓ-subgroup of an ℓ-group G. C is a *prime* subgroup of G if $g_1, g_2 \in G$ and $g_1 \wedge g_2 = e$ imply $g_1 \in C$ or $g_2 \in C$. This is equivalent to $R(C)$, the set of right cosets of C in G, is linearly ordered in the induced order (see Appendix I, Lemma 2). If C is prime and $\bigcap_{g \in G} g^{-1}Cg = \{e\}$, then C is a *representing subgroup*. In this case, $(G, R(C))$ is transitive ℓ-permutation group; conversely, if (G,Ω) is a transitive ℓ-permutation group, then G_α is a representing subgroup (cf. Theorem 1.5.1). One of the key lemmas we need is the following generalisation (due to Holland [65a]) of Theorem 4.1.1.

LEMMA 7.1.1. *Let (G,Ω) be a transitive ℓ-permutation group. Then (G,Ω) is weakly primitive if and only if G_α is a maximal representing subgroup of G for all $\alpha \in \Omega$.*

Proof: Since (G,Ω) is transitive, each G_α is a representing subgroup of G. Assume there is a representing subgroup $C \supsetneqq G_\alpha$. Define $\mathcal{C}$ by: $\gamma \mathcal{C} \delta$ if $\gamma, \delta \in \alpha C g$ for some $g \in G$. Then $\mathcal{C}$ is a congruence of (G,Ω) and $L(\mathcal{C}) \subseteq C$. Then the $\mathcal{C}$-classes are non-singletons but $L(\mathcal{C}) = \{e\}$ since C is a representing subgroup of G. Hence (G,Ω) is not weakly primitive.

Conversely, if (G,Ω) is not weakly primitive, let $\mathcal{C}$ be a congruence of (G,Ω) whose classes are non-singletons, with $L(\mathcal{C}) = \{e\}$. Let $\alpha \in \Omega$. Then $G_\alpha \subsetneqq G_{(\alpha\mathcal{C})}$ by the transitivity of (G,Ω). Moreover, $\bigcap_{g \in G} g^{-1}G_{(\alpha\mathcal{C})}g = L(\mathcal{C}) = \{e\}$ so $G_{(\alpha\mathcal{C})}$ is a representing subgroup of G properly containing G_α.

This algebraic identification of the stabilisers is exactly what we need.

LEMMA 7.1.2. *Let (G,Ω) and (H,T) be transitive ℓ-permutation groups and suppose that ψ is an ℓ-isomorphism from G onto H. If $G_{\bar\alpha}\psi = H_{\tau_0}$ for some $\tau_0 \in T$ and $\bar\alpha \in \bar\Omega$, define*

$\phi: \bar{\alpha}G \cong T$ by: $(\bar{\alpha}g)\phi = \tau_0(g\psi)$. Then $g\psi = \phi^{-1}g\phi$ for all $g \in G$, and (ψ,ϕ) is an ℓ-permutation group isomorphism between $(G,\bar{\alpha}G)$ and (H,T).

Proof: By the hypotheses, $\bar{\alpha}g = \bar{\alpha}f \iff gf^{-1} \in G_{\bar{\alpha}} \iff \tau_0(g\psi) = \tau_0(f\psi)$; so ϕ is well-defined and one-to-one. If $\tau \in T$, there exists $h \in H$ such that $\tau = \tau_0 h$. If $f = h\psi^{-1}$, then $(\bar{\alpha}f)\phi = \tau$ and $\tau(\phi^{-1}g\phi) = \tau(g\psi)$. Hence ϕ is onto and $\phi^{-1}g\phi = g\psi$ $(g \in G)$. If $\bar{\alpha}g_1 \leq \bar{\alpha}g_2$, let $e \leq f \in G$ be such that $\bar{\alpha}g_1 f = \bar{\alpha}g_2$. Now $(\bar{\alpha}g_1)\phi = \tau_0(g_1\psi) \leq \tau_0(g_1 f)\psi = \tau_0(g_2\psi) = (\bar{\alpha}g_2)\phi$ so ϕ preserves order. Thus (ψ,ϕ) is an ℓ-permutation group isomorphism.

LEMMA 7.1.3. *Let (G,Ω) be a transitive ℓ-permutation group, and C be a convex ℓ-subgroup of G such that $C \not\subseteq G_{\bar{\alpha}}$ for all $\bar{\alpha} \in \bar{\Omega}$. Then $B(\Omega) \cap G \subseteq C$.*

Proof: Let $e < g \in B(\Omega) \cap G$, say $\text{supp}(g) \subseteq [\alpha,\beta]$. For each $\bar{\alpha} \in \bar{\Omega}$, there exists $e < f_{\bar{\alpha}} \in C$ such that $\bar{\alpha} \in \text{supp}(f_{\bar{\alpha}})$. Now $\{\text{supp}(f_{\bar{\alpha}}): \alpha \leq \bar{\alpha} \leq \beta\}$ is an open cover of the compact subset $[\alpha,\beta]$ of $\bar{\Omega}$. Hence there are $\bar{\alpha}_1,\ldots,\bar{\alpha}_n \in [\alpha,\beta]$ such that $[\alpha,\beta] \subseteq \bigcup_{i=1}^{n} \text{supp}(f_{\bar{\alpha}_i})$. Let $f = f_{\bar{\alpha}_1} \cdots f_{\bar{\alpha}_n} \in C$. Then $[\alpha,\beta] \subseteq \text{supp}(f)$. So for some $m \in \mathbb{Z}^+$, $\beta \leq \alpha f^m$. Hence $e \leq g \leq f^m \in C$. Thus $g \in C$.

A routine modification of the above proof (cf. proof of Theorem 6C) gives:

LEMMA 7.1.4. *Let (G,Ω) be a periodic primitive ℓ-permutation group and C be a convex ℓ-subgroup of G such that $C \not\subseteq G_{\bar{\alpha}}$ for all $\bar{\alpha} \in \bar{\Omega}$. Then $C = G$.*

We know that G_α is a representing subgroup of G for all $\alpha \in \Omega$. We now extend this to show:

LEMMA 7.1.5. *If (G,Ω) is a transitive locally primitive ℓ-permutation group and $\bar{\alpha} \in \Omega^0$, then $G_{\bar{\alpha}}$ is a representing subgroup of G and either $\bar{\alpha}G$ is dense in $\bar{\Omega}$ or $\bar{\alpha}G = \Omega$.*

Proof: If Δ is a primitive segment with $\bar{\alpha} \in \bar{\Delta}$, then $\bar{\alpha}\hat{G}_{(\Delta)}$ is dense in $\bar{\Delta}$, or equal to Δ if $\Delta \cong \mathbb{Z}$ ($(\hat{G}_{(\Delta)},\Delta)$ is primitive so apply Theorem 4.1.1). Hence $\bar{\alpha}G$ is dense in Ω^0 or equal to Ω. $G_{\bar{\alpha}}$ is a prime subgroup and if $e < g \in G_{\bar{\alpha}}$, there exists $\beta \in \Omega$ such that $\beta g \neq \beta$. Therefore there exists $f \in G$ such that $\beta \leq \bar{\alpha}f < \beta g$. Thus $\bar{\alpha} < \beta g f^{-1} \leq \bar{\alpha} f g f^{-1}$; so $fgf^{-1} \notin G_{\bar{\alpha}}$. Thus $G_{\bar{\alpha}}$ contains no ℓ-ideal of G other than $\{e\}$. Consequently, $G_{\bar{\alpha}}$ is a representing subgroup of G.

We now give a partial converse of Lemma 7.1.5 which, with Lemma 7.1.1, is key to proving Theorems 7A and 7B.

LEMMA 7.1.6. *Let (G,Ω) be a transitive locally primitive ℓ-permutation group that has the support property. If (G,Ω) is not locally pathological, then every representing subgroup of G is contained in some $G_{\bar{\alpha}}$, $\bar{\alpha} \in \Omega^0$.*

Proof: Let $\mathcal{C}_0$ be the minimal non-singleton congruence of (G,Ω). Let $N = \{g \in L(\mathcal{C}_0)\colon \operatorname{supp}(g|\alpha\mathcal{C}_0) = \emptyset$ for all but a finite number of $\alpha\mathcal{C}_0\}$ and $M = \{g \in N\colon \operatorname{supp}(g|\alpha\mathcal{C}_0)$ is bounded in $\alpha\mathcal{C}_0$ for all $\alpha \in \Omega\}$. By the fundamental triviality, M and N are ℓ-ideals of G. Let C be a representing subgroup of G. For reductio ad absurdum, assume C is disobedient; i.e., $C \not\subseteq G_{\bar{\alpha}}$ for all $\bar{\alpha} \in \Omega^0$.

Case I. $(C \cap G_{(\alpha\mathcal{C}_0)})/G_{\alpha\mathcal{C}_0} \subseteq (\hat{G}_{(\alpha\mathcal{C}_0)})_{\bar{\alpha}}$ for some $\alpha \in \Omega$ and $\bar{\alpha} \in \overline{\alpha\mathcal{C}_0}$. Then C contains every element of G whose support is contained in $\alpha\mathcal{C}_0$ (If $C \subseteq G_{(\alpha\mathcal{C}_0)}$, then $C \subseteq G_{\bar{\alpha}}$ which contradicts the fact that C is disobedient; so there exists $e < h \in C$ such that $\alpha\mathcal{C}_0 < \alpha\mathcal{C}_0 h$. By the convexity of C, the claim follows). Hence $N \subseteq C$ if (G,Ω) is not locally doubly transitive, and $M \subseteq C$ if (G,Ω) is locally (non-pathological) doubly transitive. By the support property, $N \neq \{e\}$ (and $M \neq \{e\}$ if (G,Ω) is locally (non-pathological) doubly transitive). Since N and M are ℓ-ideals, C is not a representing subgroup of G.

Case II. $(C \cap G_{(\alpha\mathfrak{C}_0)})/G_{\alpha\mathfrak{C}_0} \not\supseteq (\hat{G}_{(\alpha\mathfrak{C}_0)})_{\bar{\alpha}}$ for all $\alpha \in \Omega$ and $\bar{\alpha} \in \overline{\alpha\mathfrak{C}_0}$.

If (G,Ω) is locally regular, $(\hat{G}_{(\alpha\mathfrak{C}_0)})_{\bar{\alpha}} = \{e\}$ and $\hat{G}_{(\alpha\mathfrak{C}_0)}$ has no proper convex subgroups for all $\alpha \in \Omega$ and $\bar{\alpha} \in \overline{\alpha\mathfrak{C}_0}$. Hence $C \supseteq L(\mathfrak{C}_0)$, a non-trivial ℓ-ideal of G. Thus C is not a representing subgroup of G.

If (G,Ω) is locally periodic, then again $C \supseteq L(\mathfrak{C}_0)$ (by Lemma 7.1.4), a contradiction.

If (G,Ω) is locally (non-pathological) doubly transitive, then $C \supseteq M$ (by Lemma 7.1.3) and again we get a contradiction.

The lemma now follows by Theorem 4A.

By Lemmas 7.1.5 and 7.1.6:

COROLLARY 7.1.7. *If Ω is homogeneous and locally primitive, the maximal representing subgroups of $A(\Omega)$ are precisely $A(\Omega)_{\bar{\alpha}}$, $\bar{\alpha} \in \Omega^0$.*

We now prove Theorem 7A.

So assume that ψ is an ℓ-embedding of G into $A(T)$ such that (G,T,ψ) is a weakly primitive transitive ℓ-permutation group. Let $\tau_0 \in T$. By Lemma 7.1.1, $(G\psi)_{\tau_0}$ is a maximal representing subgroup of $G\psi$. By Lemma 7.1.6, $(G\psi)_{\tau_0}\psi^{-1} \subseteq G_{\bar{\alpha}}$ for some $\bar{\alpha} \in \Omega^0$. Since $G_{\bar{\alpha}}$ is a representing subgroup of G (by Lemma 7.1.5), $(G\psi)_{\tau_0} = G_{\bar{\alpha}}\psi$ by the maximality of $(G\psi)_{\tau_0}$. By Lemma 7.1.2, there is $\phi\colon \bar{\alpha}G \to T$ making (ψ,ϕ) an ℓ-permutation group isomorphism between $(G,\bar{\alpha}G)$ and $(G\psi,T)$ of the desired kind.

We now prove Theorem 7B.

Let $\tau \in T$. Then $(G\psi)_\tau$ is a representing subgroup of $G\psi$, so $(G\psi)_\tau\psi^{-1} \subseteq G_{\bar{\alpha}}$ for some $\bar{\alpha} \in \Omega^0$ (by Lemma 7.1.6). Since (G,Ω) is locally primitive and has the support property, so does $(G,\bar{\alpha}G)$. Hence $(G,\bar{\alpha}G)$ is weakly primitive. By Lemma 7.1.1, $G_{\bar{\alpha}}$ is a maximal representing subgroup of G. Now there is a one-to-one containment preserving correspondence between the congruences of $(G\psi,T)$ and the convex ℓ-subgroups of $G\psi$ which contain $(G\psi)_\tau$ (Theorems 1.6.1 and 1.6.2). Let $\mathcal{C}$ be the congruence corresponding to $G_{\bar{\alpha}}\psi$. Then $(G,T/\mathcal{C},\psi)$ is weakly primitive by the converse direction of Lemma 7.1.1 and the proof is complete.

We conclude with an example to show that the conclusions of Theorem 7A are false if (G,Ω) is pathological. The example is due to Glass and McCleary [76].

<u>EXAMPLE 7.1.8</u>. Let $G = \{g \in A(\mathbb{R}): (\exists m \in \mathbb{Z}^+)(\forall \alpha \in \mathbb{R})\ (\alpha + m)g = \alpha g + m\}$ (Example 1.10.3). So $\bar{\Omega} = \Omega^0 = \Omega = \mathbb{R}$. By Lemma 7.1.1, it is enough to find a maximal representing subgroup C of G with $C \neq G_\alpha$ $(\alpha \in \mathbb{R})$, for then $(G,R(C))$ is weakly primitive with $C = G_C$. Let $\beta_i = (4^i - 1)/3$ and $\Lambda_i = \beta_i + 4^i\mathbb{Z}$ $(i \in \mathbb{Z}^+)$; i.e., for $i > 1$, β_i is the second smallest positive element of Λ_{i-1}. Then $\Lambda_1 \supsetneqq \Lambda_2 \supsetneqq \cdots$ and $\bigcap_{i=1}^{\infty} \Lambda_i = \emptyset$. Hence there exists an ultrafilter $\mathcal{F}$ on $\mathbb{R}$ such that each $\Lambda_i \in \mathcal{F}$. Let $P = \{g \in G: Fx(g) \in \mathcal{F}\}$. Then P is a prime subgroup of G (If $g_1 \wedge g_2 = e$, $Fx(g_1) \cup Fx(g_2) = \mathbb{R} \in \mathcal{F}$; so $Fx(g_1) \in \mathcal{F}$ or $Fx(g_2) \in \mathcal{F}$). Let $\alpha \in \mathbb{R}$. Choose $i \in \mathbb{Z}^+$ so that $\alpha \notin \Lambda_i$ and let $f \in G_{\Lambda_i} \setminus G_\alpha$ with period 4^i (i.e., $(\beta + 4^i)f = \beta f + 4^i$ $(\beta \in \mathbb{R})$). Then $f \in P$ as $Fx(f) \supseteq \Lambda_i \in \mathcal{F}$. Thus $P \not\subseteq G_\alpha$ for all $\alpha \in \mathbb{R}$.

Let $C \supseteq P$ be a prime subgroup of G maximal with respect to not containing $z: \alpha \mapsto \alpha + 1$. Since G is ℓ-simple (Theorem 6.5), C is a representing subgroup. Let $e < g \in G$ have period m and let $n \in \mathbb{Z}^+$ be such that $0g \leq n$. Then $g \leq z^{m+n}$. Hence a prime subgroup of G is equal to G only if it contains z. So C is the desired maximal representing subgroup of G. Note that, by Theorem 4.1.1, $(G,R(C))$ is primitive and hence pathological by Corollary 7C.

7.2. ORDERED PERMUTATION GROUPS.

Throughout this section, which is devoted to proving Theorem 7F, (G,Ω) and (H,T) will be triply transitive ordered permutation groups containing strictly positive elements of bounded support. Thus (G,Ω) and (H,T) are m-transitive for all $m \in \mathbb{Z}^+$ (Lemma 1.10.4) and if $\psi: G \cong H$ (as groups), then $g \in G^+ \cup G^-$ implies $g\psi \in H^+ \cup H^-$ by Corollary 1.10.8 ($\underline{G}^- = \{g \in G: g < e\}$). Moreover, by Corollary 1.10.8, if $e < g_0 \in G$ and $e < g_0\psi$ $(e > g_0\psi)$ for *some* $g_0 \in G$, then ψ preserves (reverses) order.

Again we will use stabilisers to identify Ω and T (The techniques of Section 2.1 do not necessarily apply here since $\{\bar{\alpha} \in \bar{\Omega}: (\exists g \in G)\sup(\text{supp}(g)) = \bar{\alpha}\}$ may not even include Ω).

A convex subgroup C of a p.o. group G is called *prime* if whenever $g_1,g_2 \in G$ and $g_1 \wedge g_2 = e$, then $g_1 \in C$ or $g_2 \in C$. This is a generalisation of the definition for ℓ-groups except that C is not necessarily a sublattice. Note $g_1 \wedge g_2$ may not always exist in G. If C is a prime subgroup of p.o. group G and $\bigcap_{g \in G} g^{-1}Cg = \{e\}$, we say that C is a *representing subgroup* of G. Note that if (G,Ω) is a triply transitive ordered permutation group containing a strictly positive element of bounded support, then G_α is a maximal subgroup of G for each $\alpha \in \Omega$ (Theorem 4.1.5).

LEMMA 7.2.1. *Let (G,Ω) be a triply transitive ordered permutation group having strictly positive elements of bounded support. Then the maximal representing subgroups of G are precisely the stabilisers $G_{\bar{\alpha}}$ for which $G_{\bar{\alpha}}$ is a maximal subgroup of G.*

Proof: Let $\bar{\alpha} \in \bar{\Omega}$. Clearly $G_{\bar{\alpha}}$ is convex. If $\bar{\alpha} < \bar{\alpha}f,\bar{\alpha}g$, let $\bar{\beta} = \min(\bar{\alpha}f,\bar{\alpha}g) > \bar{\alpha}$. By Corollary 1.10.6, there exists $e < h \in G$ such that $\text{supp}(h) \subseteq (\bar{\alpha},\bar{\beta})$. Hence $e < h < f,g$ by Lemma 1.9.4. So $G_{\bar{\alpha}}$ is prime. Moreover, since $\bar{\alpha}G$ is dense in

$\bar{\Omega}$ (by Theorem 4.1.3), $\bigcap_{g \in G} g^{-1}G_{\bar{\alpha}}g = \bigcap_{g \in G} G_{\bar{\alpha}g} = \{e\}$. Therefore each $G_{\bar{\alpha}}$ is a representing subgroup.

We now show that if C is a prime subgroup of G and for all $\bar{\alpha} \in \bar{\Omega}$ $C \not\subseteq G_{\bar{\alpha}}$, then $B(\Omega) \cap G \subseteq C$ (cf. Lemma 7.1.3). Since $B(\Omega) \cap G \lhd G$, the lemma will follow.

Assume C is a prime subgroup of G and $C \not\subseteq G_{\bar{\alpha}}$ for all $\bar{\alpha} \in \bar{\Omega}$. Let $b \in G \cap B(\Omega)$, say $\text{supp}(b) \subseteq [\alpha,\beta]$. For each $\bar{\delta} \in [\alpha,\beta]$, there is $c_{\bar{\delta}} \in C$ such that $\bar{\delta}c_{\bar{\delta}} > \bar{\delta}$. We claim we may assume $c_{\bar{\delta}} > e$ (Let $\sigma,\tau \in \Omega$ with $\bar{\delta} < \sigma < \tau < \bar{\delta}c_{\bar{\delta}}$. There are $e < f,g \in G$ such that $\bar{\delta} \in \text{supp}(f) \subseteq [\alpha,\sigma]$ and $\bar{\delta}c_{\bar{\delta}} \in \text{supp}(g) \subseteq [\tau,\tau c_{\bar{\delta}}]$. Now $f \wedge g = e$ since $\text{supp}(f) \cap \text{supp}(g) = \emptyset$. Hence $f \in C$ or $g \in C$. Thus $e < f \in C \backslash G_{\bar{\delta}}$ or $e < c_{\bar{\delta}}gc_{\bar{\delta}}^{-1} \in C \backslash G_{\bar{\delta}}$ establishing the claim). Now $\{\text{supp}(c_{\bar{\delta}}) \cap [\alpha,\beta]: \bar{\delta} \in [\alpha,\beta]\}$ is an open cover of the compact subset $[\alpha,\beta]$ of $\bar{\Omega}$. Hence there exist $\bar{\alpha}_1,\ldots,\bar{\alpha}_n \in [\alpha,\beta]$ such that $[\alpha,\beta] \subseteq \bigcup_{i=1}^{n} \text{supp}(c_{\bar{\alpha}_i})$. Let $c = c_{\bar{\alpha}_1} \ldots c_{\bar{\alpha}_n} \in C$. Then $c > e$ and fixes no point of $\bar{\Omega}$ in $[\alpha,\beta]$. Thus $c^{-m} < b < c^m$ for some $m \in \mathbb{Z}^+$, so $b \in C$ as required.

We next note that the proof of Lemma 7.1.2 shows:

<u>LEMMA</u> <u>7.2.2</u>. *Let (G,Ω) and (H,T) be transitive ordered permutation groups and suppose ψ is a group isomorphism from G onto H. If $G_{\bar{\alpha}}\psi = H_{\tau_0}$ for some $\tau_0 \in T$ and $\bar{\alpha} \in \bar{\Omega}$, define $\phi: \bar{\alpha}G \to T$ by: $(\bar{\alpha}g)\phi = \tau_0(g\psi)$. Then $g\psi = \phi^{-1}g\phi$ for all $g \in G$.*

We now prove Theorem 7F.

Let $\psi: G \cong H$ be a group isomorphism. Let $\bar{\Omega}_M = \{\bar{\alpha} \in \bar{\Omega}: G_{\bar{\alpha}}$ is a maximal subgroup of $G\}$, and $\bar{T}_M$ be defined similarly. Note $\Omega \subseteq \bar{\Omega}_M$ and $T \subseteq \bar{T}_M$. By Lemma 7.2.1 $\{G_{\bar{\alpha}}\psi: \bar{\alpha} \in \bar{\Omega}_M\} = \{H_{\bar{\tau}}: \bar{\tau} \in \bar{T}_M\}$. Moreover, distinct elements of $\bar{\Omega}$ $(\bar{T})$ have distinct stabilisers by Corollary 1.10.6. Thus we define $\phi: \Omega \to \bar{T}$ by setting $\alpha\phi = \bar{\tau}$ where $G_\alpha\psi = H_{\bar{\tau}}$. By Lemma

7.2.2, ϕ maps the set Ω one-to-one onto the set $\bar{\tau}H$ and $g\psi = \phi^{-1}g\phi$ $(g \in G)$. By Corollary 1.10.8, ψ preserves or reverses order and hence so does ϕ. Finally, if ϕ_1, ϕ_2 induce ψ and $\phi_1 \neq \phi_2$, let $\beta \in \Omega$ be such that $\beta\phi_1 \neq \beta\phi_2$. There exists $g \in G_\beta \setminus G_{\beta\phi_2\phi_1^{-1}}$. Now $(\beta\phi_2)(g\psi) = \beta g\phi_2 = \beta\phi_2 \neq (\beta\phi_2)(\phi_1^{-1}g\phi_1) = (\beta\phi_2)(g\psi)$, a contradiction. Hence the ϕ inducing ψ is unique.

CHAPTER 8
POINTWISE SUPREMA AND CLOSED SUBGROUPS

If (G,Ω) is an ℓ-permutation group and $f,g \in G$, then $\alpha(f \vee g) = \sup(\alpha f, \alpha g)$ for all $\alpha \in \Omega$. If $\{g_i : i \in I\} \subseteq G$ and $\bigvee_{i \in I} g_i$ exists in G, then is it still true that $\alpha(\bigvee_{i \in I} g_i) = \sup\{\alpha g_i : i \in I\}$ for all $\alpha \in \Omega$ even when I is infinite? If it is true whenever $\{g_i : i \in I\} \subseteq G$ and $\bigvee_{i \in I} g_i$ exists (in G), we will say that (G,Ω) has *pointwise suprema*. It is not hard to show (using $\bigwedge_{i \in I} g_i = (\bigvee_{i \in I} g_i^{-1})^{-1}$) that (G,Ω) has pointwise suprema if and only if it has pointwise infima. Unfortunately, even $(A(\Omega),\Omega)$ can fail to have pointwise suprema. For example, let $T = \{\tau_1,\tau_2\}$ with $\tau_1 < \tau_2$ and $\Omega = T \overleftarrow{\times} \mathbb{Q}$. Let $g_n \in A(\Omega)$ be defined by $g_n : (\tau,q) \mapsto (\tau, q - \frac{1}{n})$ $(\tau \in T,\ q \in \mathbb{Q})$. Then $e = \bigvee_{n \in \mathbb{Z}^+} g_n$ but $\sup\{(\tau_i,0)g_n : n \in \mathbb{Z}^+\} = \sup\{(\tau_i, -\frac{1}{n}) : n \in \mathbb{Z}^+\} = (\tau_1,0)$ $(i = 1,2)$. Hence $(\tau_2,0)(\bigvee_{n \in \mathbb{Z}^+} g_n) = (\tau_2,0)e = (\tau_2,0) \neq \sup\{(\tau_2,0)g_n : n \in \mathbb{Z}^+\}$. However,

<u>THEOREM</u> <u>8A</u>. *If Ω is homogeneous, $A(\Omega)$ has pointwise suprema.*

Let H be an ℓ-subgroup of G. H is said to be *closed* in G if whenever $\{h_i : i \in I\} \subseteq H$ and $g = \bigvee_{i \in I} h_i \in G$, then $g \in H$.

<u>THEOREM</u> <u>8B</u>. *$A(\Omega)_{\bar{\alpha}}$ is closed for all $\bar{\alpha} \in \bar{\Omega}$, for any chain Ω.*

Pointwise suprema and closed stabilizers are very closely related.

THEOREM 8C. *If (G,Ω) is an ℓ-permutation group that has pointwise suprema, then $G_{\bar{\alpha}}$ is closed for all $\bar{\alpha} \in \bar{\Omega}$. If (G,Ω) is transitive, the converse is true.*

Theorem 8A follows at once from Theorems 8B and 8C. In Section 8.1 we will establish Theorem 8B; and in Section 8.2 we will prove Theorem 8C. We will also investigate there when the converse direction in Theorem 8C holds (transitivity is not necessary). We will further examine the relationship between pointwise suprema, closed stabilisers, and the ℓ-group concepts complete ℓ-subgroup and complete distributivity. We complete Section 8.2 by examining the primitive case.

THEOREM 8D. *If (G,Ω) is a primitive ℓ-permutation group, then it fails to have pointwise suprema (closed stabilisers) if and only if it is transitively derived from a pathological ℓ-permutation group.*

In Section 8.3 we locate the closed convex ℓ-subgroups of an ℓ-permutation group (G,Ω). Recall that an ℓ-subgroup H of an ℓ-group G is prime if whenever $g_1, g_2 \in G$ and $g_1 \wedge g_2 = e$, $g_1 \in H$ or $g_2 \in H$ (see page 143). Clearly $G_{\bar{\alpha}}$ is a prime subgroup of G for all $\bar{\alpha} \in \bar{\Omega}$ whenever (G,Ω) is an ℓ-permutation group. We prove a "converse" of Theorem 8B.

THEOREM 8E. *For any chain Ω, the closed prime subgroups of $A(\Omega)$ other than $A(\Omega)$ are precisely the stabiliser subgroups $A(\Omega)_{\bar{\alpha}}$, $\bar{\alpha} \in \bar{\Omega}$.*

Let $\mathfrak{C}$ be a congruence of the ℓ-permutation group (G,Ω). Let $L(\mathfrak{C}) = \{g \in G\colon \Delta g = \Delta$ for all $\mathfrak{C}$-classes $\Delta\}$. Then $L(\mathfrak{C})$ is an ℓ-ideal of G (i.e., a normal convex ℓ-subgroup of G) and if $\mathfrak{C}_1 \subseteq \mathfrak{C}_2$, $L(\mathfrak{C}_1) \subseteq L(\mathfrak{C}_2)$. We will show:

THEOREM 8F. *Let* (G,Ω) *be an* ℓ*-permutation group and* H *a closed* ℓ*-ideal of* G. *Then* $H = L(\mathfrak{C})$ *for some congruence* $\mathfrak{C}$ *of* (G,Ω).

Finally, we generalise Theorem 8D. To do this we need an extra condition. An ordered permutation group (G,Ω) has the *support property* if for every non-singleton natural block Δ with $G_{(\Delta)} \neq \{e\}$, there is $e \neq g \in G$ such that $\mathrm{supp}(g) \subseteq \Delta$ (cf. the transitive definition given in Chapter 7).

THEOREM 8G. *Let* (G,Ω) *be an* ℓ*-permutation group having the support property. Then* $G_{\bar{\alpha}}$ *fails to be closed if and only if* $\bar{\alpha}$ *belongs to some pathologically derived primitive segment.*

The results in this chapter are due to McCleary [69], [72b], [72c], [73b] and [76].

8.1. CLOSED STABILISER SUBGROUPS.

In this section we wish to establish Theorem 8B, due originally to Lloyd [67]. Rather than give John Read's simplification of Lloyd's proof [71], we will employ Theorem 4C (or more specifically Corollary 4.4.1) after reducing the proof to the primitive case by using Theorem $3C^{\dagger}$. This approach is due to McCleary [76]. It has the advantage of being routine (given the results of Chapters 3 and 4) and, unlike Lloyd's and Read's proofs, does not require any outside results from the theory of ℓ-groups.

Let (G,Ω) be an ordered permutation group and $\mathfrak{C}$ a congruence of (G,Ω). Define $\bar{\mathfrak{C}}$ by: $\bar{\alpha}\,\bar{\mathfrak{C}}\,\bar{\beta}$ if $\bar{\alpha} = \bar{\beta}$ or $\bar{\alpha}$ and $\bar{\beta}$ belong to the Dedekind completion of the same $\mathfrak{C}$-class $(\bar{\alpha},\bar{\beta} \in \bar{\Omega})$. This yields a congruence of $(G,\bar{\Omega})$.

The following lemma is of central importance in this chapter and is of a fundamental nature.

LEMMA 8.1.1. *If* (G,Ω) *is an* ℓ*-permutation group,* $\{g_i : i \in I\} \subseteq G$ *and* $g = \bigvee_{i \in I} g_i$ *(in* G*), then for each* $\alpha \in \Omega$, *let* $\bar{\alpha} = sup\{\alpha g_i : i \in I\} \in \bar{\Omega}$. *If* $\mathfrak{C}$ *is a congruence of* (G,Ω), $\bar{\alpha} \notin \alpha g\bar{\mathfrak{C}}$ *and* $e \neq f \in G$, *then* $supp(f) \not\subseteq [inf(\bar{\alpha}\bar{\mathfrak{C}}), inf(\alpha g\bar{\mathfrak{C}}))$. *In particular, no element of* G *has support contained in* $[\bar{\alpha}, \alpha g)$.

Proof: We prove only the special case when each $\mathfrak{C}$-class is a singleton--the general case is proved similarly. Suppose $e \neq f \in G$ and $supp(f) \subseteq [\bar{\alpha}, \alpha g)$. Since $f \vee f^{-1}$ and f have the same support, we may assume $f > e$. If $\gamma \notin [\bar{\alpha}g^{-1}, \alpha)$, then $\gamma g f^{-1} = \gamma g \geq \gamma g_i$ for all $i \in I$, and if $\delta \in [\bar{\alpha}g^{-1}, \alpha)$, then $\delta g f^{-1} \geq \bar{\alpha} \geq \alpha g_i > \delta g_i$ for all $i \in I$. Hence $g_i \leq gf^{-1} < g$ for all $i \in I$. Thus $g = \bigvee_{i \in I} g_i \leq gf^{-1} < g$, a contradiction.

Let (G,Ω) be an ordered permutation group. (G,Ω) is said to be *depressible* if $g \in G$ implies that every bump of g belongs to G; i.e., if $\Delta = Conv_{\Omega}\{\alpha g^n : n \in \mathbb{Z}\}$ for some $\alpha \in \Omega$, then

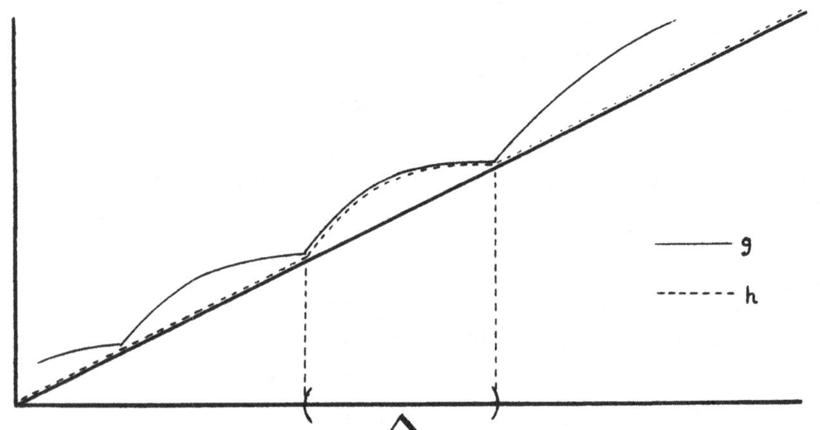

$h \in G$ where $\beta h = \begin{cases} \beta g & \text{if } \beta \in \Delta \\ \beta & \text{if } \beta \notin \Delta \end{cases}$. For example, $(A(\Omega),\Omega)$ is depressible for any chain Ω, but Examples 1.9.2 and 1.10.3 are not.

Note that if (G,Ω) is depressible, then so are all its primitive components. Hence if (G,Ω) is a depressible

ℓ-permutation group, its components are either static, integral, or transitively derived from "regular" or (non-pathological) doubly transitive primitive ℓ-permutation groups (Here is where we use Corollary 4.4.1).

The following theorem implies Theorem 8B (at least if $\bar{\alpha} \in \Omega$):

THEOREM 8.1.2. *If* (G, Ω) *is a depressible* ℓ*-permutation group, then* G_α *is closed for all* $\alpha \in \Omega$.

Proof: Suppose, for reductio ad absurdum, that $\{g_i : i \in I\} \subseteq G_\alpha$ and $g = \bigvee_{i \in I} g_i$ exists in G but $g \notin G_\alpha$.

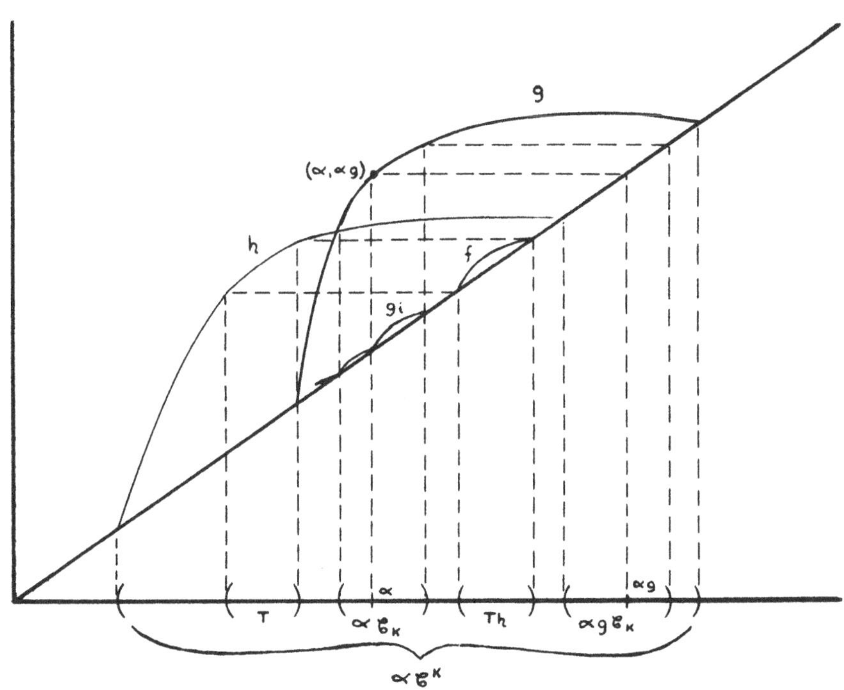

Since $\alpha g \geq \alpha g_i = \alpha$ $(i \in I)$, $\alpha g > \alpha$. Let $K = \mathrm{Val}(\alpha,\alpha g)$ (Here we use Theorem 3C[†]). Clearly K is not static. Without loss of generality, we may assume that $\beta g_i = \beta$ and $\beta g = \beta$ for all $i \in I$, $\beta \notin \alpha\mathfrak{C}^K$. Let T be a $\mathfrak{C}_K$-class contained in $\alpha\mathfrak{C}^K$. If K is integral, then there exists $h \in G_{(\alpha\mathfrak{C}^K)}$ such that $Th \subseteq [\alpha\mathfrak{C}_K,\alpha g\mathfrak{C}_K)$. The same is true if K <u>is not</u> integral since then the orbits of $\hat{G}_{(\alpha\mathfrak{C}^K)}$ are dense in $\alpha\mathfrak{C}^K/\mathfrak{C}_K$ (Theorem 4.1.1). We claim that $G_{(Th)} = \{e\}$. Let $e < f \in G_{(Th)}$. By depressibility, we may assume that $\mathrm{supp}(f) \subseteq Th \subseteq [\alpha\mathfrak{C}_K,\alpha g\mathfrak{C}_K)$. But by Lemma 8.1.1, no such f can exist ($\alpha g_i = \alpha$, so $\alpha\mathfrak{C}_K g_i = \alpha\mathfrak{C}_K$ for all $i \in I$). Thus $G_{(Th)} = \{e\}$, and so $G_{(T)} = \{e\}$ by the fundamental triviality. But this was true for any $\mathfrak{C}_K$-class $T \subseteq \alpha\mathfrak{C}^K$. Since any component of a depressible ordered permutation group is depressible, we have reduced the problem to the primitive case.

So assume (G,Ω) is a primitive depressible ℓ-permutation group with G_α not closed. If G were an o-group, it would be a subgroup of $\mathbb{R}$. Hence its only convex subgroups are G and $\{e\}$ which are closed. Hence (G,Ω) is transitively derived from a (non-pathological) doubly transitive ℓ-permutation group. Let $e < f \in G$ have bounded support, say $\mathrm{supp}(f) \subseteq (\beta_1,\beta_2)$, where $\beta_1,\beta_2 \in \alpha G$. There is $h \in G$ such that $\beta_1 h = \alpha$ and $\beta_2 h = \alpha g$. Hence $e < h^{-1}fh$ and $\mathrm{supp}(h^{-1}fh) \subseteq (\alpha,\alpha g)$. This is the desired contradiction (by Lemma 8.1.1) and the theorem is proved.

Note that if (G,Ω) is primitive and depressible, it clearly has closed stabilisers (by Corollary 4.4.1 and Lemma 8.1.1). Hence, in the proof of Theorem 8.1.2, our goal was to reduce the general case to the primitive. This we accomplished by using the structure theory of Chapter 3. We will adopt this strategy elsewhere--especially in the proof of Theorem 11A.

Let H be an ℓ-subgroup of an ℓ-group G. If whenever $\{h_i : i \in I\} \subseteq H$ has a supremum in H, this is the supremum in G, we say that H is a <u>*complete*</u> <u>*ℓ-subgroup*</u> of G.

<u>COROLLARY</u> <u>8.1.3</u>. *If G is a complete ℓ-subgroup of $A(\Omega)$, then G_α is closed for all $\alpha \in \Omega$.*

<u>*Proof:*</u> Let $\{g_i : i \in I\} \subseteq G_\alpha$ and $g = \bigvee_{i \in I} g_i$ (in G). Then $g = \bigvee_{i \in I} g_i$ (in $A(\Omega)$). Since $\{g_i : i \in I\} \subseteq A(\Omega)_\alpha$, $g \in A(\Omega)_\alpha$ by Theorem 8.1.2. Hence $g \in G \cap A(\Omega)_\alpha = G_\alpha$; i.e., G_α is closed.

<u>8.2</u>. <u>POINTWISE</u> <u>SUPREMA</u>.

In this section we wish to complete the proof of Theorem 8B as well as prove Theorems 8C and 8D (due to McCleary [72b], [72c] and [73b]). We will also thoroughly investigate the relationship between pointwise suprema, closed stabilisers, complete ℓ-subgroups and complete distributivity:

Let <u>^IJ</u> denote the set of functions from the index set I into the index set J. An ℓ-group G is said to be <u>*completely*</u> <u>*distributive*</u> if $\bigwedge_{i \in I} \bigvee_{j \in J} g_{ij} = \bigvee_{f \in {}^IJ} \bigwedge_{i \in I} g_{if(i)}$ whenever $\{g_{ij} : i \in I, j \in J\} \subseteq G$ and all the indicated suprema and infima exist (Equality does hold if I and J are finite). Byrd and Lloyd [67] have proved (Appendix II Theorem 2) an ℓ-group G is completely distributive if and only if $D(G) = \{e\}$, where <u>*$D(G)$*</u> is the intersection of all closed prime subgroups of G (see Appendix II for a proof).

We will prove the following strong version of Theorem 8C (at least for $\alpha \in \Omega$).

<u>THEOREM</u> <u>8.2.1</u>. *Let (G,Ω) be an ℓ-permutation group. Each of statements (1) to (4) implies the next. If (G,Ω) is transitive, (1) to (3) are equivalent:*

(1) (G,Ω) has pointwise suprema.

(2) G is a complete ℓ-subgroup of $A(\Omega)$.

(3) G_α *is closed for all* $\alpha \in \Omega$.

(4) G *is completely distributive.*

First we prove an easy technical lemma:

LEMMA 8.2.2. *Let* (G,Ω) *be an* ℓ*-permutation group and* $\alpha \in \Omega$ *be such that* G_α *is closed. If* $g = \bigvee_{i \in I} g_i \in G$ *and* $\bar{\alpha} = \sup\{\alpha g_i : i \in I\}$, *then there does not exist* $h \in G$ *such that* $\bar{\alpha} \leq \alpha h < \alpha g$.

Proof: Assume the hypotheses and suppose that for some $h \in G$, $\bar{\alpha} \leq \alpha h < \alpha g$. Without loss of generality, $h < g$ (take

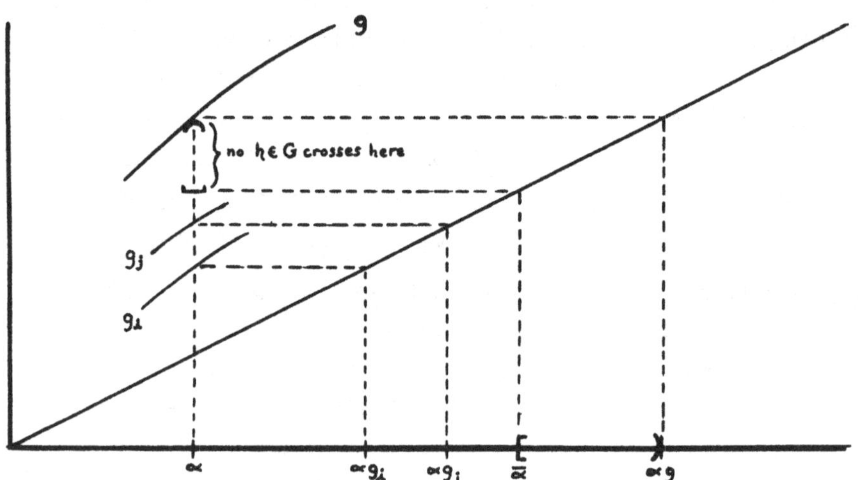

$h \wedge g$). Now $\alpha g_i \leq \bar{\alpha} \leq \alpha h$ for all $i \in I$, so $g_i h^{-1} \vee e \in G_\alpha$ for all $i \in I$. But $gh^{-1} \vee e = \bigvee_{i \in I}(g_i h^{-1} \vee e)$ and since G_α is closed, $gh^{-1} \vee e \in G_\alpha$. Hence $\alpha gh^{-1} \leq \alpha$, a contradiction.

We can now prove Theorem 8.2.1 (and hence Theorem 8C for $\bar{\alpha} \in \Omega$).

(1) → (2). If $g = \bigvee_{i \in I} g_i$ (in G) and $h \in A(\Omega)$ where $h \geq g_i$ for all $i \in I$, then $\alpha h \geq \sup\{\alpha g_i : i \in I\} = \alpha g$ for all

$\alpha \in \Omega$ (since (G,Ω) has pointwise suprema). Hence $h \geq g$ so $g = \bigvee_{i \in I} g_i$ (in $A(\Omega)$).

<u>(2) → (3)</u>. By Corollary 8.1.3.

<u>(3) → (4)</u>. $D(G) \subseteq \cap\{G_\alpha : \alpha \in \Omega\} = \{e\}$ since each G_α is closed and prime.

<u>(3) → (1)</u>. (G,Ω) transitive: If $g = \bigvee_{i \in I} g_i$ were not pointwise, there would be $\alpha \in \Omega$ such that $\alpha g > \bar{\alpha} = \sup\{\alpha g_i : i \in I\}$. Let $\bar{\alpha} \leq \beta < \alpha g$ with $\beta \in \Omega$. By transitivity, there exists $h \in G$ such that $\alpha h = \beta$, contradicting Lemma 8.2.2.

In order to complete the proof of Theorems 8B and 8C, we establish:

<u>LEMMA</u> <u>8.2.3</u>. *Let (G,Ω) be an ℓ-permutation group. If $G_{\bar{\alpha}}$ is not closed, let $\{g_i : i \in I\} \subseteq G_{\bar{\alpha}}$ with $g = \bigvee_{i \in I} g_i \notin G_{\bar{\alpha}}$. Then $G_{\bar{\beta}}$ is not closed for all $\bar{\beta} \in (\bar{\alpha}, \bar{\alpha}g)$, and no $e \neq f \in G$ has support contained in $(\bar{\alpha}, \bar{\alpha}g)$.*

<u>*Proof:*</u> Without loss of generality, $e \leq g_i$ for all $i \in I$.

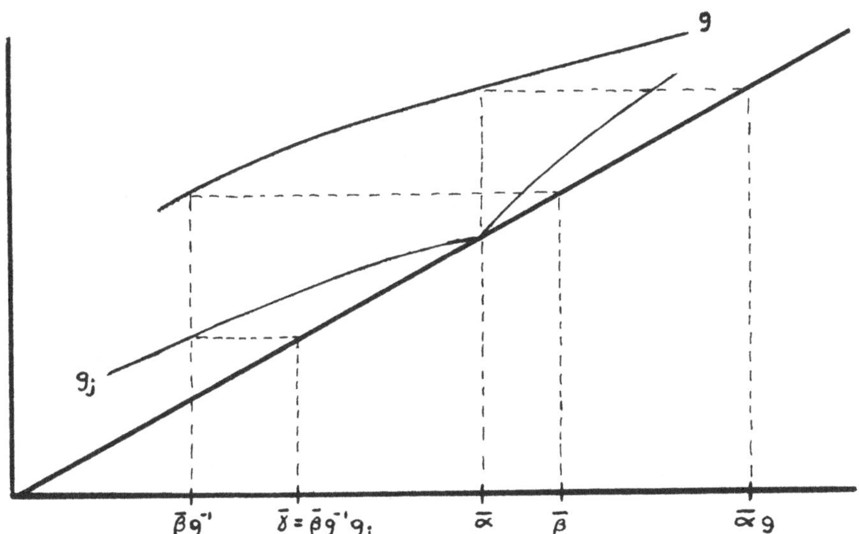

For reductio ad absurdum, suppose $G_{\bar\beta}$ is closed for some $\bar\beta \in (\bar\alpha, \bar\alpha g)$. Then $G_{\bar\beta g^{-1}} = gG_{\bar\beta}g^{-1}$ is closed and $\bar\beta g^{-1} < \bar\alpha < \bar\beta$. But $\bar\beta g^{-1} < \bar\beta = (\bar\beta g^{-1}) \bigvee_{i \in I} g_i$. Since $G_{\bar\beta g^{-1}}$ is closed, $g_j \notin G_{\bar\beta g^{-1}}$ for some $j \in I$. Let $\bar\gamma = \bar\beta g^{-1} g_j < \bar\alpha g_j = \bar\alpha$. $G_{\bar\gamma}$ is closed and $\bar\gamma g_i g^{-1} g_j < \bar\alpha g_i g^{-1} g_j = \bar\alpha g^{-1} g_j < \bar\beta g^{-1} g_j = \bar\gamma$. Hence $g_j = \bigvee_{i \in I} (g_i g^{-1} g_j \vee e) \in G_{\bar\gamma}$; i.e., $(\bar\beta g^{-1} g_j) g_j = \bar\beta g^{-1} g_j$. Thus $g_j \in G_{\bar\beta g^{-1}}$, the required contradiction.

The rest of the lemma follows from Lemma 8.1.1.

Let $\Lambda_{\mathfrak{C}} = \{\bar\alpha \in \bar\Omega : G_{\bar\alpha}$ is closed$\}$. Lemma 8.2.3 gives:

COROLLARY 8.2.4. *For any ℓ-permutation group (G,Ω), $\Lambda_{\mathfrak{C}}$ is closed with respect to the open interval topology on $\bar\Omega$.*

Theorems 8B and 8C now follow from Theorems 8.1.2 and 8.2.1 and Corollary 8.2.4.

We next wish to examine the converse directions $(3) \to (2)$ and $(2) \to (1)$ in Theorem 8.2.1. In the example we gave at the beginning of the chapter, for each $q \in \mathbb{Q}$, $\Omega_q = \{(\tau, q) : \tau \in T\}$ is a fat block of $(A(\Omega), \Omega)$ that is dead; i.e. $A(\Omega)_{(\Omega_q)} = \{e\}$. This turns out to be crucial.

Suppose that (G,Ω) is an ordered permutation group and $k(G,\Omega)$ has a minimal element K_0 that is static (i.e., for each $\mathfrak{C}^{K_0}$-class Δ, $G_{(\Delta)} = G_\Delta$). Then we will say that K_0 is *dead* and that each $\mathfrak{C}^{K_0}$ class is a *dead segment*. Define $\mathcal{D}$ by: $\alpha \mathcal{D} \beta$ if $\alpha = \beta$ or α and β belong to the same dead segment (if any). Clearly $\mathcal{D}$ is a natural congruence of (G,Ω). In the passage from

Ω to $\Omega/\mathcal{D}$, each dead segment collapses to a point of $\Omega/\mathcal{D}$ and each $\bar{\alpha} \in \bar{\Omega}$ becomes an unambiguous point of $\overline{\Omega/\mathcal{D}}$ (since distinct dead segments--being fat or singletons--cannot have a common end-point). G acts faithfully on $\Omega/\mathcal{D}$ and by Lemma 3.2.9, $(G,\Omega/\mathcal{D})$ has no dead segments. Note that $(G,\Omega/\mathcal{D})$ is just (G,Ω) stripped of dead elements and there is no change in the non-dead primitive components. Also observe that if (G,Ω) is transitive, $\mathcal{D}$ is trivial. In our example above, the $\mathcal{D}$ classes are just $\{\Omega_q : q \in \mathbb{Q}\}$. We now prove the following generalisation of (3) → (1) of Theorem 8.2.1 (transitive):

<u>THEOREM</u> <u>8.2.5</u>. *For any ℓ-permutation group (G,Ω), the following are equivalent:*

(i) $(G,\Omega/\mathcal{D})$ has pointwise suprema,

(ii) G is a complete ℓ-subgroup of $A(\Omega/\mathcal{D})$,

(iii) The stabiliser subgroups of $(G,\Omega/\mathcal{D})$ are closed,

(iv) The stabiliser subgroups of (G,Ω) are closed.

<u>Proof</u>:

<u>(i) → (ii) → (iii)</u>. By Theorem 8.2.1 and Corollary 8.2.4.

<u>(iii) ↔ (iv)</u>. If $\bar{\alpha} \in \bar{\Delta}$ for some dead segment Δ, then $G_{\bar{\alpha}} = G_{(\Delta)}$ so (G,Ω) and $(G,\Omega/\mathcal{D})$ have the same stabiliser subgroups.

<u>(iii) → (i)</u>. To save writing suppose that (G,Ω) has closed stabilisers and no dead segments. Assume, for reductio ad absurdum, that for some $\alpha \in \Omega$ and $\{g_i : i \in I\} \subseteq G$, $\bigvee_{i \in I} g_i$ exists in G but $\bar{\alpha} = \sup\{\alpha g_i : i \in I\} < \alpha g$ where $g = \bigvee_{i \in I} g_i$. Without loss of generality, we may assume $g = e$. Choose $\beta \in \Omega$ so that $\bar{\alpha} \leq \beta < \alpha g = \alpha$ and let $K = \mathrm{Val}(\alpha,\beta)$. $\alpha G \cap [\bar{\alpha},\alpha) = \emptyset$ by Lemma 8.2.2 (G_α is closed). If K were transitively derived then, by the density of orbits, there exists $h \in G$ such that $\beta\mathcal{C}_K < (\alpha\mathcal{C}_K)h < \alpha\mathcal{C}_K$, contradicting $\alpha G \cap [\bar{\alpha},\alpha) = \emptyset$. Thus K is integral or static. Hence $\alpha\mathcal{C}_K$ cannot be fat and so is extensive. Hence if $\alpha\mathcal{C}_K \neq \{\alpha\}$, then for some $h \in G$, $\alpha > \alpha h \in \alpha\mathcal{C}_K > \beta\mathcal{C}_K$ and again we obtain a contradiction. Therefore $\alpha\mathcal{C}_K = \{\alpha\}$. If K were integral, then α

would have an immediate predecessor γ in αG. Now $\gamma < \bar{\alpha}$ since $\alpha G \cap [\bar{\alpha},\alpha) = \emptyset$. But each $g_i \in G$ and so $\alpha g_i \leq \gamma$. This contradicts the fact that $\sup\{\alpha g_i : i \in I\} = \bar{\alpha} > \gamma$. It follows that K is static.

We prove that $\delta \mathcal{C}_K = \{\delta\}$ for each $\delta \in \alpha \mathcal{C}^K$ whence $\alpha \mathcal{C}^K$ is a non-trivial $\mathcal{D}$ class, the desired contradiction. Let $\delta \in \alpha \mathcal{C}^K$ with $\delta \neq \inf \alpha \mathcal{C}^K$. Since $\alpha g_i < \alpha$ for all $i \in I$ and $\alpha \mathcal{C}_K = \{\alpha\}$, $\alpha g_i \mathcal{C}_K = \{\alpha g_i\} < \alpha \mathcal{C}_K$ for all $i \in I$. Any element of $G_{(\alpha \mathcal{C}^K)}$ fixes each $\mathcal{C}_K$-class contained in $\alpha \mathcal{C}^K$ (K is static), so $g_i \notin G_{(\alpha \mathcal{C}^K)}$. Thus $\alpha \mathcal{C}^K g_i < \alpha \mathcal{C}^K$ for all $i \in I$. Now $\sup\{\delta g_i : i \in I\} \leq \inf(\alpha \mathcal{C}^K) < \delta$ and we can find $\sigma \in \alpha \mathcal{C}^K$ with $\sigma \mathcal{C}_K < \delta \mathcal{C}_K$. Then $K = \mathrm{Val}(\delta,\sigma)$. Repeating the argument of the previous paragraph with δ,σ in place of α,β, we obtain $\delta \mathcal{C}_K = \{\delta\}$. If $\inf(\alpha \mathcal{C}^K) = \delta \in \alpha \mathcal{C}^K$, then $\delta \mathcal{C}^K = \alpha \mathcal{C}^K$ is a non-extensive natural block covering $\delta \mathcal{C}_K$. Hence $\delta \mathcal{C}_K$ is extensive by Lemma 3.2.5. Since $\delta = \inf(\alpha \mathcal{C}^K)$, $\delta \mathcal{C}_K = \{\delta\}$ and the proof is complete.

<u>COROLLARY</u> <u>8.2.6</u>. *Let (G,Ω) be any ℓ-permutation group.*

(a) Any counterexample to (3) → (2) in Theorem 8.2.1 must have a dead segment Δ such that $A(\Delta) \neq \{e\}$.

(b) Any counterexample of (2) → (1) in Theorem 8.2.1 must have a dead segment Λ such that $A(\Lambda) = \{e\}$.

Proof:

(a) If (G,Ω) satisfies (3) but not (2), then G is a complete ℓ-subgroup of $A(\Omega/\mathcal{D})$ (by Theorem 8.2.5) and there exists $g = \bigvee_{i \in I} g_i \in G$ with $g > h \geq g_i$ $(i \in I)$ for some $h \in A(\Omega)$. Now for every dead segment Δ of (G,Ω), $\Delta g = \Delta h$; so $gh^{-1} \in A(\Omega)_{(\Delta)}$ for every dead segment Δ of (G,Ω). Since $gh^{-1} \neq e$, (a) follows.

(b) By Theorems 8.2.1 and 8.2.5, if (G,Ω) is a complete ℓ-subgroup of $A(\Omega)$ that doesn't have pointwise suprema, $(G,\Omega/\mathcal{D})$ has pointwise suprema. So for some dead segment Λ, $\Lambda g_i < \Lambda g$ for all $i \in I$ and $\sup\{\Lambda g_i : i \in I\} = \Lambda g$ in $(G,\Omega/\mathcal{D})$ but

$\sup\{\alpha g_i : i \in I\} \neq \alpha g$ for some $\alpha \in \Lambda$, where $g = \bigvee_{i \in I} g_i$ in G. If $A(\Lambda) \neq \{e\}$, there exists $e < h \in A(\Lambda)$. Let $f \in A(\Omega)$ be defined by $\beta f = \begin{cases} \beta h & \text{if } \beta \in \Lambda \\ \beta & \text{if } \beta \notin \Lambda \end{cases}$. Then for all $\beta \in \Lambda$, $\beta g_i \in \Lambda g_i < \Lambda g = \Lambda f^{-1} g$; so $\beta g_i < \beta f^{-1} g$. Hence $g_i < f^{-1} g < g$ for all $i \in I$. Consequently, $g \neq \bigvee_{i \in I} g_i$ (in $A(\Omega)$), a contradiction.

Note that the structure theory helped us to analyse and prove the underlying problem with the converse direction of Theorem 8.2.1. It is now easy to give counterexamples to (3) → (2) and (2) → (1).

<u>EXAMPLE</u> <u>8.2.7.</u> Let $\Omega = T \overleftarrow{\times} \mathbb{R}$ where $T \cong \mathbb{R}$, and let $G = \{g \in A(\Omega) : (\exists \bar{g} \in A(\mathbb{R}))(\forall \tau \in T)(\forall r \in \mathbb{R})((\tau,r)g = (\tau,r\bar{g}))\}$. $(G,\Omega/\mathcal{D}) \cong (A(\mathbb{R}),\mathbb{R})$ has closed stabilisers by Theorem 8B. $A(T) \neq \{e\}$. Let $g_i \in G$ be such that $(\tau,r)g_i = (\tau,r+i)$, $i \in (0,1) = I$. Then $g = \bigvee_{i \in I} g_i$ in G where $(\tau,r)g = (\tau,r+1)$ but $g > h \geq g_i$ $(i \in I)$ where $h \in A(\Omega)$ is defined by $(\tau,r)h = (\tau - 1, r + 1)$. Thus (3) holds and (2) fails in the ℓ-permutation group (G,Ω).

<u>EXAMPLE</u> <u>8.2.8.</u> Let $\Omega = T \overleftarrow{\times} \mathbb{Q}$ where $T = \{\tau_1,\tau_2\}$ with $\tau_1 < \tau_2$. We already saw that $\Omega_q = \{(\tau,q) : \tau \in T\}$ are the dead segments of $(A(\Omega),\Omega)$ $(q \in \mathbb{Q})$, $A(\Omega_q) = \{e\}$, and $(A(\Omega),\Omega)$ does not have pointwise suprema. Hence $(A(\Omega),\Omega)$ satisfies (2) but not (1).

For an example that (4) $\not\rightarrow$ (3), see McCleary [72c].

We close this section with a proof of Theorem 8D:

If (G,Ω) is static or integral, then it clearly has pointwise suprema, and so closed subgroups by Theorem 8C. So assume (G,Ω) is transitively derived from (G,T). Since (G,Ω) has no dead segments it is enough to decide if (G,Ω) has closed stabilisers by Theorem 8.2.5. Since T is dense in Ω, it is enough to decide if G_α is closed for all $\alpha \in T$ (by Corollary 8.2.4).

If (G,T) is regular, $G_\alpha = \{e\}$ which is closed.

If (G,T) is doubly transitive with $e < f \in G$ of bounded support, then by Corollary 1.10.6, Lemma 8.1.1 and Theorem 8C, G_α is closed for all $\alpha \in T$.

If (G,T) is periodic primitive with period z, and G_α were not closed, let $\{g_i : i \in I\} \subseteq G_\alpha$ with $g = \bigvee_{i \in I} g_i \notin G_\alpha$. Choose $e < h \in G_\alpha$ so that $\mathrm{supp}(h) \cap [\alpha z^{-1},\alpha] \subseteq [\alpha g^{-1},\alpha]$. Let $\tau \in [\alpha z^{-1},\alpha]$. If $\tau \in [\alpha g^{-1},\alpha]$, $\alpha g^{-1} \leq \tau h^{-1}$, so $\tau g_i \leq \alpha \leq \tau h^{-1} g$; and if $\tau \notin [\alpha g^{-1},\alpha]$, $\tau g_i \leq \tau g = \tau h^{-1} g$. Thus $h^{-1}g$ exceeds each g_i on $[\alpha z^{-1},\alpha]$ and so on T. Therefore $g_i \leq h^{-1}g < g$ for all $i \in I$, a contradiction. Consequently, G_α is closed.

Now suppose (G,T) is pathological. Let $\alpha \in T$ and $e < g \in G$ be such that $\alpha g > \alpha$. For all $\tau \notin [\alpha g^{-1},\alpha]$, we can find $g_\tau \in G_\alpha$ such that $\tau g = \tau g_\tau$. Now $g \geq g \wedge g_\tau$ for all $\tau \notin [\alpha g^{-1},\alpha]$. If $g \geq h \geq g_\tau \wedge g$ for all $\tau \notin [\alpha g^{-1},\alpha]$, then $\tau g \geq \tau h \geq \tau g$ if $\tau \notin [\alpha g^{-1},\alpha]$. So $\mathrm{supp}(gh^{-1}) \subseteq [\alpha g^{-1},\alpha]$ and, consequently, $gh^{-1} = e$. Thus $g = \bigvee_{\tau \notin [\alpha g^{-1},\alpha]} (g \wedge g_\tau)$. But $g \notin G_\alpha$ and $g \wedge g_\tau \in G_\alpha$ for all τ. Therefore G_α is not closed.

8.3. CLOSED SUBGROUPS OF ℓ-GROUPS.

In this section we wish to locate the closed convex ℓ-subgroups of an ℓ-group G. In particular, we will prove Theorems 8E, 8F and 8G. The main result (a strong form of which we will use throughout the section) is:

THEOREM 8.3.1. *If (G,Ω) is an ℓ-permutation group and C is a closed convex ℓ-subgroup of G, then C is an intersection of stabilisers.*

From this we will deduce, besides Theorems 8E, 8F and 8G:

THEOREM 8.3.2. *If G is a completely distributive ℓ-group, then every closed convex ℓ-subgroup of G is an intersection of closed prime subgroups.*

We first set up some notation. Let C be a convex ℓ-subgroup of an ℓ-group G. Since the intersection of a family of closed convex ℓ-subgroups of G is a closed convex ℓ-subgroup, there is a least closed convex ℓ-subgroup of G containing C. We call this the *closure of C* and denote it by C^*. Let $\overline{Fx}(C) = \{\bar{\alpha} \in \bar{\Omega}: (\forall g \in C)(\bar{\alpha}g = \bar{\alpha})\}$ and $G(\overline{Fx}(C)) = \{g \in G: (\forall \bar{\alpha} \in \overline{Fx}(C))(\bar{\alpha}g = \bar{\alpha})\}$. Then $G(\overline{Fx}(C)) = \bigcap\{G_{\bar{\alpha}}: \bar{\alpha} \in \overline{Fx}(C)\}$. Observe that $G_{\bar{\beta}} = G(\overline{Fx}(G_{\bar{\beta}}))$ for all $\bar{\beta} \in \bar{\Omega}$.

The following theorem immediately implies Theorem 8.3.1 and is the key to this section.

THEOREM 8.3.3. *Let (G,Ω) be an ℓ-permutation group and C a convex ℓ-subgroup of G. Then $C \subseteq G(\overline{Fx}(C)) \subseteq C^*$. Hence $C = G(\overline{Fx}(C))$ if C is closed.*

Proof: Let $H = G(\overline{Fx}(C))$. If $\mathrm{Conv}_\Omega(\beta H) \not\subseteq \mathrm{Conv}_\Omega(\beta C^*)$ for some $\beta \in \Omega$, then (without loss of generality), $\bar{\sigma} = \sup(\mathrm{Conv}_\Omega(\beta C^*)) \in \mathrm{Conv}_{\bar{\Omega}}(\beta H)$. So $\beta \leq \bar{\sigma} < \beta h$ for some $h \in H$. But $\bar{\sigma} \in \overline{Fx}(C^*) \subseteq \overline{Fx}(C)$ and $h \in H = G(\overline{Fx}(C))$; thus $\bar{\sigma}h = \bar{\sigma}$, a contradiction. Hence $\mathrm{Conv}_\Omega(\beta H) \subseteq \mathrm{Conv}_\Omega(\beta C^*)$ for all $\beta \in \Omega$.

Let $e \leq g \in H$ and $\alpha \in \Omega$. There exists $e \leq f_\alpha \in C^*$ such that $\alpha g \leq \alpha f_\alpha$. Since $e \leq g \wedge f_\alpha \leq f_\alpha$, $g \wedge f_\alpha \in C^*$. Now $g = \bigvee_{\alpha \in \Omega} (g \wedge f_\alpha) \in C^*$ as C^* closed. Thus $H \subseteq C^*$.

We will frequently use the following (Appendix II Theorem 1) due to Byrd and Lloyd [67]:

If P is a closed prime subgroup of an ℓ-group G and Q is a convex ℓ-subgroup of G containing P, then Q is closed.

For a proof, see Appendix II. One immediate consequence of this and Theorem 8.3.1 is:

COROLLARY 8.3.4. *If* (G,Ω) *is an* ℓ*-permutation group and* C *is a closed prime subgroup, then* C *is an intersection of closed stabilisers.*

By Theorems 8D and 8.2.1, if (G,Ω) is a transitive primitive ℓ-permutation group that is not pathological, then G is completely distributive.

COROLLARY 8.3.5. *If* (G,Ω) *is a pathological* ℓ*-permutation group, then* $D(G) = G$. *Hence* G *is not completely distributive.*

Proof: In the proof of Theorem 8D we saw that $G_{\bar{\alpha}}$ is not closed $(\bar{\alpha} \in \bar{\Omega})$. Hence $D(G)$, the intersection of all closed prime subgroups of G, is G (by Corollary 8.3.4).

COROLLARY 8.3.6. *Let* (G,Ω) *be an* ℓ*-permutation group.* (G,Ω) *has closed stabilisers if and only if* $C^* = G(\overline{Fx}(C))$ *for all convex* ℓ*-subgroups* C *of* G.

Proof: If (G,Ω) has closed stabilisers, then $G(\overline{Fx}(C)) = \bigcap\{G_{\bar{\alpha}}: \bar{\alpha} \in \overline{Fx}(C)\}$ is a closed convex ℓ-subgroup of G. Since $C \subseteq G(\overline{Fx}(C)) \subseteq C^*$, $G(\overline{Fx}(C)) = C^*$.

Conversely, $G^*_{\bar{\beta}} = G(\overline{Fx}(G_{\bar{\beta}})) = G_{\bar{\beta}}$ so $G_{\bar{\beta}}$ is closed for all $\bar{\beta} \in \bar{\Omega}$.

Let (G,Ω) be an ℓ-permutation group and $\mathcal{C}$ a partial congruence of (G,Ω). $\mathcal{C}$ is said to be *closed* if $G_{(\Delta)}$ is closed for every $\mathcal{C}$-class Δ. If $\mathcal{C}$ is a (full) congruence of (G,Ω), recall $L(\mathcal{C}) = \{g \in G: \Delta g = \Delta$ for all $\mathcal{C}$-classes $\Delta\} = \bigcap\{G_{(\Delta)}: \Delta$ is a $\mathcal{C}$-class$\}$ is an ℓ-ideal of G. So $L(\mathcal{C})$ is closed if $\mathcal{C}$ is.

We can now prove the following generalisation of Theorem 8F:

COROLLARY 8.3.7. *Let* (G,Ω) *be an* ℓ*-permutation group and* H *be a closed* ℓ*-ideal of* G. *Let* $\mathcal{C}$ *be the congruence of* (G,Ω) *whose classes are the* H*-orbits. Then* $H = L(\mathcal{C})$. *Thus* $(G/H,\Omega/\mathcal{C})$ *is an* ℓ*-permutation group. Hence there is a standard*

ℓ-embedding of (G,Ω) into $(\bar{H},\alpha H)$ Wr $(G/H,\Omega/\mathcal{C})$ if (G,Ω) is transitive, where $\bar{H}$ denotes the closure (in $\hat{G}_{(\alpha H)}$) of the restriction of H to αH.

Proof: Since H is an ℓ-ideal, $\mathcal{C}$ is a congruence of (G,Ω). Clearly, it is natural and $L(\mathcal{C}) = G(\overline{Fx}(H))$; so $L(\mathcal{C}) = H$ by Theorem 8.3.3. Since $L(\mathcal{C}) = \bigcap\{G_{(\Delta)}: \Delta \text{ is a } \mathcal{C}\text{-class}\}$, the lazy subgroup of the action of G on $\Omega/\mathcal{C}$ is just $L(\mathcal{C}) = H$.

We next prove the following generalisation of Theorem 8E.

THEOREM 8.3.8. *Let (G,Ω) be a depressible ℓ-permutation group. Then the closed prime subgroups of G (other than G) are precisely the stabiliser subgroups $G_{\bar{\alpha}}$, $\bar{\alpha} \in \bar{\Omega}$.*

Proof: By Theorem 8.1.2 and Corollary 8.2.4, each $G_{\bar{\alpha}}$ $(\bar{\alpha} \in \bar{\Omega})$ is a closed prime subgroup of G.

Conversely, let C be a closed prime subgroup of G, $C \neq G$. We must associate C with a point $\bar{\xi} \in \bar{\Omega}$ such that $C = G_{\bar{\xi}}$. By Theorem 8.3.3, $C = \bigcap\{G_{\bar{\alpha}}: \bar{\alpha} \in \overline{Fx}(C)\}$. If $\overline{Fx}(C) = \emptyset$, then $C = G$; so $\overline{Fx}(C) \neq \emptyset$. Since C is prime, $\{G_{\bar{\alpha}}: \bar{\alpha} \in \overline{Fx}(C)\}$ is a tower. We must show this tower has a least element, $G_{\bar{\xi}}$ (Then $C = G_{\bar{\xi}}$). Suppose, for reductio ad absurdum, that the tower has no least element. If $G_{\bar{\alpha}} \subsetneqq G_{\bar{\beta}}$, then $\Lambda(\bar{\alpha},\bar{\beta}) = \mathrm{Conv}_{\Omega}(\bar{\alpha}G_{\bar{\beta}})$ is a non-trivial extensive block of (G,Ω) (cf. Theorem 1.6.2). We now show that if $G_{\bar{\alpha}_2} \subsetneqq G_{\bar{\alpha}_1} \subsetneqq G_{\bar{\beta}}$, $\Lambda(\bar{\alpha}_1,\bar{\beta}) = \Lambda(\bar{\alpha}_2,\bar{\beta})$.

Since $G_{\bar{\alpha}_2} \subsetneqq G_{\bar{\alpha}_1}$, $\bar{\alpha}_2 \in \bar{\Lambda}(\bar{\alpha}_1,\bar{\beta})$ (Otherwise, we could choose $g \in G_{\bar{\beta}}$ such that $\bar{\alpha}_1 g \neq \bar{\alpha}_1$ and g is the identity off the natural block $\Lambda(\bar{\alpha}_1,\bar{\beta})$--here is where we use depressibility. Then $g \in G_{\bar{\alpha}_2} \setminus G_{\bar{\alpha}_1}$, a contradiction). Thus there are $f,g \in G_{\bar{\beta}}$ such that $\bar{\alpha}_1 f \leq \bar{\alpha}_2 \leq \bar{\alpha}_1 g$. Therefore $\bar{\alpha}_2 g^{-1} \leq \bar{\alpha}_1 \leq \bar{\alpha}_2 f^{-1}$ and, as $f^{-1},g^{-1} \in G_{\bar{\beta}}$, $\bar{\alpha}_1 \in \bar{\Lambda}(\bar{\alpha}_2,\bar{\beta})$. This establishes the claim.

Consequently, we will now write $\Lambda(\bar{\beta})$ for $\Lambda(\bar{\alpha},\bar{\beta})$ where $\bar{\alpha}$ is any point such that $\bar{\alpha} \in \overline{Fx}(C)$ and $G_{\bar{\alpha}} \subsetneqq G_{\bar{\beta}}$. If $\bar{\alpha},\bar{\beta} \in \overline{Fx}(C)$ and $G_{\bar{\alpha}} \subsetneqq G_{\bar{\beta}}$, let $\bar{\gamma} \in \overline{Fx}(C)$ with $G_{\bar{\gamma}} \subsetneqq G_{\bar{\alpha}} \subsetneqq G_{\bar{\beta}}$. Then $\Lambda(\bar{\alpha}) = \Lambda(\bar{\gamma},\bar{\alpha}) \subseteq \Lambda(\bar{\gamma},\bar{\beta}) = \Lambda(\bar{\beta})$ and as $\bar{\alpha} \in \bar{\Lambda}(\bar{\beta})\setminus\bar{\Lambda}(\bar{\alpha})$,

$\mathcal{J} = \{\bar{\Lambda}(\bar{\alpha}): \bar{\alpha} \in \overline{Fx}(C)\}$ is a strictly decreasing tower of (non-empty) extensive blocks of $(G,\bar{\Omega})$ having no least element. By Lemma 3.2.5, $\sup(\Lambda(\bar{\alpha})) < \sup(\Lambda(\bar{\beta}))$ if $\bar{\Lambda}(\bar{\alpha}) \subsetneq \bar{\Lambda}(\bar{\beta})$ in $\mathcal{J}$. Hence if $\bar{\xi} = \inf\{\sup(\Lambda(\bar{\alpha})): \bar{\alpha} \in \overline{Fx}(C)\}$, $\bar{\xi} \in \bar{\Lambda}(\bar{\alpha})$ for all $\bar{\alpha} \in \overline{Fx}(C)$. But since $C \subseteq G_{\bar{\alpha}}$ for all $\bar{\alpha} \in \overline{Fx}(C)$, $\sup(\Lambda(\bar{\alpha}))g = \sup(\Lambda(\bar{\alpha}))$ for all $g \in C$. Consequently, $C \subseteq G_{\bar{\xi}}$; i.e., $\bar{\xi} \in \overline{Fx}(C)$. Thus $\bar{\xi} \in \bar{\Lambda}(\bar{\xi})$ and as $\Lambda(\bar{\xi})$ is a $G_{\bar{\xi}}$ orbit, $\bar{\Lambda}(\bar{\xi}) = \{\bar{\xi}\}$. Therefore $\mathcal{J}$ has a minimal element, the desired contradiction.

We next consider Theorem 8.3.2.

Let (G,Ω) be an ℓ-permutation group. Define $\underline{\mathfrak{C}}$ by: $\alpha \mathfrak{C} \beta$ if $\alpha = \beta$ or $G_{\bar{\gamma}}$ is not closed for any $\bar{\gamma}$ between α and β. $\mathfrak{C}$ is a congruence of (G,Ω) since $G_{\bar{\gamma}}$ is closed if and only if $G_{\bar{\gamma}g}$ is. Recall $\Lambda_{\mathfrak{C}} = \{\bar{\beta} \in \bar{\Omega}: G_{\bar{\beta}}$ is closed$\}$ is closed (Corollary 8.2.4).

We now use Appendix II Theorem 1 to identify the closed congruences of (G,Ω).

<u>LEMMA</u> <u>8.3.9</u>. *Let (G,Ω) be any ℓ-permutation group. Then $\mathfrak{C}$ is natural and the smallest closed congruence of (G,Ω). All $\mathfrak{C}$ classes are extensive. A partial congruence $\mathfrak{G}$ of (G,Ω) is closed if and only if each $\mathfrak{G}$-class is a union of $\mathfrak{C}$ classes.*

<u>*Proof:*</u> If Δ is a $\mathfrak{C}$-class, then for all $\alpha \in \Omega$, if $\alpha > \sup(\Delta) = \bar{\delta}$, $[\bar{\delta},\alpha) \cap \Lambda_{\mathfrak{C}} \neq \emptyset$. Hence $G_{(\Delta)} = G_{\bar{\delta}}$ is closed by Corollary 8.2.4. Therefore $\mathfrak{C}$ is closed. Moreover, Δ is extensive (For if $\delta \in \Delta$ and $A = \mathrm{Conv}_{\Omega}(\delta G_{(\Delta)})$, let $\bar{\alpha} = \sup(A)$. Then $G_{\bar{\alpha}} \supseteq G_{(\Delta)} = G_{\bar{\delta}}$ and so is closed by Appendix II Theorem 1. Hence $\bar{\alpha} \notin \bar{\Delta}$. Thus $A = \Delta$).

If each $\mathfrak{G}$-class is a union of $\mathfrak{C}$-classes, then $\mathfrak{G}$ is closed by Appendix II Theorem 1.

If $\mathfrak{G}$ is closed and Λ is a $\mathfrak{G}$-class, $G_{\bar{\beta}} = G_{(\Lambda)}$ is closed where $\bar{\beta} = \sup(\Lambda)$. Hence $\bar{\beta} \notin \bar{\Delta}$ for any non-singleton $\mathfrak{C}$-class Δ. Thus the $\mathfrak{G}$-classes are unions of $\mathfrak{C}$-classes.

COROLLARY 8.3.10. *Let* (G,Ω) *be an* ℓ*-permutation group and* P *a prime subgroup of* G. *Then* $\overline{Fx}(P^*) = \overline{Fx}(P) \cap \Lambda_{\mathfrak{C}}$ *and* $P^* = G(\overline{Fx}(P) \cap \Lambda_{\mathfrak{C}})$.

Proof: If $\bar\beta \in \overline{Fx}(P^*)$, then $P^* \subseteq G_{\bar\beta}$ so $G_{\bar\beta}$ is closed. Hence $\bar\beta \in \overline{Fx}(P) \cap \Lambda_{\mathfrak{C}}$. Conversely, since $G_{\bar\alpha}$ is closed for all $\bar\alpha \in \overline{Fx}(P) \cap \Lambda_{\mathfrak{C}}$, $H = \cap\{G_{\bar\alpha}: \bar\alpha \in \overline{Fx}(P) \cap \Lambda_{\mathfrak{C}}\} \supseteq P$ and is closed. Hence $H \supseteq P^*$ so $\overline{Fx}(P) \cap \Lambda_{\mathfrak{C}} \subseteq \overline{Fx}(P^*)$. The rest of the corollary follows from Theorem 8.3.3.

By Corollary 8.3.4 and Lemma 8.3.9:

COROLLARY 8.3.11. *Let* (G,Ω) *be an* ℓ*-permutation group. Then* $D(G) = L(\mathfrak{C})$. *Hence, if* G *is completely distributive,* $L(\mathcal{C})$ *is closed for every natural congruence* $\mathcal{C}$ *of* (G,Ω).

We can now prove Theorem 8.3.2.

If G is completely distributive, $D(G) = \{e\}$ so $(G,\Omega/\mathfrak{C})$ is an ℓ-permutation group with closed stabilisers by Lemma 8.3.9 and Corollary 8.3.11. Let $\mathcal{D}'$ be the "dead" congruence for $(G,\Omega/\mathfrak{C})$. Then $(G,(\Omega/\mathfrak{C})/\mathcal{D}')$ is an ℓ-permutation group with closed stabilisers by Theorem 8.2.5. Since stabiliser subgroups are prime, Theorem 8.3.2 now follows from Theorem 8.3.1. //

Finally, we prove Theorem 8G which generalises Theorem 8D. Recall that an ordered permutation group (G,Ω) satisfies the support property if for each non-dead segment Δ with $|\Delta| > 1$, there exists $e \neq g \in G$ such that $\text{supp}(g) \subseteq \Delta$. We must prove that $\mathfrak{C}$ has as its non-singleton classes pathologically derived segments. We may assume that (G,Ω) has no dead segments. If Δ is a non-singleton natural block, $G_{(\Delta)} = G_{\inf(\Delta)}$ is closed by Lemma 8.2.3 (Here we use the support property). Hence $\mathcal{C} = \{\Delta g: g \in G\}$ is a closed natural partial congruence of (G,Ω). By Lemma 8.3.9, $\mathfrak{C}$ is trivial (in which case all $G_{\bar\alpha}$ are closed) or $\mathfrak{C}$ has primitive segments and is a minimal non-trivial full natural congruence of (G,Ω). So it is enough to show that if Δ is a primitive segment and $\alpha \in \Delta$, then G_α fails to be closed if and only if Δ is a pathologically derived segment.

Let $K \in k(G,\Omega)$ be such that $\Delta = \alpha \mathfrak{C}^K$. Note that $(G_K)_\alpha$ is not closed if and only if (G_K,Ω_K) is transitively derived from a pathological ℓ-permutation group (by Theorem 8D). Let $L_K = \{g|\alpha\mathfrak{C}^K: \operatorname{supp}(g) \subseteq \alpha\mathfrak{C}^K\}$, an ℓ-ideal of G_K.

If (G_K,Ω_K) is pathological, so is (L_K,Ω_K) by Lemma 6.3. Therefore there exist $\hat{g}_i \in (L_K)_\alpha$ and $\hat{g} \in L_K\backslash(L_K)_\alpha$ with $\hat{g} = \bigvee_{i\in I} \hat{g}_i$. By the support property, we can find $g_i, g \in G$ $(i \in I)$ such that $\operatorname{supp}(g), \operatorname{supp}(g_i) \subseteq \Delta$ $(i \in I)$, and $g|\Delta = \hat{g}$, $g_i|\Delta = \hat{g}_i$ $(i \in I)$. If $f \in G$ and $g_i \le f < g$ for all $i \in I$, then $\operatorname{supp}(f) \subseteq \Delta$ so $\bigvee_{i\in I} \hat{g}_i \le \hat{f} < \hat{g}$, a contradiction. Thus $g = \bigvee_{i\in I} g_i$ and $g_i \in G_\alpha$ $(i \in I)$ but $g \in G\backslash G_\alpha$. Hence G_α is not closed.

Conversely, if G_α is not closed, let $g_i \in G_\alpha$ $(i \in I)$, $g = \bigvee_{i\in I} g_i \in G$ with $g \notin G_\alpha$. Since $G_{(\Delta)}$ is closed, $\alpha g \in \Delta$. If (G_K,Ω_K) were not pathologically derived, $(G_K)_\alpha$ would be closed in G_K. If (G_K,Ω_K) is not doubly transitively derived, G_K is ℓ-simple (Theorem 6C); so $G_K = L_K$. If (G_K,Ω_K) is doubly transitively derived, $L_K \supseteq G_K \cap B(\Omega_K)$ by Lemma 6.3. Let $\hat{g} = g|\Delta$, $\hat{g}_i = g_i|\Delta$. Then $\hat{g}_i \in (G_K)_\alpha$ and $\hat{g} \notin (G_K)_\alpha$ so $\hat{g} \neq \bigvee_{i\in I} \hat{g}_i$. Hence there would exist $e < \hat{h} \in L_K$ such that $\hat{g}_i \le \hat{g}\hat{h}^{-1} < \hat{g}$ for all $i \in I$. By the support property, there would exist $e < h \in G$ such that $h|\Delta = \hat{h}$ and $\operatorname{supp}(h) \subseteq \Delta$. Then $g_i \le gh^{-1} < g$ for all $i \in I$, the desired contradiction. Consequently, (G_K,Ω_K) is pathologically derived. //

The support property is crucial for Theorem 8G as the following example shows.

<u>EXAMPLE</u> <u>8.3.12</u>. Let $(W,\Omega) = (\mathbb{Z},\mathbb{Z})\ \mathrm{Wr}\ (\mathbb{Z},\mathbb{Z})$ and let $G = \{(\{g_n\},\bar{g}) \in W: \exists m \in \mathbb{Z}^+(g_k = g_n \text{ if } k \equiv n \pmod m)\}$. Then (G,Ω) is a transitive locally primitive ℓ-permutation group whose local component is integral. Define $g = (\{g_n\},\bar{g}) \in G$ and $g^{(i)} = (\{g_n^{(i)}\},\bar{g}^{(i)}) \in G$ $(i \in \mathbb{Z}^+)$ by: $\alpha\bar{g} = \alpha\bar{g}^{(i)} = \alpha$ $(\alpha \in \mathbb{Z}$, $i \in \mathbb{Z}^+)$, $\alpha g_n = \alpha + 1$, $\alpha g_n^{(i)} = \begin{cases} \alpha & \text{if } n \equiv 0 \pmod i \\ \alpha + 1 & \text{otherwise} \end{cases}$

$(\alpha, n \in \mathbb{Z};\ i \in \mathbb{Z}^+)$. Then $g = \bigvee_{i \in \mathbb{Z}^+} g^{(i)}$ and $g^{(i)} \in G_{(0,0)}$ for all $i \in \mathbb{Z}^+$, but $g \notin G_{(0,0)}$. Hence $G_{(0,0)}$ is not closed.

This example also has the disturbing property that it has no lateral completion in (W, Ω)--see the list of open problems.

For further results concerning complete distributivity, etc., see McCleary [72b], [72c], [73b] and [76].

CHAPTER 9
AUTOMORPHISMS OF $A(\Omega)$

In Section 2.4 we showed that if Ω is doubly homogeneous, then every ℓ-automorphism of $A(\Omega)$ is conjugation by an element of $A(\bar{\Omega})$ (Theorem 2.4.1). In Chapter 7 we extended the result to show that if Ω is homogeneous and locally primitive, every ℓ-automorphism of $A(\Omega)$ is conjugation by an element of $A(\Omega^0)$ (Corollary 7E). In this chapter we wish to pursue two directions. The first is to try to extend Corollary 7E to the non-locally primitive case. The second is to try to improve on Corollary 7E by replacing Ω^0 by Ω; i.e., to attempt to prove that every ℓ-automorphism of $A(\Omega)$ is inner (cf. Corollary 2.4.4). This latter can only be hoped for if Ω is homogeneous as the following easy example (1.1.3) shows.

Let $\mathbb{Z}_1,\mathbb{Z}_2$ be disjoint copies of $\mathbb{Z}$ and $\Omega = \mathbb{Z}_1 \overleftarrow{\cup} \mathbb{Z}_2$. Then we saw that Ω is not homogeneous and $A(\Omega)$ is ℓ-isomorphic to $\mathbb{Z} \oplus \mathbb{Z}$ ordered by: $(m,n) \geq 0$ if $m,n \geq 0$. Hence $A(\Omega)$ is Abelian and has only one inner automorphism, the identity. But ψ is an ℓ-automorphism of $\mathbb{Z} \oplus \mathbb{Z}$ where $(m,n)\psi = (n,m)$ $(m,n \in \mathbb{Z})$. Consequently, $A(\Omega)$ has an outer ℓ-automorphism.

We will prove the following theorem due to Holland [75a]:

<u>THEOREM 9A</u>. *There exists a doubly homogeneous chain Ω such that $A(\Omega)$ has an outer ℓ-automorphism.*

The proof relies only on the results established in Section 2.4.

We return to generalising Corollary 7E. Let $k = k(\Omega)$ be the collection of covering pairs of congruences of the transitive ℓ-permutation group $(A(\Omega),\Omega)$. Let <u>Ω^0</u> $= \{\bar{\alpha} \in \bar{\Omega} : \bar{\alpha} \in \bigcap_{K \in k'} \bar{\Delta}_K$

where k' is coinitial in k, Δ_K is a $\mathcal{C}^K$ class and $\Delta_K \subseteq \Delta_{K'}$ if $K \leq K'\}$. Note that $\Omega^0 \supseteq \Omega$ and if Ω is locally primitive, the definition of Ω^0 coincides with the one given in Chapter 7. We will continue to call Ω^0 the *local completion* of Ω.

Using stabilisers in a more subtle way than we did in Chapter 7, we will prove the following result due to McCleary [69] (It is an improvement of results by Holland [65a] and Lloyd [64]):

THEOREM 9B. *If Ω is homogeneous, every ℓ-automorphism of $A(\Omega)$ is conjugation by an element of $A(\Omega^0)$.*

The proof incorporates some of the ideas of Sections 7.1 and 8.3.

Finally, we will prove the following theorem about ℓ-characteristic subgroups (A convex ℓ-subgroup H of an ℓ-group G is *ℓ-characteristic* if $H\psi = H$ for every ℓ-automorphism ψ of G):

THEOREM 9C.

(1) If Ω is doubly homogeneous, every normal subgroup of $A(\Omega)$ is ℓ-characteristic.

(2) If Ω is homogeneous, every closed ℓ-ideal of $A(\Omega)$ is ℓ-characteristic.

Note that if every ℓ-automorphism of $A(\Omega)$ is inner, every ℓ-ideal of $A(\Omega)$ is automatically ℓ-characteristic. Hence Theorem 9C is a weakening of our previous intention. The proof of Theorem 9C (1) relies on Theorems 2.4.1, 2.3.1, Lemma 2.3.5 and the patching lemma. Not surprisingly, the proof of Theorem 9C (2) is similar to that of Theorem 9B. Theorem 9C is due to Holland [75a].

We commence by constructing a chain Ω satisfying Theorem 9A.

By Theorem 2.4.1, if ψ is an ℓ-automorphism of $A(\Omega)$, there is $\hat{\psi} \in A(\bar{\Omega})$ such that $g\psi = \hat{\psi}^{-1}g\hat{\psi}$ $(g \in A(\Omega))$. Now $\Omega\hat{\psi}$ is an orbit of $A(\Omega)$ ordermorphic to Ω. Thus $A(\Omega)$ has an outer ℓ-automorphism only if Ω is ordermorphic to another orbit of $A(\Omega)$ in $\bar{\Omega}$. We construct such an Ω.

Let $T = \overrightarrow{\prod_{\mu \in \omega_1}} \mathbb{Z}$; i.e., $\tau < \sigma$ if and only if there exists $\mu \in \omega_1$ such that $\tau(\mu) < \sigma(\mu)$ and $\tau(\nu) = \sigma(\nu)$ for all $\nu < \mu$. Let $\mathcal{F}$ be an ultrafilter on ω_1 that contains all final segments of ω_1 (i.e., if $\mu \in \omega_1$, $\{\nu \in \omega_1 : \nu \geq \mu\} \in \mathcal{F}$). For each $\tau \in T$, let $X(\tau) = \{\mu \in \omega_1 : \tau(\mu)$ is even$\}$. Let $\Omega = \{\beta \in T : X(\beta) \in \mathcal{F}\}$, a subchain of T. We now show that Ω satisfies Theorem 9A.

(1) $\underline{\Omega}$ __is__ __homogeneous__: Let $\alpha,\beta \in \Omega$. Define $g \in A(T)$ by: $(\tau g)(\mu) = \tau(\mu) + \beta(\mu) - \alpha(\mu)$ $(\mu \in \omega_1)$. Then $\alpha g = \beta$, and $g|\Omega \in A(\Omega)$ since $X(\tau g) \supseteq X(\tau) \cap X(\alpha) \cap X(\beta)$.

(2) $\underline{\Omega}$ __is__ __doubly__ __homogeneous__: For each $\mu \in \omega_1$, define $\alpha \equiv_\mu \beta$ if $\alpha(\nu) = \beta(\nu)$ for all $\nu \leq \mu$ $(\alpha,\beta \in \Omega)$. For any $\alpha \in \Omega$, write α^μ for the equivalence class of α under $\equiv_\mu$. Observe that α^μ is ordermorphic to Ω via $\beta \mapsto \beta'$ where $\beta'(\nu) = \beta(\mu + \nu)$ $(\nu \in \omega_1)$ (Since $\mu + \nu = \nu$ if $\nu \geq \mu^2 + \omega$, $\beta \in \alpha^\mu$ implies $\beta' \in \Omega$ by the definition of $\mathcal{F}$). Consequently, if $\alpha,\beta \in \Omega$, then $\{\sigma \in \alpha^\mu : \sigma \leq \alpha\}$ is ordermorphic to $\{\tau \in \beta^\mu : \tau \leq \beta\}$.

Let $\gamma < \alpha < \beta$ (in Ω). By Lemma 1.10.2, it is enough to find $g \in A(\Omega)_\gamma$ such that $\alpha g = \beta$. Choose $\mu \in \omega_1$ least such that $\gamma(\mu) \neq \alpha(\mu)$. Let $A = \{\delta \in \Omega : \delta \leq \sigma$ for all $\sigma \in \alpha^\mu\}$ and $B = \Omega \backslash A$. By the previous paragraph, B is ordermorphic to Ω and hence is homogeneous. But $\alpha,\beta \in B$ so there exists $g' \in A(B)$ such that $\alpha g' = \beta$. Define $g \in A(\Omega)$ by: $\tau g = \begin{cases} \tau g' & \text{if } \tau \in B \\ \tau & \text{if } \tau \in A \end{cases}$.

Then $\alpha g = \beta$ and $\gamma g = \gamma$ (since $\gamma \in A$).

(3) $\underline{T \subseteq \bar{\Omega}}$: Let $\tau \in T$. Then $\tau = \sup\{\sigma_\nu : \nu \in \omega_1\} = \inf\{\sigma'_\nu : \nu \in \omega_1\}$ where $\sigma_\nu, \sigma'_\nu \in \Omega$ are defined by:

$$\sigma_\nu(\mu) = \begin{cases} \tau(\mu) & \text{if } \mu \le \nu \\ \tau(\mu) - 1 & \text{if } \mu > \nu \text{ and } \tau(\mu) \text{ odd} \\ \tau(\mu) - 2 & \text{if } \mu > \nu \text{ and } \tau(\mu) \text{ even} \end{cases}$$

$$\sigma'_\nu(\mu) = \begin{cases} \tau(\mu) & \text{if } \mu \le \nu \\ \tau(\mu) + 1 & \text{if } \mu > \nu \text{ and } \tau(\mu) \text{ odd} \\ \tau(\mu) + 2 & \text{if } \mu > \nu \text{ and } \tau(\mu) \text{ even} \end{cases}$$

Note that $\sigma_\nu < \sigma_{\nu'} < \sigma'_{\nu'} < \sigma'_\nu$ if $\nu < \nu'$. Thus $T \subseteq \bar{\Omega}$.

Since $|\Omega| = \aleph_1$, every point of $\bar{\Omega}$ has character c_{00}, c_{01}, c_{10} or c_{11}.

(4) $T = \{\bar{\alpha} \in \bar{\Omega}: \bar{\alpha}$ has character $c_{11}\}$: We first show that every point of T has character c_{11}. Suppose $\tau \in T$ and $\tau = \sup\{\sigma_n: n \in \omega\}$ where $\tau > \sigma_n \in \Omega$ and $\sigma_n < \sigma_m$ if $n, m \in \omega$, $n < m$. There exists a least $n_0 \in \omega$ such that $\sigma_{n_0}(0) = \tau(0)$. For all $n > n_0$, $\sigma_n(0) = \tau(0)$. There exists a least $\mu_1 \in \omega_1$ such that $\sigma_{n_0}(\mu_1) < \tau(\mu_1)$. Let $n_1 \in \omega$ be least such that $n_1 > n_0$ and $\sigma_{n_1}(\mu_1) = \tau(\mu_1)$. So $\sigma_n(\mu) = \tau(\mu)$ for all $n \ge n_1$ and $\mu \le \mu_1$. Continuing in this way we obtain $\{n_i: i \in \omega\}$ and $\{\mu_i: i \in \omega\}$ strictly increasing sequences in ω and ω_1 respectively (where $\mu_0 = 0$). Let $\nu = \sup\{\mu_i: i \in \omega\} < \omega_1$ and define $\rho \in T$ by: $\rho(\mu) = \begin{cases} \tau(\mu) & \text{if } \mu \ne \nu \\ \tau(\mu) - 1 & \text{if } \mu = \nu \end{cases}$. Let $m \in \omega$. There exists $j \in \omega$ such that $m < n_j$. Now $\sigma_m < \sigma_{n_j}$ so $\sigma_m(\mu) \le \rho(\mu)$ if $\mu < \mu_j$ by construction. But $\sigma_m(\mu_j) < \sigma_{n_j}(\mu_j) = \tau(\mu_j) = \rho(\mu_j)$ so $\sigma_m < \rho$ for all $m \in \omega$. Since $\rho < \tau$, $\tau \ne \sup\{\sigma_m: m \in \omega\}$, the desired contradiction. So no element of T is the supremum of a countable strictly increasing sequence of elements of Ω. Similarly for infimum. Thus each point of T has character c_{11}.

Now if $\{\alpha_\nu: \nu \in \omega_1\}$ is a strictly increasing set of points of Ω, then as $\{\alpha_\nu(0): \nu \in \omega_1\}$ is an increasing set of integers, there exists $\lambda(0) \in \omega_1$ such that $\alpha_\mu(0) = \alpha_\nu(0)$ if $\mu, \nu \ge \lambda(0)$.

Suppose $\lambda(\xi) \in \omega_1$ has been defined for all $\xi < \zeta < \omega_1$ so that $\lambda(\xi) \leq \lambda(\eta)$ whenever $\xi \leq \eta < \zeta$ and $\alpha_\mu(\xi) = \alpha_\nu(\xi)$ if $\mu,\nu \geq \lambda(\xi)$. Now $\{\alpha_\nu: \nu \in \omega_1 \ \& \ \nu \geq \sup\{\lambda(\xi): \xi < \zeta\}\}$ is uncountable so there exists $\lambda(\zeta) \in \omega_1$ such that $\lambda(\zeta) \geq \lambda(\xi)$ for all $\xi < \zeta$ and $\alpha_\mu(\zeta) = \alpha_\nu(\zeta)$ if $\mu,\nu \geq \lambda(\zeta)$. By induction on $\zeta \in \omega_1$, if $\{\alpha_\nu: \nu \in \omega_1\}$ has an upper bound in $\bar{\Omega}$, then $\sup\{\alpha_\nu: \nu \in \omega_1\} = \tau \in T$ where $\tau(\zeta) = \alpha_{\lambda(\zeta)}(\zeta)$ $(\zeta \in \omega_1)$. Thus all points of character c_{11} belong to T (Observe that the proof just given shows that the points of $\bar{\Omega}\setminus T$ have character c_{00}).

Now let $\hat{\psi} \in A(T)$ be defined by: $(\tau\hat{\psi})(\mu) = \tau(\mu) + 1$ $(\tau \in T;\ \mu \in \omega_1)$. Then $\Omega\hat{\psi} = T\setminus\Omega$ and $(T\setminus\Omega)\hat{\psi} = \Omega$. If $g \in A(\Omega)$, then $\hat{\psi}^{-1}g\hat{\psi} \in A(\Omega)$ since $T\setminus\Omega$ is an orbit of $A(\Omega)$ by (4). Thus conjugation by $\hat{\psi}$ is indeed an ℓ-automorphism of $A(\Omega)$ that is outer. This proves Theorem 9A.

We next turn to Theorem 9C (1) since its proof (like that of Theorem 9A) depends only on Chapter 2. The key to the proof is Lemma 2.3.5. To apply that lemma, we need doubly homogeneous chains of countable coterminality. We obtain such chains as subchains of Ω and extend from them to Ω via the patching lemma.

So assume that Ω is doubly homogeneous and ψ is an ℓ-automorphism of $A(\Omega)$. By Theorem 2.4.1, there is $\hat{\psi} \in A(\bar{\Omega})$ such that $f\psi = \hat{\psi}^{-1}f\hat{\psi}$ $(f \in A(\Omega))$. Let $e < g \in A(\Omega)$ and $T = \{\bar{\alpha} \in \bar{\Omega}: \bar{\alpha}g = \bar{\alpha} \ \& \ \bar{\alpha}\hat{\psi} = \bar{\alpha}\}$. Let $(\bar{\alpha},\bar{\beta})$ be any maximal interval in $\bar{\Omega}\setminus T$ and $\Omega_1 = \Omega \cap (\bar{\alpha},\bar{\beta})$. So Ω_1 is doubly homogeneous. We claim it has countable coterminality. If not, say Ω_1 does not have cofinality $\aleph_0$. Then $(\bar{\alpha},\bar{\beta})$ does not have cofinality $\aleph_0$. Let $\bar{\delta} \in (\bar{\alpha},\bar{\beta})$. Without loss of generality, $\bar{\delta}g \neq \bar{\delta}$. So $\bar{\delta} < \bar{\delta}g < \bar{\delta}g^2 < \ldots$. Let $\bar{\delta}_0 = \sup\{\bar{\delta}g^n: n \in \omega\}$. Then $\bar{\delta}_0 \in (\bar{\alpha},\bar{\beta})$ since $(\bar{\alpha},\bar{\beta})$ has cofinality greater than $\aleph_0$. But $\bar{\delta}_0 g = \bar{\delta}_0$ so $\bar{\delta}_0\hat{\psi} \neq \bar{\delta}_0$. Hence $\bar{\delta}_0|\hat{\psi}| > \bar{\delta}_0$ where $|\hat{\psi}| = \hat{\psi} \vee \hat{\psi}^{-1}$. Let $\bar{\delta}_1 = \sup\{\bar{\delta}_0|\hat{\psi}|^n: n \in \omega\} < \bar{\beta}$. So $\bar{\delta}_1\hat{\psi} = \bar{\delta}_1$ whence $\bar{\delta}_1 g \neq \bar{\delta}_1$. Continuing we obtain $\{\bar{\delta}_n: n \in \omega\} \subseteq (\bar{\alpha},\bar{\beta})$ such that $\bar{\delta}_{2n}g = \bar{\delta}_{2n}$ and $\bar{\delta}_{2n+1}\hat{\psi} = \bar{\delta}_{2n+1}$ $(n \in \omega)$. But if $\bar{\gamma} = \sup\{\bar{\delta}_n: n \in \omega\}$, then

$\bar{\gamma}g = \bar{\gamma} = \bar{\gamma}\hat{\psi}$. Thus $\bar{\gamma} = \bar{\beta}$. Consequently, $(\bar{\alpha},\bar{\beta})$ has cofinality $\aleph_0$.

Now $g|\Omega_1 \in A(\Omega_1)$ and $\hat{\psi}|(\bar{\alpha},\bar{\beta}) \in A((\bar{\alpha},\bar{\beta}))$. By Lemma 2.3.5, $g\psi|\Omega_1$ is dominated by the supremum of two conjugates (in $A(\Omega_1)$) of $g|\Omega_1$. Therefore, by the patching lemma (as Ω_1 ranges through all such maximal intervals), $g\psi \leq f_1^{-1}gf_1 \vee f_2^{-1}gf_2$ for some $f_1,f_2 \in A(\Omega)$. Hence if L is any ℓ-ideal of $A(\Omega)$, $L\psi \subseteq L$. This is true for all ℓ-automorphisms of $A(\Omega)$. Thus every ℓ-ideal of $A(\Omega)$ is ℓ-characteristic. Theorem 9C (1) now follows from Theorem 2.3.1.

If Ω is rigidly homogeneous, the only ℓ-automorphism of $A(\Omega)$ is the identity. Hence

<u>COROLLARY</u> <u>9.1</u>. *If Ω is homogeneous and primitive, every ℓ-ideal of $A(\Omega)$ is ℓ-characteristic.*

We now turn our attention to Theorem 9B. Recall that (G,Ω) has the interval support property if for each $\bar{\alpha},\bar{\beta} \in \bar{\Omega}$ with $\bar{\alpha} < \bar{\beta}$, there exists $e \neq g \in G$ such that $\text{supp}(g) \subseteq (\bar{\alpha},\bar{\beta})$. We break the proof of Theorem 9B into two cases:

<u>THEOREM</u> <u>9.2</u>. *If Ω is homogeneous, it is locally regular or has the interval support property.*

<u>*Proof:*</u> Let $\mathfrak{C}$ be the intersection of all the non-singleton congruences of $(A(\Omega),\Omega)$--a tower by Theorem 3A. By Corollary 3B, $\mathfrak{C}$ is a congruence.

<u>Case</u> <u>1</u>. $\mathfrak{C}$ is not the singleton congruence. Then it is the unique minimal congruence and Ω is locally primitive. If the primitive segments are regular, the theorem is proved; if they are not, they are doubly homogeneous (by Theorem 4B) and thus Ω enjoys the interval support property.

<u>Case</u> <u>2</u>. $\mathfrak{C}$ is the singleton congruence. So Ω is not locally primitive. Also Ω is dense (otherwise $\mathfrak{F}$ is the minimal non-trivial congruence where $\alpha \mathfrak{F} \beta$ if there are only finitely many

elements of Ω between α and β). Let Δ be an interval of Ω with $|\Delta| > 1$. Let $\delta \in \Delta$ with δ not an endpoint of Δ. Since $\mathcal{C}$ is trivial, there is a non-singleton congruence $\mathcal{K}$ such that $\delta\mathcal{K} \subseteq \Delta$. Let $\alpha \in \delta\mathcal{K}$ with $\alpha \neq \delta$. Since Ω is homogeneous, there exists $g \in A(\Omega)$ with $\alpha g = \delta$ and we may assume $\mathrm{supp}(g) \subseteq \delta\mathcal{K} \subseteq \Delta$ (Take g_0 the bump of g with α in its support). Thus Ω has the interval support property.

In the homogeneous locally regular case we already know every ℓ-automorphism of $A(\Omega)$ is conjugation by an element of $A(\Omega^0)$. We can do better:

<u>THEOREM</u> <u>9.3</u>. *If Ω is homogeneous and locally regular, every ℓ-automorphism of $A(\Omega)$ is inner.*

<u>Proof</u>: By Corollary 7.1.7, $\{A(\Omega)_{\bar{\alpha}} : \bar{\alpha} \in \Omega^0\}$ is the set of maximal representing subgroups of $A(\Omega)$. But if $\bar{\alpha} \in \Omega^0\backslash\Omega$, $\bar{\alpha}$ is a cut in some primitive segment, say Δ. Since Ω is locally regular, $A(\Omega)_{\bar{\alpha}} = A(\Omega)_\delta$ where $\delta \in \Delta$. Hence $\{A(\Omega)_\alpha : \alpha \in \Omega\}$ is the set of maximal representing subgroups of $A(\Omega)$. Let $\alpha_0 \in \Omega$ and ψ be an ℓ-automorphism of $A(\Omega)$. So $A(\Omega)_{\alpha_0}\psi = A(\Omega)_\beta$ for some $\beta \in \Omega$. By Lemma 7.1.2, $\phi \in A(\Omega)$ where $(\alpha_0 g)\phi = \beta(g\psi)$ $(g \in A(\Omega))$, and $g\psi = \phi^{-1}g\phi$. Thus ψ is an inner automorphism of $A(\Omega)$.

We must now consider the interval support case. By Theorem 8E, we know that the proper closed prime subgroups of $A(\Omega)$ are precisely the stabilisers $A(\Omega)_{\bar{\alpha}}$ $(\bar{\alpha} \in \bar{\Omega})$. We now wish to distinguish the points of Ω^0 from arbitrary points of $\bar{\Omega}$. The next lemma is the key to proving Theorem 9B.

<u>LEMMA</u> <u>9.4</u>. *If Ω is homogeneous and has the interval support property, then the minimal closed prime subgroups of $A(\Omega)$ are precisely the stabilisers $A(\Omega)_{\bar{\alpha}}$ ($\bar{\alpha} \in \Omega^0$ or $\bar{\alpha}$ is a cut in some doubly homogeneous primitive component).*

Proof: Let $\bar\alpha \in \bar\Omega$ be the supremum or infimum of some non-singleton block Δ, and let $\delta \in \Delta$. Then $A(\Omega)_\delta \subsetneqq A(\Omega)_{\bar\alpha}$ so $A(\Omega)_{\bar\alpha}$ is not a minimal prime. If $\bar\alpha \in \bar\Omega$ is a cut in some regular $\beta\mathcal{C}^K/\mathcal{C}_K$ (K non-minimal, $\beta \in \Omega$), then $A(\Omega)_\beta \subsetneqq A(\Omega)_{(\beta\mathcal{C}_K)} = A(\Omega)_{\bar\alpha}$ and again $A(\Omega)_{\bar\alpha}$ is not minimal. Hence $A(\Omega)_{\bar\alpha}$ is minimal only if $\bar\alpha \in \Omega^0$ or $\bar\alpha$ is a cut in some doubly homogeneous $\beta\mathcal{C}^K/\mathcal{C}_K$. Since Ω enjoys the interval support property, if $\bar\alpha, \bar\beta$ fit the above description, $A(\Omega)_{\bar\alpha} \subseteq A(\Omega)_{\bar\beta}$ only if $\bar\alpha = \bar\beta$. By Theorem 8E, $A(\Omega)_{\bar\alpha}$ is minimal if $\bar\alpha \in \Omega^0$ or is a cut in some doubly homogeneous primitive component.

We can now prove Theorem 9B. Firstly, observe that if $\bar\alpha \in \Omega^0$, then $A(\Omega)_{\bar\alpha}$ is a representing subgroup of $A(\Omega)$. However, if $\bar\alpha$ is a cut in some non-minimal doubly homogeneous primitive component $\beta\mathcal{C}^K/\mathcal{C}_K$, then $A(\Omega)_{\bar\alpha} \supseteq L(\mathcal{C}_K) \neq \{e\}$ and so is not a representing subgroup. By Lemma 9.4, if ψ is an ℓ-automorphism of $A(\Omega)$, $\{A(\Omega)_{\bar\alpha}\psi : \bar\alpha \in \Omega^0\} = \{A(\Omega)_{\bar\alpha} : \bar\alpha \in \Omega^0\}$. Theorem 9B now follows from Lemma 7.1.2.

We can now prove Theorem 9C (2). Suppose that the theorem is false. Then there is a closed ℓ-ideal H of $A(\Omega)$ and an ℓ-automorphism ψ of $A(\Omega)$ such that $H\psi \neq H$. Clearly, ψ cannot be inner. By Theorems 9.2 and 9.3, Ω has the interval support property. By Theorems 8F and 3A, the closed ℓ-ideals of $A(\Omega)$ form a tower. Hence, without loss of generality, $H \subsetneqq H\psi$. Let $M = \bigcup_{m\in\omega} H\psi^m$ and $N = \bigcap_{m\in\omega} H\psi^{-m}$. Then N is a closed ℓ-ideal of $A(\Omega)$ and hence $N = L(\mathcal{C})$ for some congruence $\mathcal{C}$ of $(A(\Omega),\Omega)$. Now $\Omega/\mathcal{C}$ is homogeneous and H/N is an ℓ-ideal of $A(\Omega/\mathcal{C})$ that is not fixed by the ℓ-automorphism of $A(\Omega/\mathcal{C})$ induced by ψ. Hence we may assume that $N = \{e\}$. Let $\mathcal{C}_n$ be the congruence such that $L(\mathcal{C}_n) = H\psi^n$ ($n \in \mathbb{Z}$). So the $\mathcal{C}_n$ classes are the orbits of points of Ω by $H\psi^n$. Let $\bar{\mathcal{C}}_n$ be the natural congruence of $(A(\Omega),\bar\Omega)$ whose non-singleton classes are the convexifications (in $\bar\Omega$) of the orbits of $H\psi^n$. If $\alpha\mathcal{C}_n\beta$, then $\beta = \alpha g$ for some $g \in H\psi^n$. Hence $\beta\hat\psi = (\alpha g)\hat\psi = (\alpha\hat\psi)(g\psi)$, where ψ is conjugation by $\hat{\psi} \in A(\Omega^0)$ (by Theorem 9B). Thus

$$\alpha \mathcal{C}_n \beta \text{ implies } (\alpha\hat{\psi})\bar{\mathcal{C}}_{n+1}(\beta\hat{\psi}). \qquad (\dagger)$$

Let $\bar{\Lambda}$ be a non-trivial $\bar{\mathcal{C}}_0$-class. By transitivity, there exists $g \in A(\Omega)$ such that $\bar{\Lambda} \subseteq \bar{\Lambda}\hat{\psi}g$. Replacing ψ by conjugation by $\hat{\psi}g$, we may assume $\bar{\Lambda} \subseteq \bar{\Lambda}\hat{\psi} \subseteq \bar{\Lambda}\hat{\psi}^2 \subseteq \ldots$. Let $\bar{\Delta} = \bigcup_{m \in \omega} \bar{\Lambda}\hat{\psi}^m$. Then $\bar{\Delta}\hat{\psi} = \bar{\Delta}$ so $\hat{\psi}$ induces an ℓ-automorphism $\tilde{\psi}$ of $A(\bar{\Delta} \cap \Omega)$ under which the closed ℓ-ideal $\tilde{H}$ corresponding to H is not fixed, and $\bigcup_{m \in \omega} \tilde{H}\tilde{\psi}^m = A(\bar{\Delta} \cap \Omega)$. Since $\bar{\Delta} \cap \Omega$ is homogeneous, we may assume that $M = A(\Omega)$.

We next show there is an orbit T of $A(\Omega)$ such that $T\hat{\psi} = T$. Let $\alpha, \beta \in \Omega$ with $\alpha < \beta$. There exists $n \in \mathbb{Z}$ such that $\alpha \mathcal{C}_{n+1} \beta$ but $\beta \notin \alpha\mathcal{C}_n$. So $\alpha\hat{\psi}^{-1}\bar{\mathcal{C}}_n\beta\hat{\psi}^{-1}$ by $(\dagger)$. Since $\bar{\mathcal{C}}_n$ is extensive (for $(A(\Omega),\bar{\Omega})$), there exists $\sigma \in \Omega$ such that $\sigma < \alpha\hat{\psi}^{-1}$ and $\sigma\bar{\mathcal{C}}_n\alpha\hat{\psi}^{-1}$. By homogeneity, there exists $g \in A(\Omega)$ such that $\sigma g = \alpha$. Then $\alpha = \sigma g < \alpha\hat{\psi}^{-1}g < \beta\hat{\psi}^{-1}g$. But $\sigma\bar{\mathcal{C}}_n\beta\hat{\psi}^{-1}$ so $\sigma g\bar{\mathcal{C}}_n\beta\hat{\psi}^{-1}g$; i.e., $\alpha\bar{\mathcal{C}}_n\beta\hat{\psi}^{-1}g$. Since $\beta \notin \alpha\mathcal{C}_n$ and $\alpha < \beta$, $\beta\hat{\psi}^{-1}g < \beta$. As $\alpha = \sigma g < \alpha\hat{\psi}^{-1}g$, $\hat{\psi}^{-1}g$ fixes some $\bar{\tau} \in \bar{\Omega}$ with $\alpha < \bar{\tau} < \beta$. Thus $\bar{\tau}\hat{\psi}^{-1} = \bar{\tau}g^{-1}$. So $\bar{\tau}A(\Omega)\hat{\psi}^{-1} = \tau(A(\Omega)\psi)\hat{\psi}^{-1} = \bar{\tau}g^{-1}A(\Omega) = \bar{\tau}A(\Omega)$ and $T = \bar{\tau}A(\Omega)$ is the desired orbit.

Now if T is any orbit of $A(\Omega)$ and $T\hat{\psi} = T$, then $\bar{\tau}A(\Omega)$ is dense in $\bar{\Omega}$ for any $\bar{\tau} \in T$ (If $\mathcal{C}$ is defined by: $\alpha\mathcal{C}\beta$ if $\alpha = \beta$ or no point of $\bar{\tau}A(\Omega)$ lies between α and β, then $\mathcal{C}$ is a congruence of $(A(\Omega),\Omega)$. By Theorems 8B, 8.2.1 and Corollary 8.3.11, $L(\mathcal{C})$ is a closed ℓ-ideal of $A(\Omega)$. But $N = \{e\}$ and $M = A(\Omega)$ so $L(\mathcal{C}_n) \subseteq L(\mathcal{C}) \subsetneq L(\mathcal{C}_{n+1})$ for some $n \in \mathbb{Z}$ or $\mathcal{C}$ is trivial. Since $L(\mathcal{C}_n) = H\psi^n$ and $L(\mathcal{C}_{n+1}) = H\psi^{n+1}$, $L(\mathcal{C}) \subsetneq L(\mathcal{C})\psi$ if $\mathcal{C}$ is not trivial. By definition, $L(\mathcal{C})\psi = L(\mathcal{C})$ so $\mathcal{C}$ is trivial. Hence $\bar{\tau}A(\Omega)$ is dense in $\bar{\Omega}$).

Let $\Lambda = \{\bar{\alpha} \in \bar{\Omega}\backslash\Omega\colon \bar{\alpha}A(\Omega) \text{ is dense in } \bar{\Omega}\}$. By the previous two paragraphs, $\Lambda \neq \emptyset$. Clearly $\Lambda A(\Omega) = \Lambda$. If we can show that Λ is an orbit of $A(\Omega)$ it will follow that Λ and Ω are the <u>only</u> dense orbits of $A(\Omega)$. But $T\hat{\psi} = T$ for some orbit T which is necessarily dense. Hence we will have $\Lambda\hat{\psi} = \Lambda$ or $\Omega\hat{\psi} = \Omega$. Since $\Omega\hat{\psi}$ is dense, in either case we will have $\Omega\hat{\psi} = \Omega$. Thus ψ

is inner and consequently $H\psi = H$ the desired contradiction. So to complete the proof of Theorem 9C (2), we show that Λ is an orbit of $A(\Omega)$.

Let $\bar{\alpha}, \bar{\beta} \in \Lambda$ and $n \in \mathbb{Z}$. Let $\alpha \in \Omega$ and pick $\beta \in \alpha\mathcal{C}_n$ so that $\beta > \alpha$. Since $\bar{\alpha}A(\Omega)$ is dense in $\bar{\Omega}$, there is $g \in A(\Omega)$ such that $\alpha < \bar{\alpha}g < \beta$. Hence $\bar{\alpha} \in \alpha\bar{\mathcal{C}}_n g^{-1} = \bar{\Pi}_n$, a non-singleton $\bar{\mathcal{C}}_n$ class. Similarly, $\bar{\beta} \in \bar{\Sigma}_n$ a non-singleton $\bar{\mathcal{C}}_n$ class. By transitivity, there exists $g_n \in A(\Omega)$ such that $\bar{\Pi}_n g_n = \bar{\Sigma}_n$. Let $\Lambda_0 = \Omega \setminus \bar{\Pi}_0$ and $\Lambda_{-m} = \bar{\Pi}_{1-m} \setminus \bar{\Pi}_{-m}$ $(m \in \mathbb{Z}^+)$. Then $\{\Lambda_{-m} : m \in \omega\}$ is

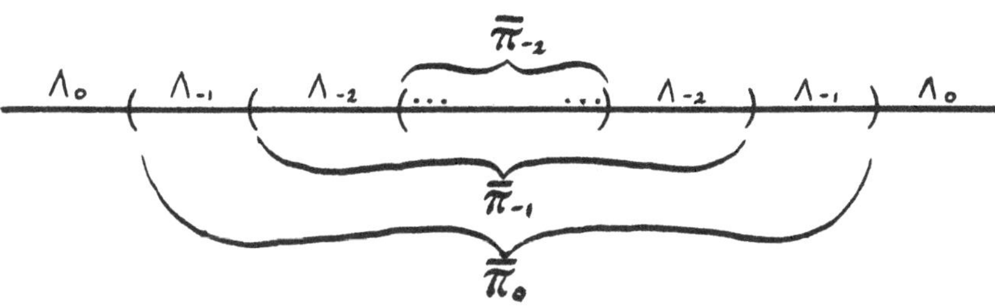

a partition of Ω. Define f by: $\sigma f = \sigma g_{-m}$ if $\sigma \in \Lambda_{-m}$ $(m \in \omega)$. Since each $\bar{\Pi}_{-m}$ and $\bar{\Sigma}_{-m}$ is a block of $A(\Omega)$ and $\bigcap_{n\in\mathbb{Z}} \bar{\Pi}_n = \{\bar{\alpha}\}$, $\bigcap_{n\in\mathbb{Z}} \bar{\Sigma}_n = \{\bar{\beta}\}$, it follows that $f \in A(\Omega)$ and $\bar{\alpha}f = \bar{\beta}$. Hence Λ is an orbit of $A(\Omega)$ and the proof is complete.

For further results about ℓ-characteristic subgroups of $A(\Omega)$ see Holland [75a]. In particular, an example is given there to show:

THEOREM 9.5. *There exists a homogeneous chain Ω such that $A(\Omega)$ has an ℓ-ideal that is not ℓ-characteristic.*

We outline the proof. For each $n \in \mathbb{Z}^+$, let Λ_n be a doubly homogeneous chain with $|\Lambda_n| = \aleph_n$ and coteriminality

$\aleph_{n-1}$. Let $T_n = \mathbb{Z} \overleftarrow{\times} \Lambda_n$ and $T = \overrightarrow{\prod}_{n\in\mathbb{Z}^+} T_n$. Let $X_n = \{(k,\alpha) \in T_n: k \text{ is even}\}$ and $\mathcal{F}$ be a non-principal ultrafilter on $\mathbb{Z}^+$. For each $\tau \in T$, let $Y(\tau) = \{n \in \mathbb{Z}^+: \tau_n \in X_n\}$ and $\Omega = \{\tau \in T: Y(\tau) \in \mathcal{F}\}$. It is easily seen that Ω is homogeneous.

Define natural equivalence relations on Ω by: $\sigma \mathcal{C}_n \tau$ if $\sigma_m = \tau_m$ for all $m \leq n \in \mathbb{Z}^+$. Then $\mathcal{C}_m \subseteq \mathcal{C}_n$ if $n \leq m$. Moreover, each $\mathcal{C}_n$ is a congruence of $(A(\Omega),\Omega)$ if the Λ_n's are sufficiently dissimilar. Hence, by Theorem 5A, we can ℓ-embed $(A(\Omega),\Omega)$ in $\mathrm{Wr}\{(A(T_n),T_n): n \in (\mathbb{Z}^+)^*\}$, where $(\mathbb{Z}^+)^*$ is the set $\mathbb{Z}^+$ ordered by $0 > 1 > 2 > \ldots$. That is, each $g \in A(\Omega)$ is determined by a $\mathbb{Z}^+\times\Omega$ matrix $\{g(n,\sigma)\}_{n\in\mathbb{Z}^+,\sigma\in\Omega}$, where $g(n,\sigma) \in A(T_n)$, $(\sigma g)_n = \sigma_n g(n,\sigma)$, and $g(n,\sigma) = g(n,\tau)$ if $\sigma \mathcal{C}_{n-1} \tau$.

For each $m \in \mathbb{Z}^+$, choose $\alpha_m \in \Lambda_m$ and define $f_m \in A(T_m)$ by:

$$(k,\beta)f_m = \begin{cases} (k+2,\beta), & \text{if } \beta = \alpha_m \\ (k,\beta) & \text{, otherwise} \end{cases} \qquad (k \in \mathbb{Z},\ \beta \in \Lambda_m).$$

Then $X_m f_m = X_m$. Define $g \in A(T)$ by

$$g(n,\sigma) = \begin{cases} e & \text{if } n = 1 \text{ or } \sigma_{n-1} \in X_{n-1} \\ f_n & \text{otherwise} \end{cases} \qquad (n \in \mathbb{Z}^+,\ \sigma \in T).$$

Then $g|\Omega \in A(\Omega)$. Let L be the ℓ-ideal of $A(\Omega)$ generated by $g|\Omega$.

For each $m \in \mathbb{Z}^+$, let $\hat{\psi}_m \in A(T_m)$ be defined by: $(k,\beta)\hat{\psi}_m = (k+1,\beta)$, and let $\hat{\psi} \in A(T)$ be defined by $(\tau\hat{\psi})_m = \tau_m\hat{\psi}_m$. Then $T\hat{\psi} = T$, $\Omega\hat{\psi} = T\backslash\Omega$ and conjugation by $\hat{\psi}$ is an outer ℓ-automorphism. Moreover, it can be shown that $g\psi = \hat{\psi}^{-1}g\hat{\psi} \in L$; so L is not ℓ-characteristic.

PART IV

APPLICATIONS TO LATTICE-ORDERED GROUPS

CHAPTER 10
EMBEDDING THEOREMS FOR LATTICE-ORDERED GROUPS

In Chapter 2 we proved that every ℓ-group can be ℓ-embedded in a divisible ℓ-group. This result is analogous to the group theory result. In this chapter we wish to prove two further analogues of embedding theorems (as well as giving another proof of Corollary 2K cited above).

<u>THEOREM</u> <u>10A</u>. *Every countable ℓ-group can be ℓ-embedded in a two generator ℓ-group.*

<u>THEOREM</u> <u>10B</u>. *Every ℓ-group can be ℓ-embedded in an ℓ-group in which any two strictly positive elements are conjugate.*

The standard tool for proving the group theoretic analogues of Theorems 2K, 10A and 10B (where "strictly positive" is replaced by "of infinite order") is the free product with amalgamation. In order for this creature to exist, we need the amalgamation property: A class $\mathcal{K}$ of algebras has the *amalgamation property* if whenever $G, H_1, H_2 \in \mathcal{K}$ and $\theta_i: G \to H_i$ are monomorphisms $(i = 1,2)$, there is

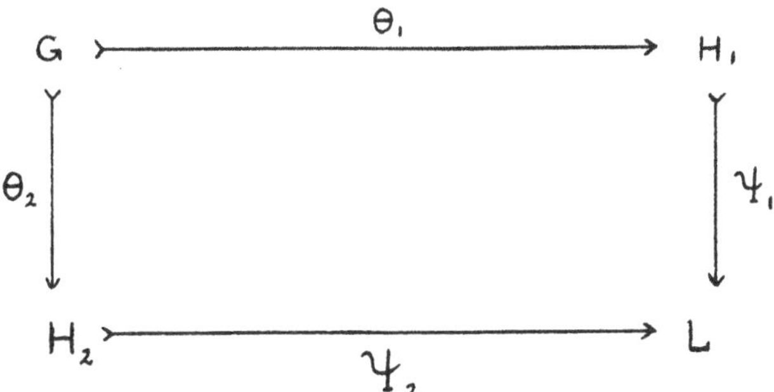

an algebra $L \in \mathcal{K}$ and monomorphisms $\psi_i: H_i \to L$ $(i = 1,2)$ such that the diagram commutes; i.e., for all $g \in G$, $g\theta_1\psi_1 = g\theta_2\psi_2$. Both the class of groups and the class of distributive lattices enjoy the amalgamation property. However, surprisingly,

<u>THEOREM</u> <u>10C</u>. *The class of all ℓ-groups does not enjoy the amalgamation property.*

The repercussions of Theorem 10C are frightful and justify extensive study of ℓ-permutation groups for anyone interested in ℓ-groups. Anyone doubting this should examine the proofs of the group-theoretic analogues of theorems in this chapter and Chapters 12 and 13, and compare them with the devices we are forced to use. In especial, we do not know if each recursively presented ℓ-group can be ℓ-embedded in a finitely presented ℓ-group. (This is the ℓ-group analogue of Graham Higman's famous embedding theorem for groups. All proofs of Higman's theorem for groups (or other classes of algebras) rely heavily on the amalgamation property or on facts that imply the amalgamation property.)

We begin with the proof of Theorem 10C. Let $H_1 = (\mathbb{Z} \oplus \mathbb{Z}) \overleftarrow{\rtimes} \mathbb{Z}$ where $(m,n,0)$ conjugated by $(0,0,1)$ is $(n,m,0)$. We write $a = (1,0,0)$, $b = (0,1,0)$ and $c = (0,0,1)$. Then H_1 is a (metabelian) ℓ-group. Let $G = (\langle a\rangle \oplus \langle b\rangle) \overleftarrow{\times} \langle c^2\rangle$, an Abelian ℓ-subgroup of H_1. $A = \langle a\rangle$ and $B = \langle b\rangle$ are prime subgroups of G so $\Omega = R(A) \overleftarrow{\cup} R(B)$ is a chain (Appendix I Lemma 2) where $R(X)$ is the set of right cosets of X in G ($X = A$ or B); i.e., $\Omega = (\mathbb{Z} \overleftarrow{\otimes} \mathbb{Z}) \overleftarrow{\cup} (\mathbb{Z} \overleftarrow{\otimes} \mathbb{Z})$, two copies of $\mathbb{Z} \overleftarrow{\otimes} \mathbb{Z}$. Since $A \cap B = \{e\}$, $(G\theta_2,\Omega)$ is an ℓ-permutation group where $\theta_2: G \to A(\Omega)$ is given by: $(Xf)(g\theta_2) = X(fg)$ ($X = A$ or B; $f,g \in G$). Let $h \in A(\Omega)$ be given by: $(Af)h = Af$ and $(Bf)h = Bfc^2$ ($f \in G$). Let H_2 be the ℓ-subgroup of $A(\Omega)$ generated by $G\theta_2$ and h, and let $\theta_1: G \to H_1$ be the inclusion map. We now prove, by reductio ad absurdum, that there does not exist an ℓ-group L with ℓ-embeddings $\psi_i: H_i \to L$ ($i = 1,2$) such that $g\theta_1\psi_1 = g\theta_2\psi_2$ for all $g \in G$. So suppose such L, ψ_1, ψ_2 exist. Let C be the convex ℓ-subgroup of H_2 generated by $b\theta_2$ and $k = h(c\theta_2)^{-2}$. Since $B(b\theta_2) = B$ and

$Bk = B$, $C \subseteq H_2 \cap A(\Omega)_B$. Now $B(a\theta_2) = Ba \neq B$ so $a\theta_2 \notin C$. By Zorn's lemma, there is a convex ℓ-subgroup P of L containing $C\psi_2$ which is maximal with respect to not containing $a\theta_2\psi_2$ $(= a\theta_1\psi_1)$. Since $c^{-1}bc = a$ and $b\theta_1\psi_1 = b\theta_2\psi_2 \in C\psi_2$, $c\psi_1 \notin P$. By Appendix I Lemma 3, P is a prime subgroup of L; so T, the set of right cosets of P in L, is a chain. Let $\chi: L \to A(T)$ be defined by: $(Px)(y\chi) = Pxy$ $(x,y \in L)$. Then χ is an ℓ-homomorphism and $(L\chi,T)$ is a transitive ℓ-permutation group. Since G is Abelian, $Bfb = Bbf = Bf$ for all $f \in G$. Thus $\text{supp}(b\theta_2) \subseteq R(A)$. Hence $b\theta_2 \wedge h = e$ (in $A(\Omega)$). Therefore $b\theta_2\psi_2\chi \wedge h\psi_2\chi = e$ (in $A(T)$). Now $(Pc\psi_1)(b\theta_2\psi_2\chi) = (Pc\psi_1)(b\theta_1\psi_1\chi) = P(cb)\psi_1 = P(ac)\psi_1 = P(a\theta_1\psi_1)(c\psi_1) \neq Pc\psi_1$, as $a\theta_1\psi_1 = a\theta_2\psi_2 \notin P$. Hence $(Pc\psi_1)(h\psi_2\chi) = Pc\psi_1$. Consequently, $Pc\psi_1 = P(c\psi_1)(h\psi_2\chi) > Ph\psi_2\chi = P(kc^2\theta_2)\psi_2\chi = Pc^2\theta_2\psi_2\chi = P(c\psi_1)^2$, as $c\psi_1 \notin P$ and $k\psi_2 \in C\psi_2 \subseteq P$. Thus $Pc\psi_1 > P(c\psi_1)^2$. But $c\psi_1 > e$, the desired contradiction.

We next give a proof of Theorem 10A. The proof mirrors one given for groups by B. H. and Hanna Neumann [59]. For the definition of $[G,G]$ see page 129. We need the following lemma:

<u>LEMMA</u> <u>10.1</u>. *Let* (G,Ω) *be an* ℓ*-permutation group and* $(W,R) = (G,\Omega)\ Wr\ (\mathbb{Z},\mathbb{Z})$. *Then there is an* ℓ*-embedding* $\psi: G \to W$ *so that* $G\psi \subseteq [W,W]$.

<u>Proof</u>: Let $g \in G$. Define $g\psi \in W$ by: $g\psi = (\{g_n\},0)$, where

$$g_n = \begin{cases} g & \text{if } n = 0 \\ e & \text{if } n \neq 0 \end{cases}.$$

Then $\psi: g \mapsto g\psi$ is clearly an ℓ-isomorphism.

Let $g \in G$ and define $f = f(g) = (\{f_n\},0)$ where

$$f_n = \begin{cases} g^{-1} & \text{if } n \geq 0 \\ e & \text{if } n < 0 \end{cases}.$$

Let $z = (\{e\},1)$. Then $z^{-1}fz = (\{h_n\},0)$

where $h_n = \begin{cases} g^{-1} & \text{if } n \geq 1 \\ e & \text{if } n < 0 \end{cases}$. So $g\psi = [f,z] \in [W,W]$.

Proof of Theorem 10A: Let $\{g(i): i \in \mathbb{Z}^+\}$ be an enumeration of the countable ℓ-group G and let $f(i) = (\{f(i)_n\},0) \in W$ be defined by: $f(i)_n = \begin{cases} g(i)^{-1} & \text{if } n \geq 0 \\ e & \text{if } n < 0 \end{cases}$. So if z and ψ are defined as in the proof of Lemma 10.1, $g(i)\psi = [f(i),z]$ for all $i \in \mathbb{Z}^+$. Let $(H,T) = (W,R)\ \mathrm{Wr}\ (\mathbb{Z},\mathbb{Z})$ and define $\theta: W \to H$ by:

$$w\theta = (\{h_n\},0) \text{ where } h_n = \begin{cases} w & \text{if } n = 0 \\ e & \text{if } n \neq 0 \end{cases}.$$

Then θ is clearly an ℓ-embedding and $g(i)\psi\theta = (\{h(i)_n\},0)$ where $h(i)_0 = g(i)\psi$. Let $b = (\{e\},1) \in H$, and $a = (\{a_n\},0) \in H$ be defined by:

$$a_n = \begin{cases} z & \text{if } n = 0 \\ f(i) & \text{if } n = 2i - 1 \quad (i \in \mathbb{Z}^+) \\ e & \text{otherwise} \end{cases}.$$

We complete the proof by showing that $G\psi\theta$ is contained in the ℓ-subgroup of H generated by a and b. First observe that $b^{2i-1}ab^{-(2i-1)} = (\{c(i)_n\},0)$

$$\text{where } c(i)_n = \begin{cases} f(i + j) & \text{if } n = 2j \ \&\ -i < j \in \mathbb{Z} \\ z & \text{if } n = 1 - 2i \\ e & \text{otherwise} \end{cases} \quad (i \in \mathbb{Z}^+).$$

Hence $b^{2i-1}ab^{-(2i-1)}a = (\{d(i)_n\},0)$ where

$$d(i)_n = \begin{cases} f(i)z & \text{if } n = 0 \\ f(i + j) & \text{if } n = 2j \neq 0 \ \&\ -i < j \in \mathbb{Z} \\ z & \text{if } n = 1 - 2i \\ f(j) & \text{if } n = 2j - 1\ (j \in \mathbb{Z}^+) \\ e & \text{otherwise} \end{cases}.$$

Therefore

$[b^{2i-1}ab^{-(2i-1)},a] = (\{k(i)_n\},0),$

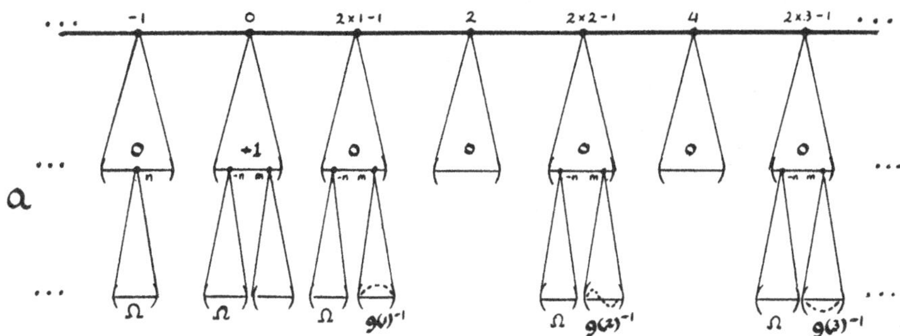

where $k(i)_n = \begin{cases} [f(i),z] & \text{if } n = 0 \\ e & \text{if } n \neq 0 \end{cases}$; i.e.,

$[b^{2i-1}ab^{-(2i-1)},a] = g(i)\psi\theta$, as desired.

We next provide a different proof of Corollary 2K which illustrates both a new technique, and a new class of chains.

Let α be an ordinal number. A chain Ω is said to be an <u>η_α-*set*</u> if whenever $A, B \subseteq \Omega$ with $A < B$ and $|A|, |B| < \aleph_\alpha$, there exists $\sigma \in \Omega$ such that $A < \sigma < B$. Any η_α-set of cardinality $\aleph_\alpha$ is called an <u>α-*set*</u>; e.g., a 0-set is just $\mathbb{Q}$. η_α-sets are precisely the $\aleph_\alpha$-saturated models of the theory of dense linear ordering without endpoints (Chang and Keisler [73], Proposition 5.4.2). Consequently, we obtain the following (see Chang and Keisler, op. cit.):

(a) Any two α-sets are ordermorphic (Theorem 5.1.3).

(b) Any α-set is <u>$\aleph_\alpha$-*homogeneous*</u> (i.e., any ordermorphism between subsets (of the chain) of cardinality $< \aleph_\alpha$ can be extended to an automorphism of the chain) (Theorem 5.1.14).

(c) If Ω is an η_α-set and T is a chain with $|T| \leq \aleph_\alpha$, then T can be order embedded in Ω (Theorem 5.1.12).

(d) Assuming GCH (Generalised Continuum Hypothesis) α-sets exist whenever $\aleph_\alpha$ is a <u>*regular cardinal*</u> (i.e., if $(\mathrm{cf}(\aleph_\alpha) = \aleph_\alpha)$ (Proposition 5.1.5).

(e) If $\aleph_\alpha$ is regular and Ω is an α-set, then $|\bar{\Omega}| = 2^{\aleph_\alpha}$ (Gilman and Jerison [60], pp. 185-189).

It follows immediately that

(f) If $\aleph_\alpha$ is regular, then any point of an α-set has character $c_{\alpha\alpha}$.

(g) If $\aleph_\alpha$ is regular and Ω is an α-set, then $\mathrm{cf}(\Omega) = \mathrm{ci}(\Omega) = \aleph_\alpha$ (For (g), for $\mathrm{cf}(\Omega) = \aleph_\alpha$, take $B = \emptyset$; for $\mathrm{ci}(\Omega) = \aleph_\alpha$, take $A = \emptyset$).

For each cardinal κ, $\underline{\kappa^+}$ (the successor cardinal of κ) is regular. Hence the class of regular cardinals is cofinal in the class of all cardinals. In view of (b), Theorem 2E, Holland's theorem and Lemma 2.5.2, to prove Corollary 2K it is enough to show:

<u>THEOREM</u> <u>10.2</u>. *If $\aleph_\alpha$ is regular and T is a chain with $|T| \leq \aleph_\alpha$, then T can be order-embedded in an α-set Ω so that each element of $A(T)$ extends to a member of $A(\Omega)$.*

<u>*Proof:*</u> Suppose $A,B \subseteq T$ with $A < B$ and $|A|,|B| < \aleph_\alpha$. If no point of T lies between A and B adjoin an α-set between A and B. Since the number of subsets of T of cardinality less than $\aleph_\alpha$ is at most $\aleph_\alpha$, we will have adjoined at most $\aleph_\alpha$ α-sets to T as A and B run through all possible candidates. Hence the resulting set Ω has cardinality $\aleph_\alpha$.

We now show that Ω is an α-set. Let $A',B' \subseteq \Omega$ with $A' < B'$ and $|A'|,|B'| < \aleph_\alpha$. If there exist $A,B \subseteq T$ with $\sup A = \sup A'$ and $\inf B = \inf B'$, then $|A|,|B| < \aleph_\alpha$ without loss of generality. Hence there exists $\sigma \in \Omega$ with $A < \sigma < B$; and so $A' < \sigma < B'$. If no such A and B exist, then (without loss of generality) there is $\sigma \in A'$ such that $\emptyset \neq A = \{\sigma' \in A' : \sigma' > \sigma\}$ is contained in an α-set Δ_0 adjoined to T in constructing Ω. Then $|A|,|B| < \aleph_\alpha$ where $B = B' \cap \Delta_0$ (Of course, B may be empty). Now $A < B$ and $A,B \subseteq \Delta_0$, so there is $\delta \in \Delta_0$ such that $A < \delta < B$. But Δ_0 is a convex subset of Ω. Therefore $A' < \delta < B'$ and Ω is an α-set.

We next define an equivalence relation $\mathcal{R}$ on Ω by: $\sigma_1 \mathcal{R} \sigma_2$ if $\sigma_1 = \sigma_2$ or $(\sigma_1 < \sigma_2$ & $T \cap [\sigma_1, \sigma_2) = \emptyset)$ or $(\sigma_2 < \sigma_1$ & $T \cap [\sigma_2, \sigma_1) = \emptyset)$. The equivalence classes of $\mathcal{R}$ are of three types:

(I) $\{\tau\}$ for some $\tau \in T$ (such a τ has initial character $\aleph_\alpha$).

(II) $\Delta \cup \{\tau\}$ for some $\tau \in T$ (i.e., Δ is a "tagged" α-set); such a τ has initial character $< \aleph_\alpha$ (in T).

(III) An α-set we adjoined to T to obtain Ω.

Clearly any two equivalence classes of the same type are ordermorphic (use (a) for (II) and (III)), and no two equivalence classes of different types are ordermorphic. For each $\Lambda_1 \cup \{\tau_1\}, \Lambda_2 \cup \{\tau_2\}$ equivalence classes of type II, choose an ordermorphism $\phi_{12}: \Lambda_1 \cong \Lambda_2$; and for each Δ_1, Δ_2 equivalence classes of type III, choose an ordermorphism $\phi_{12}: \Delta_1 \cong \Delta_2$. Note that each equivalence class of type III is obtained by a cut in equivalence classes of types I and II; viz: Δ_1 is inserted between $\sup\{\tau_\mu \mathcal{R}: \mu \in M\}$ and $\inf\{\tau_\nu \mathcal{R}: \nu \in N\}$ where $\tau_\mu, \tau_\nu \in T$ $(\mu \in M, \nu \in N)$, $|M|, |N| < \aleph_\alpha$ and, in T, $\sup\{\tau_\mu: \mu \in M\} = \inf\{\tau_\nu: \nu \in N\} \in \bar{T} \backslash T$.

Let $f \in A(T)$. Define $\tilde{f} \in A(\Omega)$ as follows: If $\tau \in T$, $\tau\tilde{f} = \tau f$. If $\sigma\mathcal{R}$ is of type II, say $\sigma\mathcal{R} = \Lambda_1 \overset{\leftarrow}{\cup} \{\tau\}$, let $\tau f \mathcal{R} = \Lambda_2 \overset{\leftarrow}{\cup} \{\tau f\}$. For $\delta_1 \in \Lambda_1$, let $\delta_1 \tilde{f} = \delta_1 \phi_{12}$. Let Δ_1 be a type III class, say Δ_1 is inserted between $\{\tau_\mu \mathcal{R}: \mu \in M\}$ and $\{\tau_\nu \mathcal{R}: \nu \in N\}$. Let Δ_2 be the type III class inserted between $\{\tau_\mu f \mathcal{R}: \mu \in M\}$ and $\{\tau_\nu f \mathcal{R}: \nu \in N\}$. For $\delta \in \Delta_1$, let $\delta\tilde{f} = \delta\phi_{12}$. Clearly $\tilde{f} \in A(\Omega)$ and extends f. This completes the proof of Theorem 10.2.

Now suppose $\aleph_\alpha$ is regular and Ω is an α-set. Then clearly no point of $\bar{\Omega}$ has character $c_{\beta\gamma}$ where $\beta, \gamma < \alpha$. Let $\aleph_\beta$ be regular with $\beta < \alpha$. Let $\{\sigma_\nu: \nu \in \omega_\beta\}$ be an increasing sequence of points in Ω. Then $\bar{\sigma} = \sup\{\sigma_\nu: \nu \in \omega_\beta\} \in \bar{\Omega}$ and hence

$\bar{\sigma}$ has character $c_{\beta\alpha}$. Similarly, there are points of $\bar{\Omega}$ of character $c_{\alpha\beta}$. But there are only $\aleph_\alpha$ subsets of Ω of cardinality less than $\aleph_\alpha$. Since $|\bar{\Omega}| = 2^{\aleph_\alpha}$ (by (e)), there are $2^{\aleph_\alpha}$ points in $\bar{\Omega}$ of character $c_{\alpha\alpha}$. By (b), if $\aleph_\beta$ is regular with $\beta < \alpha$, then any two points of $\bar{\Omega}$ of character $c_{\beta\alpha}$ ($c_{\alpha\beta}$) lie in the same orbit of $A(\Omega)$. Now if $\bar{\sigma}, \bar{\tau} \in \bar{\Omega}\backslash\Omega$ have character $c_{\alpha\alpha}$, then $\{\sigma \in \Omega: \sigma < \bar{\sigma}\}$ and $\{\sigma \in \Omega: \sigma < \bar{\tau}\}$ are α-sets and hence ordermorphic by (a), say by g_1. Similarly there is an ordermorphism $g_2: \{\sigma \in \Omega: \sigma > \bar{\sigma}\} \cong \{\sigma \in \Omega: \sigma > \bar{\tau}\}$. Define

$$g \in A(\Omega) \text{ by } \sigma g = \begin{cases} \sigma g_1 & \text{if } \sigma < \bar{\sigma} \\ \sigma g_2 & \text{if } \sigma > \bar{\sigma} \end{cases}.$$

Then $\bar{\sigma} g = \bar{\tau}$ so any holes in Ω of character $c_{\alpha\alpha}$ lie in the same orbit T of $A(\Omega)$. If ψ is an ℓ-automorphism of $A(\Omega)$, it is conjugation by some $\hat{\psi} \in A(\bar{\Omega})$ by Theorem 2.4.1. Now $\hat{\psi}$ permutes the orbits of $A(\Omega)$. Since T and Ω are the only orbits of $A(\Omega)$ of character $c_{\alpha\alpha}$, $\Omega\hat{\psi} = T$ or $\Omega\hat{\psi} = \Omega$. But $|T| = 2^{\aleph_\alpha} > \aleph_\alpha = |\Omega|$. Hence $\Omega\hat{\psi} = \Omega$ and ψ is an inner automorphism of $A(\Omega)$. We have thus established:

<u>LEMMA</u> <u>10.3</u>. *If* $\aleph_\alpha$ *is regular and* Ω *is an* α*-set, then the orbits of* $A(\Omega)$ *comprise* Ω, *all points of* $\bar{\Omega}\backslash\Omega$ *of character* $c_{\alpha\alpha}$, *and all points of* $\bar{\Omega}$ *of character* $c_{\beta\alpha}$ $(c_{\alpha\beta})$ $(\beta < \alpha$ *and* $\aleph_\beta$ *regular). Therefore, any* ℓ*-automorphism of* $A(\Omega)$ *is inner.*

<u>COROLLARY</u> <u>10.4</u>. *If* $\aleph_\alpha$ *is regular and* G *is an* ℓ*-group with* $|G| \leq \aleph_\alpha$, *then* G *can be* ℓ*-embedded in an* ℓ*-group in which every* ℓ*-automorphism is inner, namely* $A(\Omega)$ *where* Ω *is an* α*-set.*

Theorem 10.2 is due to Holland [63], Lemma 10.3 and Corollary 10.4 to McCleary [73a].

In the proof of Theorem 10.2 we did a "weak tagging" to show that any element of $A(T)$ can be lifted to an element of $A(\Omega)$. Before proving Theorem 10B (again using α-sets), we digress to give a further (and truer) application of tagging due to Scrimger [70].

THEOREM 10.5. *Let* (G,Ω) *be an ordered permutation group. Assume that whenever* (H,Ω) *is an ordered permutation group with* $\alpha G = \alpha H$ *for all* $\alpha \in \Omega$, *then* $H \subseteq G$. *Then there is a chain* $T \supseteq \Omega$ *such that* G *is* ℓ*-isomorphic to* $A(T)$.

Proof: Let $\{\Omega_i : i \in I\}$ be the set of G-orbits of points of Ω (So if (G,Ω) is transitive, $|I| = 1$). The idea is to enlarge Ω by adding to each element of Ω_i a tag, so that points of Ω_j receive an identifiably different tag from those of Ω_i if $\Omega_i \neq \Omega_j$. This will ensure that $A(T)$ will respect the original orbits of (G,Ω). However, we must be careful to ensure that no point of Ω_i gets mapped by an element of $A(T)$ onto a point on a tag.

For the purposes of this proof, we consider $\mathbb{Z}^+$ to be the set of non-zero finite ordinals. So $n = \{0,1,\ldots,n-1\}$. Since $\mathbb{Q}$ is countable, there is a one-to-one correspondence between it and $\mathbb{Z}^+$. Denote this correspondence by $q \mapsto n_q$. Let $X = \{(m,q): q \in \mathbb{Q}\ \&\ m \in n_q\}$ ordered antilexicographically $((m,q) < (m',q')$ if $q < q'$ or $(q = q'\ \&\ m < m'))$. Since

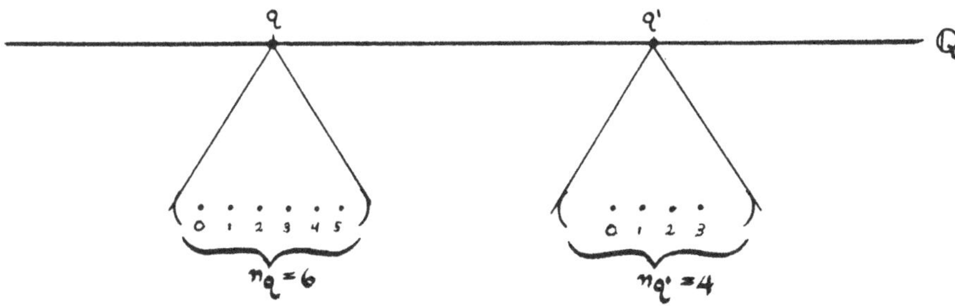

$\mathbb{Q}$ is dense, there are infinitely many points of X between (m,q) and (m',q') if $q \neq q'$. Hence if $f \in A(X)$ and $(m,q)f = (m',q')$, $n_q \leq n_{q'}$. But $(m',q')f^{-1} = (m,q)$, so $n_{q'} \leq n_q$. Thus $n_q = n_{q'}$.

Therefore $q = q'$ and $f = e$. Consequently, $A(X) = \{e\}$.

For any ordinal μ, let $X(\mu) = \mu^* \overleftarrow{\cup} X \overleftarrow{\cup} \mu$. If $f \in A(X(\mu))$, then as X contains no final well-ordered segment, $Xf = X$. Hence $\mu f = \mu$ and $\mu^* f = \mu^*$. Since $A(X) = \{e\}$ and (μ^*) μ is (inversely) well-ordered, $f = e$; i.e., $A(X(\mu)) = \{e\}$.

Well order I (the set indexing the G-orbits of points of Ω) and let $\{\mu_i : i \in I\}$ be a collection of limit ordinals with $|\Omega| < \mu_i < \mu_j$ if $i,j \in I$ with $i < j$. For $\alpha \in \Omega_i$, let $T(\alpha) = \{\alpha\} \overleftarrow{\cup} X_\alpha(\mu_i)$ where $X_\alpha(\mu_i)$ is a copy of $X(\mu_i)$ such that $X_\alpha(\mu_i) \cap X_\beta(\mu_j) = \emptyset$ unless $i = j$ and $\alpha = \beta$; i.e., we have tagged α with a copy of $X(\mu_i)$. So if $\alpha, \beta \in \Omega$ belong to different G-orbits, they receive an essentially different tag. Let $T = \overleftarrow{\bigcup_{\alpha \in \Omega}} T(\alpha)$; i.e., $T(\alpha) < T(\beta)$ if $\alpha < \beta$ (in Ω). Clearly $\Omega \subseteq T$. Let $f \in A(T)$ and suppose $\alpha \in \Omega_i$ with $\alpha \in \mathrm{supp}(f)$. So $\alpha f \in T(\beta)$ for some $\beta \in \Omega_j$ (some $j \in I$). Now $\alpha f \notin \mu_j$ (otherwise μ_j contains an infinite initial segment of $\mu_i^* f$; i.e., a non well-ordered subset). If $\alpha f \in X$, then X contains an infinite initial segment of $\mu_i^* f$ which is impossible. Also $\alpha f \notin \mu_j^*$ (Otherwise μ_j^* contains $\mu_i^* f$ as a final segment whence $Xf = X$ and $\mu_i f$ is an initial segment of μ_j. Hence $\mu_i < \mu_j$. By applying f^{-1} we get $\mu_j < \mu_i$, the desired contradiction). Therefore $\alpha f = \beta$, and by the above argument $\mu_i = \mu_j$. Hence $f|\Omega$ preserves the G-orbits of points of Ω; i.e., $A(T)|\Omega$ has the same orbits as G (for points of Ω). By hypothesis, $A(T)|\Omega \subseteq G$. But every element of G naturally extends to an element of $A(T)$. Consequently, $A(T)$ is ℓ-isomorphic to G (via restriction to Ω).

COROLLARY 10.6. *If $\{\Omega_i : i \in I\}$ is a family of pairwise disjoint chains, there is a chain T such that $A(T) \cong \prod_{i \in I} A(T_i)$.*

Chang and Ehrenfeucht [62] used tagging to prove:

THEOREM 10.7. *Every proper subgroup of $\mathbb{R}$ is isomorphic to $A(T)$ for some chain T.*

The idea is to give different tags to different cosets of G in $\mathbb{R}$ in the same way as in the proof of Theorem 10.5. The details are straightforward.

We conclude the chapter with a proof of Theorem 10B. Specifically, we will prove:

THEOREM 10.8. *Let $\aleph_\alpha > \aleph_0$ be a regular cardinal and G be an ℓ-group with $|G| \leq \aleph_\alpha$. Then there is an ℓ-group $H \supseteq G$ such that $|H| \leq \aleph_\alpha$, and if $e < h_1, h_2 \in H$, $\gamma_0 \in supp(h_1)$ and $\delta_0 \in supp(h_2)$, there is $f \in H$ such that $\gamma_0 f = \delta_0$ and $f^{-1}h_1 f = h_2$.*

The key to the proof is Theorem 2.2.5. If (G,T) is an ℓ-permutation group and $e < g \in G$ has one bump, g is not conjugate in $A(T)$ to any $e < f \in G$ which has more than one bump. Indeed, as we saw in both the proof of Corollary 2.2.3 and the illustrations prior to Theorem 2.2.5, two elements may have the same number of bumps and yet not be conjugate in $A(\bar{T})$. So we will have to enlarge T to Ω, say, so each $e < g \in G$ gets mapped to an element of $A(\Omega)$ which has a lot of supporting intervals with whole intervals between any two distinct bumps of the image of g. Then, hopefully, any two strictly positive elements of (the image of) G will be conjugate in $G^{(1)} \subseteq A(\Omega)$. If we can achieve this we will be home since we can then proceed to spread the disease, by induction, up a countable sequence of ℓ-groups $G = G^{(0)} \subseteq G^{(1)} \subseteq \ldots$ where any two strictly positive elements of $G^{(m)}$ are conjugate in $G^{(m+1)}$ $(m \in \omega)$. Then $H = \bigcup_{m \in \omega} G^{(m)}$ will satisfy the theorem (if $|G^{(m)}| \leq \aleph_\alpha$ all m).

So we will start with (G,T) an ℓ-permutation group and obtain such an Ω. By Corollary 2L, we may assume T is doubly homogeneous and $G \subseteq B(T)$. An obvious candidate for Ω is an α-set since each chain of size not exceeding $\aleph_\alpha$ can be order-embedded in an α-set. As we want each $e < g \in G$ to extend to an element $g\psi$ of $A(\Omega)$ such that $g\psi$ has lots of bumps and whole segments of Ω between any two distinct bumps, it would be rather nice if we could arrange that the set of bumps of $g\psi$ is an α-set and similarly for the set of "fixed point intervals." More precisely, if (G,T) is an ordered permutation group and $g \in G$, a *fixed point interval* of g is a convex subset Δ of T maximal with respect to $g \in G_\Delta$. Recall that the set of supporting intervals of an element g is just the set of supports of bumps of g.

Our proof will involve transfinite induction for $\aleph_\alpha$ steps.

In the first lemma, we will ensure two things: (a) If $\tau \in T$ is the infimum or supremum of a set of bumps of $g \in G$, then we can squeeze a bump of the image of g between them and τ. This will help obtain density of the set of supporting intervals of the image of g (if we repeat the process enough times). (b) If $\tau \in T$ and $\{\tau\}$ is a fixed point interval of $g \in G$, then there is a whole segment (in the extension of T) about τ that is a fixed point interval of the image of g. Even though the image of g may still have fixed point intervals comprising just one point, we will be able to remove this problem if we repeat the process sufficiently often.

LEMMA 10.9. *Let (G,T) be an ℓ-permutation group. Then there exists a $T^\flat$ containing T such that $|T^\flat| \leq max(|G|,|T|)$ and an ℓ-embedding $\psi: G \to A(T^\flat)$ satisfying:*

(i) For each $\tau \in T$, there exist $\tau_1^\flat, \tau_2^\flat \in T^\flat$ such that $\{\tau' \in T: \tau' < \tau\} < \tau_1^\flat < \tau < \tau_2^\flat < \{\tau' \in T: \tau' > \tau\}$;

(ii) For each $g \in G$, $g\psi|T = g$;

(iii) If I is a non-empty collection of supporting intervals of g and $inf_{\bar{T}}(\cup I) = \sigma \in T$, then there is a supporting interval Λ of $g\psi$ such that $\sigma < \Lambda < \cup I$ (and dually for $sup_{\bar{T}}(\cup I) = \tau \in T$); and

(iv) If Δ is a fixed point interval of g, then Δ is a subset of a fixed point interval (of cardinality greater than 1) of $g\psi$.

As a good guide to the proof of this lemma, keep in mind $(B(\mathbb{Q}),\mathbb{R})$ which has two orbits $\mathbb{Q}$ and $\mathbb{R}\setminus\mathbb{Q}$. Consider what happens to g and h where g has support $(-1,0) \cup (1,2) \cup (2,3)$ and h has support $(0,\sqrt{2})$. We will sketch what happens to g in the proof.

Proof: Let k be the set of G-orbits of T. For each $K \in k$, choose $0_K \in K$. For each $\beta \in K$, pick $c_\beta \in G$ such that $0_K c_\beta = \beta$, insisting that $c_{0_K} = e$. If $\beta,\gamma \in K$, let $c_{\beta\gamma} = c_\beta^{-1} c_\gamma$. Consequently, $\beta c_{\beta\gamma} = \gamma$, $c_{\gamma\beta} = c_{\beta\gamma}^{-1}$ and $c_{\beta\gamma} c_{\gamma\delta} = c_{\beta\delta}$ if $\beta,\gamma,\delta \in K$. Define ℓ-ideals L_K and R_K of G_{0_K} by $L_K = \{g \in G\colon \operatorname{supp}(g) \cap [\beta,0_K] = \emptyset$ for some $\beta < 0_K\}$, and R_K dually. Using Holland's theorem (Appendix I), we can find $T_1(K)$

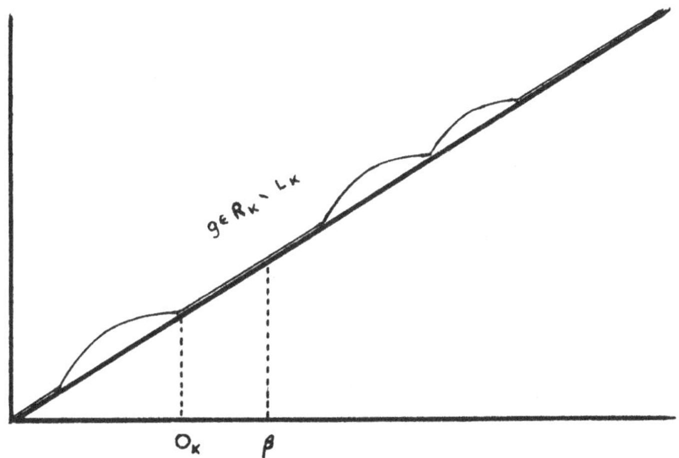

and $T_2(K)$, non-empty disjoint chains such that $(G_{0_K}/L_K, T_1(K))$ and $(G_{0_K}/R_K, T_2(K))$ are ℓ-permutation groups where $|T_1(K)|, |T_2(K)| \le |G|$. Moreover, by Corollary 2L, we may assume that all elements of these ℓ-permutation groups have bounded support and that $T_i(K)$ is doubly homogeneous $(i = 1,2)$. Now let $T(K) = T_1(K) \overleftarrow{\cup} \{\xi_K\} \overleftarrow{\cup} T_2(K)$ where $\{\xi_K\colon K \in k\}$ is a set of new symbols. Define $\psi_K\colon G_{0_K} \to A(T(K))$ by:

$$\tau(g\psi_K) = \begin{cases} \tau(gL_K) & \text{if} \quad \tau \in T_1(K) \\ \xi_K & \text{if} \quad \tau = \xi_K \\ \tau(gR_K) & \text{if} \quad \tau \in T_2(K) \end{cases}$$

. We have made sure that if $\{O_K\}$ was a fixed point interval of some $g \in G$, then there is a whole interval around O_K in $T(K)$ that is a fixed point interval of $g\psi_K$ (since every element of G_{O_K}/L_K (G_{O_K}/R_K) has bounded support in $T_1(K)$ $(T_2(K))$). This avoids (2) and (3) on pages 61,62 not being conjugate--provided that O_K is strategically placed. Clearly ψ_K is an ℓ-homomorphism.

We now generalise the Wreath product to form the *orbit Wreath product*. If $\beta \in T$, let $K(\beta) \in k$ be that G-orbit to which β belongs. Let $T^b = \{(\tau,\beta): \beta \in T \ \& \ \tau \in T(K(\beta))\}$. So $|T^b| \leq \max(|G|,|T|)$. T^b is totally ordered by: $(\tau_1,\beta_1) < (\tau_2,\beta_2)$ if $\beta_1 < \beta_2$ (in T) or $\beta_1 = \beta_2$ and $\tau_1 < \tau_2$ (in $T(K(\beta_1))$).

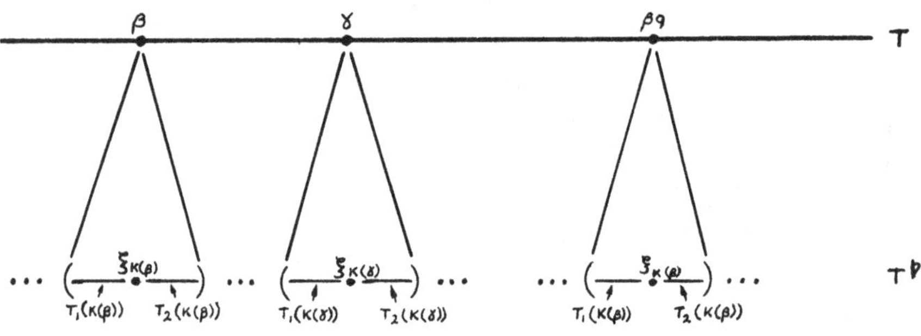

We can order-embed T in T^b by: $\beta \mapsto (\xi_{K(\beta)},\beta)$. So (i) holds. Under this identification, we have tagged each point of T by a tag dependent on the G-orbit (cf. proof of Theorem 10.5). Now let $W = \{(\hat{h},g): g \in G \ \& \ \hat{h} \in \prod_{\beta \in T} G\psi_{K(\beta)}\}$. (W,T^b) is an ℓ-permutation group where $(\tau,\beta)(\hat{h},g) = (\tau\hat{h}_{K(\beta)},\beta g)$ and the group and lattice operations are the same as for the Wreath product of two ℓ-permutation groups (see page 117). The only difference between the Wreath product of two ℓ-permutation groups and what we have described here is that the global component is not transitive and the local components of the orbit Wreath product depend on the G-orbit of the global part.

Define $\psi: G \to W$ by: $g\psi = (\hat{g},g)$ where $\hat{g}_\beta = (c_\beta g c_{\beta g}^{-1})\psi_{K(\beta)}$ (cf. page 120). (Note that $0_K c_\beta g c_{\beta g}^{-1} = 0_K$ so ψ is well-defined.)

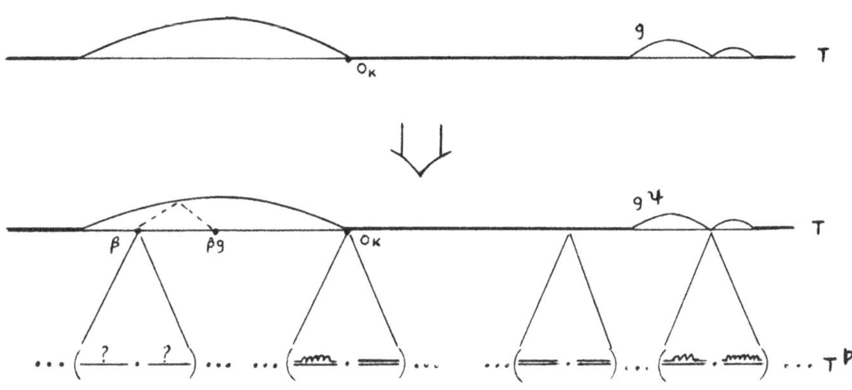

If $g,h \in G$ and for some $\gamma_1 < \beta < \gamma_2$ (in T), $g|[\gamma_1,\gamma_2] = h|[\gamma_1,\gamma_2]$, then $\beta g = \beta h$ and for all $\gamma \in [\gamma_1 c_\beta^{-1}, \gamma_2 c_\beta^{-1}]$, $\gamma c_\beta g = \gamma c_\beta h$. Hence $\gamma c_\beta g c_{\beta g}^{-1} = \gamma c_\beta h c_{\beta h}^{-1}$. But $\gamma_1 c_\beta^{-1} < 0_K < \gamma_2 c_\beta^{-1}$ so $(c_\beta g c_{\beta g}^{-1})(c_\beta h c_{\beta h}^{-1})^{-1} \in L_{K(\beta)} \cap R_{K(\beta)}$. Thus $\hat{g}_\beta = (c_\beta g c_{\beta g}^{-1})\psi_{K(\beta)} = (c_\beta h c_{\beta h}^{-1})\psi_{K(\beta)} = \hat{h}_\beta$. That is, $\hat{g}_\beta$ is determined by the action of g on arbitrarily small neighbourhoods of β.

We now show that ψ is an ℓ-embedding. If $g,h \in G$, then $(\widehat{gh})_\beta = (c_\beta g h c_{\beta gh}^{-1})\psi_{K(\beta)} = (c_\beta g c_{\beta g}^{-1} \cdot c_{\beta g} h c_{\beta gh}^{-1})\psi_{K(\beta)} = \hat{g}_\beta \hat{h}_{\beta g}$ since $\psi_{K(\beta)}$ is a homomorphism. Thus $(gh)\psi = g\psi h\psi$ and so ψ is a homomorphism. Moreover, if $\beta g < \beta$, then $(g \vee e)|[\beta g, \beta g^{-1}] = e|[\beta g, \beta g^{-1}]$; so $(\widehat{g \vee e})_\beta = \hat{e}_\beta$. If $\beta < \beta g$, then $(g \vee e)|[\beta g^{-1}, \beta g] = g|[\beta g^{-1}, \beta g]$; so $(\widehat{g \vee e})_\beta = \hat{g}_\beta$. If $\beta g = \beta$, then $\beta(g \vee e) = \beta g = \beta$ and hence $(\widehat{g \vee e})_\beta = c_\beta (g \vee e) c^{-1}_{\beta(g \vee e)} \psi_{K(\beta)}$ $= c_\beta g c^{-1}_{\beta g} \psi_{K(\beta)} \vee c_\beta e c^{-1}_\beta \psi_{K(\beta)} = \hat{g}_\beta \vee e$. Therefore

$$(g \vee e)\psi = (\hat{h}, g \vee e) \quad \text{where} \quad \hat{h} = \begin{cases} e & \text{if } \beta g < \beta \\ \hat{g}_\beta & \text{if } \beta g > \beta \\ \hat{g}_\beta \vee e & \text{if } \beta g = \beta \end{cases}.$$ But this is precisely $g\psi \vee e$. Consequently, ψ is an ℓ-homomorphism. Furthermore, $\beta(g\psi) = (\xi_{K(\beta)}, \beta)(g\psi) = (\xi_{K(\beta)}\hat{g}_\beta, \beta g) = (\xi_{K(\beta)}, \beta g) =$ $(\xi_{K(\beta g)}, \beta g) = \beta g$, so ψ extends the inclusion map. Thus ψ is one-to-one and satisfies (ii).

It remains to prove that (iii) and (iv) hold.

(iii) Let I be a non-empty collection of supporting intervals of g and let $\sigma = \inf_T(\cup I)$. Then $\sigma g = \sigma$ and for each $\beta > \sigma$, there exists $\tau \in T$ such that $\tau \in [\sigma, \beta]$ and $\tau g \neq \tau$. If $K = K(\sigma)$, then $c_\sigma g c^{-1}_{\sigma g} \in G_{O_K} \backslash R_K$. Hence $g\psi$ has a supporting interval contained in $(T_2(K), \sigma)$. Since $\sigma < (T_2(K), \sigma) < \cup I$, (iii) follows.

(iv) Let Δ be a fixed point interval of g. If $\Delta = \{\delta\}$, then $\delta(g\psi) = \delta$, so Δ is contained in a fixed point interval of $g\psi$. Since each element of $G_{O_{K(\delta)}} \psi_{K(\delta)} | T_j(K(\delta))$ has bounded support in $T_j(K(\delta))$ $(j = 1,2)$, there is an interval of T^{b} properly containing $\{\delta\}$ on which $g\psi$ is the identity; so we have expanded the singleton fixed point interval to a non-singleton one. If $|\Delta| > 1$, let $\Delta_{T^b} = \mathrm{Conv}_{T^b}(\Delta)$. We show that if $(\tau, \delta) \in \Delta_{T^b}$, then $(\tau, \delta) g\psi = (\tau, \delta)$ (whence Δ is contained in a fixed point interval of $g\psi$). If δ is the least element of Δ, let $\delta_1 \in \Delta$ be such that $\delta < \delta_1$. Now $\tau \in T_2(K(\delta))$ and as $g|[\delta, \delta_1] = e|[\delta, \delta_1]$, $c_\delta g c^{-1}_{\delta g} \in R_{K(\delta)}$; so $\tau \hat{g}_{K(\delta)} = \tau$. Thus $(\tau, \delta)(g\psi) = (\tau \hat{g}_{K(\delta)}, \delta g) = (\tau, \delta)$. Similarly, if δ is the largest element of Δ or there exist $\delta_1, \delta_2 \in \Delta$ with $\delta_1 < \delta < \delta_2$. //

We have still some problems remaining. Firstly, if we look at (2) on page 61, we are not guaranteed that any two elements fitting that picture are conjugate--$\bar{\sigma}$ may belong to T or be a hole in T. Also, if we have a non-empty set I of supporting intervals of g and $\inf_{\bar{T}}(\bigcup I) \in \bar{T}\backslash T$, we haven't necessarily increased the density (this is just a more general setting of the previous sentence). Indeed, if we took $(G,T) = (A(\mathbb{Q}),\mathbb{Q})$ and $e < g,h \in G$ have supports $(0,1) \cup (1,2)$ and $(0,\sqrt{2}) \cup (\sqrt{2},2)$, then in passing to $(G,T^{\flat},\psi)$, there is a fixed point interval of $g\psi$ around 1 but still only a hole for a "fixed point interval" of $h\psi$ around $\sqrt{2}$. So it seems that we may be further from our goal than before. If we had just passed to $(G,\bar{T})$ first, we might have $|\bar{T}| = 2^{\aleph_\alpha} > \aleph_\alpha$ and $\aleph_\alpha$ steps might have been insufficient--we quickly get into trouble and the method of proof fails. Hence we must insert some cuts into $T^{\flat}$ before proceeding. Since we seek α-sets, we must get by with very few cuts. This will be our thrust in Lemma 10.11.

We first require some more background.

We now generalise the concept of an α-set. Let Ω be a chain. $\Gamma \subseteq \Omega$ is said to be an *α-splitting subset* of Ω if whenever $A,B \subseteq \Omega$, $|A|,|B| < \aleph_\alpha$ and $A < B$, there exists $\gamma \in \Gamma$ such that $A < \gamma < B$. Let κ be a cardinal number. Ω is an *α-set of degree* κ if $|\Omega| = \aleph_\alpha$ and $\Omega = \bigcup_{\mu<\kappa} \Omega_\mu$ where each Ω_μ is an α-splitting subset of cardinality $\aleph_\alpha$ and $\Omega_\mu \cap \Omega_{\mu'} = \emptyset$ if $\mu \neq \mu'$. Note that if Ω is an α-set of degree κ, then $\kappa \leq \aleph_\alpha$ and each Ω_μ is an α-set that is dense in $\bar{\Omega}$ and Ω is itself an α-set. Conversely, any α-set is an α-set of degree 1. We now generalise (a) of page 187.

LEMMA 10.10. *If $\aleph_\alpha$ is a regular cardinal and κ is a cardinal not exceeding $\aleph_\alpha$ then any two α-sets of degree κ are ordermorphic. In fact, if $\Omega = \bigcup_{\mu<\kappa} \Omega_\mu$ and $\Delta = \bigcup_{\mu<\kappa} \Delta_\mu$ where each Ω_μ (Δ_μ) is an α-splitting subset of Ω (Δ) of cardinality $\aleph_\alpha$, and $\gamma_0 \in \Omega_0$, $\delta_0 \in \Delta_0$, then there exists an ordermorphism*

$\eta\colon \Omega \to \Delta$ *such that* $\Omega_\mu \eta = \Delta_\mu$ *for all* $\mu < \kappa$ *and* $\gamma_0\eta = \delta_0$.

Proof: Let $\{\gamma_\nu : \nu < \aleph_\alpha\}$ and $\{\delta_\nu : \nu < \aleph_\alpha\}$ be well-orderings of Ω and Δ respectively. We construct a sequence $\{\eta_\nu : \nu < \aleph_\alpha\}$ of functions (with $\gamma_0\eta_0 = \delta_0$) satisfying, for each $\nu < \aleph_\alpha$:

(1) $\{\gamma_{\nu'} : \nu' < \nu\} \subseteq \mathrm{dom}(\eta_\nu) \subseteq \Omega$.

(2) $\{\delta_{\nu'} : \nu' < \nu\} \subseteq \mathrm{image}(\eta_\nu) \subseteq \Delta$.

(3) η_ν is one-to-one and order-preserving.

(4) If $\nu' < \nu$, then $\eta_{\nu'} \subseteq \eta_\nu$.

(5) $\gamma \in \Omega_\mu \cap \mathrm{dom}(\eta_\nu)$ if and only if $\gamma\eta_\nu \in \Delta_\mu$.

Then $\eta = \bigcup_{\nu < \aleph_\alpha} \eta_\nu$ will be required ordermorphism, the onto-ness being guaranteed by (2).

By (a) and (b) on page 187, we may assume $\gamma_0\eta_0 = \delta_0$; i.e., $\eta_0 = \{(\gamma_0, \delta_0)\}$.

Assume $\{\eta_\xi : \xi < \nu\}$ has already been constructed satisfying (1)-(5) as well as

(6) $|\eta_\xi| < \aleph_\alpha$ for all $\xi < \nu$.

If ν is a limit ordinal, define $\eta_\nu = \bigcup_{\xi<\nu} \eta_\xi$. This clearly satisfies (1)-(6).

If ν is a successor ordinal, say $\nu = \zeta + 1$, then $\mathrm{dom}(\eta_\zeta) \neq \Omega$, by (6). Let ξ be the least ordinal such that $\gamma_\xi \notin \mathrm{dom}(\eta_\zeta)$. By hypothesis, $\zeta \leq \xi$. Then for a unique $\mu_0 < \kappa$, $\gamma_\xi \in \Omega_{\mu_0}$. Let $A = \{\gamma \in \mathrm{dom}(\eta_\zeta) : \gamma < \gamma_\xi\}$ and $B = \{\gamma \in \mathrm{dom}(\eta_\zeta) : \gamma_\xi < \gamma\}$. Then $A < \gamma_\xi < B$, so $A\eta_\zeta < B\eta_\zeta$. But $|A\eta_\zeta|, |B\eta_\zeta| < \aleph_\alpha$. Hence there exists a least ordinal ξ' such that $\delta_{\xi'} \in \Delta_{\mu_0}$ and $A\eta_\zeta < \delta_{\xi'} < B\eta_\zeta$. Note that since $A \cup B = \mathrm{dom}(\eta_\zeta)$, $\delta_{\xi'} \notin \mathrm{image}(\eta_\zeta)$. If $\delta_\zeta \in \mathrm{image}(\eta_\zeta)$, let $\gamma_{\zeta'} = \delta_\zeta\eta_\zeta^{-1}$. If $\delta_\zeta = \delta_{\xi'}$, let $\gamma_{\zeta'} = \gamma_\xi$. If $\delta_\zeta \notin \mathrm{image}(\eta_\zeta) \cup \{\delta_{\xi'}\}$, then $\delta_\zeta \in \Delta_{\mu_1}$, say. We can obtain $\gamma_{\zeta'} \in \Omega_{\mu_1}$ such that $\Theta\eta_\zeta^{-1} < \gamma_{\zeta'} < \Phi\eta_\zeta^{-1}$, where $\Theta = \{\delta \in \mathrm{image}(\eta_\zeta) : \delta < \delta_\zeta\}$ and $\Phi = \{\delta \in \mathrm{image}(\eta_\zeta) : \delta_\zeta < \delta\}$. Again, $\gamma_{\zeta'} \notin \mathrm{dom}(\eta_\zeta)$. Now define $\eta_\nu = \eta_\zeta \cup \{(\gamma_\xi, \delta_{\xi'}), (\gamma_{\zeta'}, \delta_\zeta)\}$. Then η_ν satisfies (1)-(6), completing the proof. //

Let $\aleph_\alpha$ be a regular cardinal greater than $\aleph_0$. Let $\underline{T^\#}$ be the union of T together with all cuts (A,B), where either $cf(A) < \aleph_\alpha$ or $ci(B) < \aleph_\alpha$. $T^\# \subseteq \bar{T}$ so we can extend the order on T to one on $T^\#$.

LEMMA 10.11. *If (G,T) is an ℓ-permutation group with $|T| \le \aleph_\alpha$, then $(G,T^\#)$ is an ℓ-permutation group satisfying:*

(i) $|T^\#| \le \aleph_\alpha$.

(ii) If $\Lambda \subseteq T$ and $|\Lambda| < \aleph_\alpha$, then $\sup \Lambda \in T^\#$ ($\inf \Lambda \in T^\#$) provided Λ is bounded above (below) in T.

(iii) Each fixed point interval of $g \in G$ is contained in a fixed point interval of the extension of g.

Proof: Since $T^\# \subseteq \bar{T}$, we can extend each $g \in G$ to an element of $A(T^\#)$. Specifically, if $(A,B) \in T^\#$, $(A,B)g = (Ag,Bg) \in T^\#$.

Now the number of subsets of T of cardinality $\aleph_\mu < \aleph_\alpha$ is $|T|^{\aleph_\mu} \le \aleph_\alpha^{\aleph_\mu}$. Hence $|T^\#| \le \max(\aleph_\alpha, \sup\{\aleph_\alpha^{\aleph_\mu}: \mu < \alpha\})$. But $\sup\{\aleph_\alpha^{\aleph_\mu}: \mu < \alpha\} = \sup\{\aleph_\alpha: \mu < \alpha\} = \aleph_\alpha$ (we are assuming G.C.H.). Thus (i) is proved.

To prove (ii), let $A_1 = \{\tau \in T: \tau \le \lambda$ for some $\lambda \in \Lambda\}$, $B_1 = \{\tau \in T: \tau > \Lambda\}$, $A_2 = \{\tau \in T: \tau < \Lambda\}$ and $B_2 = \{\tau \in T: \tau \ge \lambda$ for some $\lambda \in \Lambda\}$. Λ is cofinal in A_1 and coinitial in B_2. Hence $(A_1,B_1),(A_2,B_2) \in T^\#$ and are the infimum and supremum of Λ in $T^\#$. (iii) is immediate.

Passing from T to $T^\#$ will again permit us more freedom to expand the supporting and fixed point intervals in accordance with our plan. Moreover, every element of G is bounded (in $T^\#$).

In order to prove Theorem 10.8 we will proceed by induction interlacing $\#$ and $\flat$ as mentioned on page 199. It should be added that if (G,T) is our original ℓ-permutation group and $\bar{\tau} \in \bar{T}\setminus T$ has character $c_{\alpha\alpha}$ and there is $e < g \in G$ such that $\bar{\tau}$ is a supremum and infimum of collections of $\aleph_\alpha$ supporting intervals of g, then in enlarging G and T through subsequent $\#$s and $\flat$s we will never remove this phenomenon. It seems, therefore, that our program is doomed. However, the "final" enlargement of any $e < f \in G$ (even if f has but one supporting interval originally) will also enjoy this phenomenon as we will see.

Proof of Theorem 10.8: Let (G,T) be an ℓ-permutation group with T doubly homogeneous, $|T| \leq |G|$ and $G \subseteq B(T)$ (Corollary 2L). Let $\psi_0 = e$ and $\Omega_0 = T$. Assume that the sequences $\{\Omega_\mu : \mu < \mu_0\}$ and $\{\psi_\mu : \mu < \mu_0\}$ have been constructed satisfying for each $\mu < \mu_0$,

(I) $|\Omega_\mu| \leq \aleph_\alpha$.

(II) (G,Ω_μ,ψ_μ) is an ℓ-permutation group with $G\psi_\mu \subseteq B(\Omega_\mu)$.

(III) If μ is a limit ordinal, $\Omega_\mu = \bigcup_{\nu<\mu} \Omega_\nu$.

(IV) If $\mu = \lambda + 2n + 1$ for λ a limit ordinal and n finite, then $\Omega_\mu = (\Omega_{\lambda+2n})^{\#}$.

(V) If $\mu = \lambda + 2n + 2$ for λ a limit ordinal and n finite, then $\Omega_\mu = (\Omega_{\lambda+2n+1})^{\flat}$.

(VI) If $\mu' < \mu$, then $G\psi_{\mu'} = G\psi_\mu | \Omega_{\mu'}$.

(VII) If $\mu' < \mu$ and $\Delta_{\mu'}$ is a fixed point interval of $g\psi_{\mu'}$, then $\mathrm{Conv}_{\Omega_\mu}(\Delta_{\mu'})$ is contained in a fixed point interval Δ_μ of $g\psi_\mu$.

If μ_0 is a limit ordinal, let $\Omega_{\mu_0} = \bigcup_{\mu<\mu_0} \Omega_\mu$ and $\psi_{\mu_0} = \bigcup_{\mu<\mu_0} \psi_\mu$. Clearly (I)-(VI) are satisfied. If Δ_μ is a fixed point interval of $g\psi_\mu$ with $\mu < \mu_0$, and $\beta \in \mathrm{Conv}_{\Omega_{\mu_0}}(\Delta_\mu)$, then $\beta \in \Omega_{\mu'}$ for some μ' such that $\mu \leq \mu' < \mu_0$. Hence, by the induction hypothesis, $\beta(g\psi_{\mu_0}) = \beta(g\psi_{\mu'}) = \beta$. Thus (VII) is satisfied.

If $\mu_0 = \lambda + 2n + 1$, let $\Omega_{\mu_0} = (\Omega_{\lambda+2n})^{\#}$ and let ψ_{μ_0} be the obvious embedding given in Lemma 10.11. Then (I)-(VII) are satisfied. Similarly if $\mu_0 = \lambda + 2n + 2$ (using Lemma 10.9).

So we can construct sequences $\{\Omega_\mu : \mu < \aleph_\alpha\}$ and $\{\psi_\mu : \mu < \aleph_\alpha\}$ satisfying (I)-(VII). Let $\Omega = \bigcup_{\mu<\aleph_\alpha} \Omega_\mu$ and $\psi = \bigcup_{\mu<\aleph_\alpha} \psi_\mu$.

We now prove:

(1) Ω is an α-set.

(2) (G,Ω,ψ) is an ℓ-permutation group.

(3) If $e \neq g \in G$, the set of supporting intervals of $g\psi$ <u>$(I(g\psi))$</u> together with the fixed point intervals of $g\psi$ (omitting the greatest and least) <u>$(F^*(g\psi))$</u> form an α-set of degree 2.

(4) If $e \neq g \in G$ and $\Delta \in F^*(g\psi)$, then Δ is an α-set.

Proof: (1) $|\Omega| \leq \aleph_\alpha^2 = \aleph_\alpha$. Let $A,B \subseteq \Omega$, $A < B$ and $|A|,|B| < \aleph_\alpha$. Then there exists $\mu < \aleph_\alpha$ such that $A,B \subseteq \Omega_\mu$. So $A,B \subseteq \Omega_{\lambda+2n}$ for some limit ordinal λ and finite n. By (IV) and Lemma 10.11, $\sup A, \inf B \in (\Omega_{\lambda+2n})^{\#} = \Omega_{\lambda+2n+1} \subseteq \Omega$. If $\sup A < \inf B$, there exists $\sigma \in (\Omega_{\lambda+2n+1})^{\flat} = \Omega_{\lambda+2n+2} \subseteq \Omega$ such that $\sup A < \sigma < \inf B$ (whence $A < \sigma < B$). If $\sup A = \sigma' = \inf B$, then $\sigma' \notin A$ without loss of generality. So there exists $\tau \in (\Omega_{\lambda+2n+1})^{\flat} = \Omega_{\lambda+2n+2} \subseteq \Omega$ such that $A < \tau < \sigma' \leq B$. Thus Ω is an α-set.

(2) Clearly $g\psi \in A(\Omega)$ for all $g \in G$. Let $g,h \in G$ and $\sigma \in \Omega$. Then $\sigma \in \Omega_\mu$ for some $\mu < \aleph_\alpha$. $\sigma(gh)\psi = \sigma(gh)\psi_\mu = (\sigma(g\psi_\mu))(h\psi_\mu) = (\sigma(g\psi))(h\psi) = \sigma((g\psi)(h\psi))$. Thus ψ is a group homomorphism. Also $(g \vee e)\psi = (g\psi \vee e)$, similarly. Therefore ψ is an ℓ-homomorphism that is one-to-one since each ψ_μ is.

(3) Note that whenever $\mu' < \mu$, $I(g\psi_{\mu'})$ can be naturally embedded in $I(g\psi_\mu)$ and similarly for $F^*(g\psi_{\mu'})$. Thus $I(g\psi) = \bigcup_{\mu<\aleph_\alpha} I(g\psi_\mu)$ and $F^*(g\psi) = \bigcup_{\mu<\aleph_\alpha} F^*(g\psi_\mu)$. So to show that $I(g\psi) \cup F^*(g\psi)$ is an α-set of degree 2, it is easily seen to be enough to prove that if $\mu < \aleph_\alpha$ and $\mathcal{A},\mathcal{B} \subseteq I(g\psi_\mu) \cup F^*(g\psi_\mu)$, $\mathcal{A} < \mathcal{B}$ and $|\mathcal{A}|,|\mathcal{B}| < \aleph_\alpha$, then for some $\mu' > \mu$ and $\Lambda \in I(g\psi_{\mu'})$ $(\Delta \in F^*(g\psi_{\mu'}))$ $\mathcal{A} < \Lambda < \mathcal{B}$ $(\mathcal{A} < \Delta < \mathcal{B})$. We may clearly assume that μ fits case (IV). If $\mathcal{B}$ is non-empty and

does not have a least element that is a fixed point interval of $g\psi_\mu$, then there is a coinitial subset $\mathcal{B}'$ of $\mathcal{B}$ such that $\mathcal{B}' \subseteq I(g\psi_\mu)$. Now $\aleph_\alpha > |\mathcal{B}| \geq ci(\mathcal{B}')$. By (IV) and Lemma 10.11, $\inf(\cup\mathcal{B}') = \sigma \in \Omega_{\mu+1}$; and by (V) and Lemma 10.9, there exists $\Lambda' \in I(g\psi_{\mu+2})$ such that $\sigma < \Lambda' < \cup\mathcal{B}'$. By (IV) and Lemma 10.11 (and $\aleph_\alpha > ci(\mathcal{B}')$) again, we find $\tau = \inf(\cup\mathcal{B}') \in \Omega_{\mu+3}$, where $\sigma < \Lambda' < \tau < \mathcal{B}'$. But then $\tau \in \Delta$ for some $\Delta \in F^*(g\psi_{\mu+3})$. Thus $\sigma < \Lambda' < \Delta < \cup\mathcal{B}'$ and so $\mathcal{A} < \Lambda < \Delta < \mathcal{B}$ where $\Lambda \in I(g\psi_{\mu+3})$ contains Λ'. If $\mathcal{A}$ is non-empty and does not have a greatest element that is a fixed point interval of $g\psi_\mu$, then we can proceed dually to the above. So assume that $\mathcal{A}$ has a greatest element that is a fixed point interval of $g\psi_\mu$ and dually for $\mathcal{B}$. By the maximality of fixed point intervals of $g\psi_\mu$, there is $\Lambda' \in I(g\psi_\mu)$ such that $\mathcal{A} < \Lambda' < \mathcal{B}$. Using $\mathcal{B} \cup \{\Lambda'\}$ in place of $\mathcal{B}$ we can proceed as in the first instance. If $\mathcal{A} = \emptyset$ and $\mathcal{B}$ has a least element Δ' that belongs to $F^*(g\psi_\mu)$, then Δ' is not a subset of the least fixed point interval of $g\psi$. So for some $\mu_1 > \mu$, there exists $\Lambda \in I(g\psi_{\mu_1})$ such that $\Lambda < \Delta'$. Working in Ω_{μ_1} instead of Ω_μ and letting $\mathcal{A} = \{\Lambda\}$, we revert to the dual of the first case. If $\mathcal{B} = \emptyset$ and $\mathcal{A}$ has a greatest element Δ' that belongs to $F^*(g\psi_\mu)$, proceed dually. Finally, if $\mathcal{A} = \emptyset = \mathcal{B}$, replace $\mathcal{B}$ with the set consisting of a single element of $I(g\psi_\mu)$ and proceed by the first case. This completes the proof of (3).

(4) Let $\Delta \in F^*(g\psi)$ and μ_0 be the least ordinal such that $\Delta \cap \Omega_{\mu_0} \neq \emptyset$. Then $\Delta_\mu = \Delta \cap \Omega_\mu$ is a fixed point interval of $g\psi_\mu$ for all $\mu \geq \mu_0$. By (VII), $\Delta = \bigcup_{\mu_0 \leq \mu < \aleph_\alpha} \Delta_\mu$. Δ is an α-set by the same proof as used for (1).

Let $e < h_1, h_2 \in G$. Then $e < h_1\psi, h_2\psi \in G\psi$. Now $I(h_i\psi) \cup F^*(h_i\psi)$ are α-sets of degree 2 by (3) $(i = 1,2)$. Let $\gamma_0 \in supp(h_1)$ and $\delta_0 \in supp(h_2)$. By Lemma 10.10, there is an ordermorphism ϕ such that $I(h_1\psi)\phi = I(h_2\psi)$ and $F(h_1\psi)\phi = F(h_2\psi)$, and the supporting interval of $h_1\psi$ containing γ_0 is mapped to the supporting interval of $h_2\psi$ containing δ_0.

Each element of $F^*(h_i\psi)$ is an α-set by (4) $(i = 1,2)$. Hence if $\Delta \in F(h_1\psi)$, there is $f_\Delta \in A(\Omega)$ such that $\Delta\phi = \Delta f_\Delta$. If $\Lambda \in I(h_1\psi)$, let $\beta_1 \in \Lambda$ and $\beta_2 \in \Lambda\phi$ (where $\beta_1 = \gamma_0$ and $\beta_2 = \delta_0$ if Λ is the interval of support of $h_1\psi$ containing γ_0).

For each $n \in \mathbb{Z}$, there is $f_n = f_n(\Lambda) \in A(\Omega)$ such that $\beta_1(h_1\psi)^j f_n = \beta_2(h_2\psi)^j$ $(j = n, n+1)$ by (b) on page 187. By the patching lemma, there is $f \in A(\Omega)$ such that $f|\Delta = f_\Delta$ for all $\Delta \in F(h_1\psi)$ and $f|[\beta_1(h_1\psi)^n, \beta_1(h_1\psi)^{n+1}] = f_n|[\beta_1(h_1\psi)^n, \beta_1(h_1\psi)^{n+1}]$ $(n \in \mathbb{Z},\ \beta_1, \beta_2 \in \Lambda,\ \Lambda \in I(h_1\psi))$. By construction $\gamma_0 f = \delta_0$ and $f^{-1}(h_1\psi)f = h_2\psi$. The proof of Theorem 10.8 is now complete.

By using Theorem 10B and the patching lemma, the following partial amalgamation result can be proved (see Pierce [72]):

<u>THEOREM</u> <u>10.12</u>. *Let G be an Archimedean o-group and H_1, H_2 be ℓ-groups. Let $\theta_i : G \to H_i$ be ℓ-embeddings $(i = 1,2)$. Then there exists an ℓ-group L and ℓ-embeddings $\psi_i : H_i \to L$ $(i = 1,2)$ such that $\theta_1\psi_1 = \theta_2\psi_2$.*

The proof of Theorem 10B relied on α-sets for $\aleph_\alpha$ regular. Unfortunately, the existence of such sets is equivalent to $2^{\aleph_\alpha} = \aleph_{\alpha+1}$. For those who, like the author, feel the nearness of the *Dies Irae* every time they invoke the Generalised Continuum Hypothesis, there is a word of consolation. By using a modification of α-sets that doesn't depend on G.C.H., all the theorems in this section can be deduced provided we relax some of the cardinality restrictions on Ω (see Weinberg [80] and [81a]); i.e., all the theorems in this chapter are independent of the Generalised Continuum Hypothesis (provided we relax some of the cardinality restrictions on Ω).

CHAPTER 11
NORMAL VALUED LATTICE-ORDERED GROUPS

Let (G,Ω) be an ℓ-permutation group and $e \neq g \in G$. If $\alpha \in \mathrm{supp}(g)$, then the convex subset of Ω swept out by the images of α under powers of g is denoted by $\underline{\Delta(\alpha,g)}$; i.e., $\Delta(\alpha,g) = \mathrm{Conv}_\Omega(\alpha\langle g\rangle) = \{\beta \in \Omega: \alpha|g|^n \leq \beta \leq \alpha|g|^{n+1}$ for some $n \in \mathbb{Z}\}$. (G,Ω) is *overlapping* if there are $\alpha \in \Omega$ and $f,g \in G$ such that $\Delta(\alpha,f)$ and $\Delta(\alpha,g)$ are incomparable (with respect to

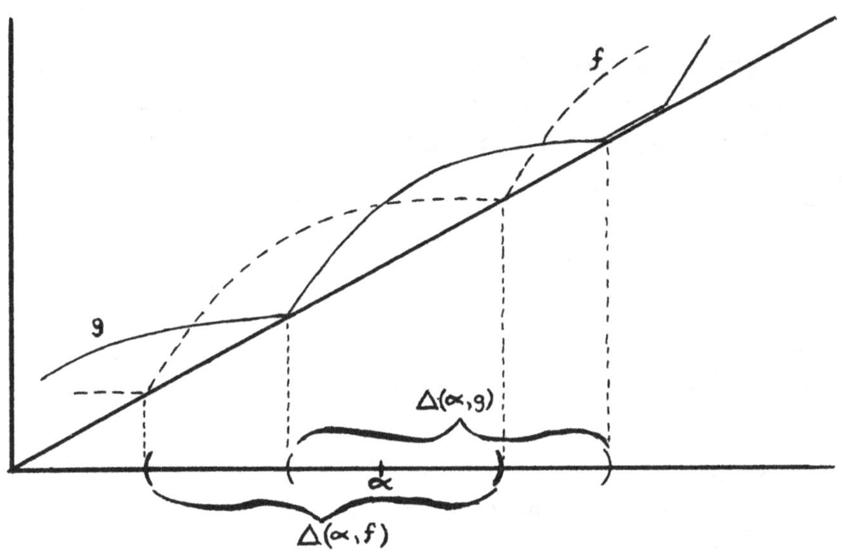

inclusion). If (G,Ω) is a primitive ℓ-permutation group, then it is overlapping if and only if it is transitively derived from a periodic or doubly transitive ℓ-permutation group; i.e., in the transitive primitive case, if and only if it is periodic or doubly transitive. We will extend this to the non-primitive case by our structure theory (Chapter 3).

A seemingly unrelated concept from classical ℓ-group theory is normal valued. Let G be an ℓ-group and H a convex ℓ-subgroup of G. H is a *value* in G if there exists $e \neq g \in G$ such that H is maximal with respect to being a convex ℓ-subgroup of G not containing g. In this case we say that H is a *value of g*. Each convex ℓ-subgroup of G that properly contains H contains g. Hence the convex ℓ-subgroup $H' \supsetneq H$, where $H' = \cap\{M: M \text{ is a convex } \ell\text{-subgroup of } G \text{ and } M \supsetneq H\}$. H' is called the *cover* of H. An ℓ-group G is said to be *normal valued* if $H \triangleleft H'$ for each value H in G. If (G,Ω) is a transitive primitive ℓ-permutation group, each G_α is a maximal convex ℓ-subgroup of G. It is a value of any $g \in G\setminus G_\alpha$ and $G'_\alpha = G$. But $f^{-1}G_\alpha f = G_{\alpha f}$ so G is normal valued only if $G_\alpha = G_{\alpha f}$ for all $f \in G$; i.e., only if (G,Ω) is regular. But if (G,Ω) is regular, G is Abelian and hence, clearly, normal valued. Thus if (G,Ω) is a primitive ℓ-permutation group, G is normal valued if and only if (G,Ω) is static, integral or regularly derived. So in the primitive case non-overlapping and normal valued are the same. We will use our structure theory of Chapter 3 to extend this and prove:

THEOREM 11A. *Let (G,Ω) be an ℓ-permutation group. The following are equivalent:*

(1) G is normal valued.

(2) (G,Ω) is non-overlapping.

(3) (G,Ω) has only static, integral and regularly derived primitive components.

(4) $fg \leq g^2f^2$ for all $e \leq f,g \in G$.

By Theorem 5A, we have as an immediate consequence:

COROLLARY 11B. *If (G,Ω) is a transitive ℓ-permutation group, then (G,Ω) can be ℓ-embedded in $Wr\{(\mathbb{R},\mathbb{R}): K \in k\}$ for $k = k(G,\Omega)$ if and only if G is normal valued.*

Note that condition (4) shows that the class $\underline{\mathcal{N}}$ of normal valued ℓ-groups forms a variety (i.e., is defined by a set of equations): $G \in \mathcal{N}$ if and only if $(\forall x \in G)(\forall y \in G)$ $[(x \vee e)(y \vee e) \wedge (y \vee e)^2(x \vee e)^2 = (x \vee e)(y \vee e)]$.

Theorem 11A is due to Wolfenstein [68] and Read [75], Corollary 11B to Read [75].

In sharp contrast to Theorem 11A, we will show the following result due to Ball (unpublished):

THEOREM 11C. *$A(\mathbb{R})$ has no normal values.*

We will use Theorem 11A to deduce the following theorem due to Holland [76b]:

THEOREM 11D. *If V is a proper variety of ℓ-groups, then $V \subseteq \mathcal{N}$.*

We give two more applications of our ideas. If U and V are varieties of ℓ-groups, define the class $\underline{U \cdot V}$ by: $G \in U \cdot V$ if there exists an ℓ-ideal N of G such that $N \in U$ and $G/N \in V$. It is easy to see that $U \cdot V$ is a variety if we use the following result of G. Birkhoff (see, e.g., Cohn [65]) which we state only for ℓ-groups:

A class $\mathcal{K}$ of ℓ-groups is a variety if and only if $\mathcal{K}$ is closed under ℓ-subgroups, ℓ-homomorphic images and products.

It is easy to see that $\cdot$ is associative. We write $\underline{V^{n+1}}$ for $V^n \cdot V$ ($n \in \mathbb{Z}^+$, V a variety of ℓ-groups), $\underline{A}$ for the variety of all abelian ℓ-groups, $\underline{\mathcal{E}}$ for the variety of all one element ℓ-groups, and $\underline{\mathcal{L}}$ for the variety of all ℓ-groups. The following facts are well-known ((a) is easy by classical ℓ-group results; e.g., Bigard, Keimel and Wolfenstein [77, Proposition 2.4.7 and Corollary 2.5.9]):

(a) $\mathcal{N} \cdot \mathcal{N} = \mathcal{N}$ (see Martinez [74]).

(b) If $V \neq \mathcal{E}$ is a variety of ℓ-groups, then $A \subseteq V$ (see Weinberg [65] or Glass, Holland and McCleary [80]).

THEOREM 11E. *$\mathcal{N}$ is the smallest variety of ℓ-groups that contains* $\bigcup_{n \in \mathbb{Z}^+} A^n$ *(and hence* $\bigcup_{n \in \mathbb{Z}^+} V^n$ *if* $V \neq \mathcal{E}, \mathcal{L}$ *).*

If H is an ℓ-group and $\{G_i : i \in I\}$ is a family of normal-valued convex ℓ-subgroups of H, then G (the convex ℓ-subgroup of H generated by $\bigcup_{i \in I} G_i$) is normal valued. This is easy using Bigard, Keimel and Wolfenstein [77, Proposition 2.4.7 and Corollary 2.5.9]. We can generalise this result by using Theorems 11A and 11D.

THEOREM 11F. *If V is a variety, H is an ℓ-group and $\{G_i : i \in I\}$ is a family of convex ℓ-subgroups of H with each $G_i \in V$, then $G \in V$ where G is the convex ℓ-subgroup of H generated by $\bigcup_{i \in I} G_i$.*

Theorem 11E is due to Glass, Holland and McCleary [80]; Theorem 11F to Holland [79].

If G is an ℓ-group and P is a prime subgroup of G, then P contains a minimal prime subgroup of G (by Zorn's lemma). Paul Conrad conjectured that if G is an ℓ-group in which each proper prime subgroup contains a unique minimal prime subgroup, then G is normal valued. We complete the chapter with a proof of the following theorem due to George M. Bergman (unpublished):

THEOREM 11G. *There is a non-normal valued ℓ-group in which each prime subgroup contains a unique minimal prime subgroup.*

Proof of Theorem 11A: We first prove that (3) is equivalent to both (2) and (4).

(2) → (3). Assume (G,Ω) has a primitive component (G_K, Ω_K) that is not static, integral or regularly derived. By Corollary 4.4.1, it is doubly transitively or periodically derived. To prove that (G,Ω) is overlapping, it is enough to find $\alpha < \beta < \gamma$ (in Ω) with $K = \mathrm{Val}(\alpha,\gamma)$, and $e < f, g \in G_{(\beta\zeta^K)}$ such

that $\gamma\mathcal{C}_K \subseteq \Delta(\beta,g)\setminus\Delta(\beta,f)$ and $\alpha\mathcal{C}_K \subseteq \Delta(\beta,f)\setminus\Delta(\beta,g)$. This reduces the problem to the primitive case. So we may assume that (G,Ω) is doubly transitively or periodically derived. In either case (using Theorem 4.3.1), there exist $\alpha < \beta < \gamma$ and $e < f,g \in G$ such that $f \in G_\gamma$, $g \in G_\alpha$, $\alpha f = \beta$ and $\beta g = \gamma$. Now $\alpha \in \Delta(\beta,f)\setminus\Delta(\beta,g)$ and $\gamma \in \Delta(\beta,g)\setminus\Delta(\beta,f)$, so (G,Ω) is overlapping.

<u>(4) → (3)</u>. Assume (G,Ω) has a primitive component (G,Ω) that is not static, integral or regularly derived. It is enough to find $\alpha \in \mathrm{dom}(K)$ and $e < f,g \in G_{(\alpha\mathcal{C}^K)}$ such that $(\alpha\mathcal{C}_K)fg > (\alpha\mathcal{C}_K)g^2f^2$. So, as in (2) → (3), we may assume that (G,Ω) is doubly transitively or periodically derived. Let $\alpha,\beta,\gamma \in \Omega$ and $e < f,g \in G$ be such that $f \in G_\gamma$, $g \in G_\alpha$, $\alpha f = \beta$ and $\beta g = \gamma$. Then $\alpha fg = \gamma = \gamma f^2 > \alpha f^2 = \alpha g^2 f^2$ as desired.

<u>(3) → (2)</u>. Suppose that (G,Ω) is overlapping. So there exist $\alpha < \beta < \gamma$ in Ω and $e < f,g \in G$ such that $\alpha \in \Delta(\beta,f)\setminus\Delta(\beta,g)$ and $\gamma \in \Delta(\beta,g)\setminus\Delta(\beta,f)$. Let $K = \mathrm{Val}(\alpha,\gamma)$. For some $n \in \mathbb{Z}^+$, $\gamma g^{-n} < \alpha f^n$. So $\gamma g^{-n} \notin \gamma\mathcal{C}_K$ or $\alpha f^n \notin \alpha\mathcal{C}_K$. Without loss of generality, we assume the latter. Then $K = \mathrm{Val}(\alpha,\alpha f)$. Hence <u>K is not</u> static. If $\bar{\delta} = \sup \Delta(\beta,f) = \sup \Delta(\alpha,f)$, $\bar{\delta}\mathcal{C}_K \in (\alpha\mathcal{C}^K/\mathcal{C}_K)$. But $f \in G_{(\bar{\delta}\mathcal{C}_K)}\setminus G_{(\alpha\mathcal{C}_K)}$. Therefore K is not regularly derived or integral.

<u>(3) → (4)</u>. Suppose $fg \not\leq g^2f^2$ for some $e \leq f,g \in G$. Then there is $\alpha \in \Omega$ such that $\alpha fg > \alpha g^2f^2$. Let $K = \mathrm{Val}(\alpha,\alpha fg)$. Since $\alpha \leq \alpha g^2 f^2 < \alpha fg$, K is not static, and as $\alpha \leq \alpha f, \alpha g \leq \alpha fg$, $f,g \in G_{(\alpha\mathcal{C}^K)}$. If (G,Ω) were integral or regularly derived, G_K would be Abelian and $\gamma\mathcal{C}_K < \gamma\mathcal{C}_K f$ or $\gamma\mathcal{C}_K < \gamma\mathcal{C}_K g$ for all $\gamma \in \alpha\mathcal{C}^K$. Thus $\alpha\mathcal{C}_K fg < \alpha\mathcal{C}_K gfgf = (\alpha\mathcal{C}_K)g^2f^2$, a contradiction.

Thus we have established (2) ↔ (3) ↔ (4).

<u>(3),(4) → (1)</u>. Assume G is not normal valued. Then there is a value H in G such that $H \ntriangleleft H'$. Let $\Omega = R(H)$, the set of right cosets of H in G. Ω is a chain by Appendix I Lemmas 2

and 3. Let $L = \bigcap_{g \in G} g^{-1}Hg$ and $M = G/L$. Then (M,Ω) is a (transitive) locally primitive ℓ-permutation group ($M_H = H$ is covered by $H' = M_{(\{Hg:g \in H'\})}$). Since M_H is not normal in its cover, the local component must be doubly transitively or periodically derived. Since (4) → (3), $fg \not\leq g^2f^2$ for some $e \leq f,g \in M = G/L$. Hence $fg \not\leq g^2f^2$ for some $e \leq f,g \in G$ (the natural map $G \to G/L$ preserves order).

<u>(1) → (4)</u>. Assume $G \in \mathcal{N}$. For reductio ad absurdum, assume (4) fails for G. Let $\mathcal{H}$ be the set of values in G. Totally order $\mathcal{H}$ and let Ω_H be the set of right cosets of the value H in G. Then Ω_H is a chain by Appendix I Lemmas 2 and 3. Let $\Omega = \bigcup_{H \in \mathcal{H}} \Omega_H$. Ω is a chain where: $\alpha < \beta$ if $\alpha \in \Omega_{H_1}$, $\beta \in \Omega_{H_2}$ and $H_1 < H_2$ (in the total order on $\mathcal{H}$), or $H_1 = H_2$ and $\alpha < \beta$ in Ω_{H_1}. There is a natural ℓ-embedding of G into $A(\Omega)$ (see Appendix I proof of Holland's theorem) so we may regard (G,Ω) as an ℓ-permutation group. Since (3) → (4), (G,Ω) has a primitive component K such that (G_K,Ω_K) is doubly transitively or periodically derived. Since each Ω_H is an orbit of G, there is $H \in \mathcal{H}$ such that $\alpha = H \in \mathrm{dom}(K)$. Thus there exist $e < f,g \in G_{(\alpha\mathcal{C}^K)}$ such that $\alpha\mathcal{C}_K g = \alpha\mathcal{C}_K$, $\alpha\mathcal{C}_K f^{-1}g^{-1} < \alpha\mathcal{C}_K f^{-1} < \alpha\mathcal{C}_K$ and $\alpha\mathcal{C}_K f^{-1}g^{-1}f = \alpha\mathcal{C}_K f^{-1}g^{-1}$. Clearly $G_{(\alpha\mathcal{C}_K)}$ is a value of f. But $g \in G_{(\alpha\mathcal{C}_K)}$; so $f^{-2}gf^2 \in G_{(\alpha\mathcal{C}_K)}$ since $G \in \mathcal{N}$. Further, $\alpha\mathcal{C}_K f^{-1}g^{-1}f^2 = \alpha\mathcal{C}_K f^{-1}g^{-1} < \alpha\mathcal{C}_K$ so $f^{-1}g^{-1}f^2 \vee e \in G_{(\alpha\mathcal{C}_K)}$. Hence $e \leq f \leq f \vee f^{-2}gf^2 = (f^{-1}g^{-1}f^2 \vee e)f^{-2}gf^2 \in G_{(\alpha\mathcal{C}_K)}$. Consequently, $f \in G_{(\alpha\mathcal{C}_K)}$, a contradiction.

<u>COROLLARY</u> <u>11.1</u>. *If (G,Ω) and (G,T) are ℓ-permutation groups, then (G,Ω) has only static, integral and regularly derived primitive components if and only if (G,T) has.*

We omit the proof of the following theorem (due to Wolfenstein [68], Read [75], and Glass and Holland [73]) since it is very similar to that of Theorem 11A. Again the crux is to use (3) and (4) of Theorem 11A (A set of values $\mathfrak{X}$ of an ℓ-group G is said to be *plenary* if (a) each $e \neq g \in G$ has a value $H \in \mathfrak{X}$ and (b) if H_2 is a value in G and $H_2 \supseteq H_1 \in \mathfrak{X}$, then $H_2 \in \mathfrak{X}$).

THEOREM 11.2. *Let* (G,Ω) *be an* ℓ-*permutation group. The following are equivalent:*

(i) $G \in \mathcal{N}$.

(ii) $[f,g]^n \leq |f| \vee |g|$ *for all* $f,g \in G$ *and* $n \in \mathbb{Z}^+$.

(iii) $[f,g]^n \leq f \vee g$ *for all* $e \leq f,g \in G$ *and* $n \in \mathbb{Z}^+$.

(iv) $[f,g]^2 \leq f \vee g$ *for all* $e \leq f,g \in G$.

(v) $e \vee (f^2g^{-2} \wedge f^{-2}g^2) \leq g$ *for all* $e \leq f,g \in G$.

(vi) G *has a plenary set* $\mathfrak{X}$ *of normal values* $(H \in \mathfrak{X} \rightarrow H \lhd H')$.

We now prove Theorem 11C. If H is a subgroup of a group G, let $N(H) = \{g \in G: g^{-1}Hg = H\}$, the *normaliser* of H in G. We will actually prove:

THEOREM 11.3. *If* P *is a prime subgroup of* $A(\mathbb{R})$, *then* $N(P) = P$.

Since every value is prime (Appendix I Lemma 3), Theorem 11C follows at once from Theorem 11.3. The proof of Theorem 11.3 hinges on the following observation: If P is a prime subgroup, $g \in N(P)$ and $f \wedge g^{-1}fg = e$, then $f \in P$ (Otherwise $g^{-1}fg \in P$, so $f \in gPg^{-1} = P$). So $g^{-1}fg \in P$ and hence $f \vee g^{-1}fg \in P$.

Proof of Theorem 11.3: Let $g \in N(P)$. We assume first $e < g$ and g has one bump. Let $\alpha \in \mathrm{supp}(g)$ and $\sigma,\tau \in \mathbb{R}$ with

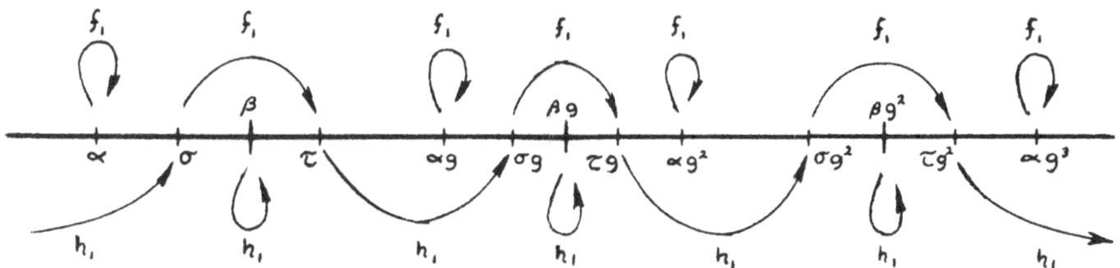

$\alpha < \sigma < \tau < \alpha g$. By doubly transitivity and the patching lemma, we can find $e < f \in A(\mathbb{R})$ such that $\text{supp}(f) \subseteq \bigcup_{n\in\mathbb{Z}} (\alpha g^{2n}, \alpha g^{2n+1})$ and $\sigma g^{2n} f = \tau g^{2n}$ $(n \in \mathbb{Z})$. Then $\text{supp}(g^{-1}fg) \subseteq \bigcup_{n\in\mathbb{Z}} (\alpha g^{2n+1}, \alpha g^{2n+2})$. So $f_1 = g^{-1}fg \vee f \in P$. Let $\beta \in (\sigma,\tau)$; so $\beta < \tau < \alpha g < \sigma g < \beta g$. Hence we may find $e < h \in A(\mathbb{R})$ such that $\text{supp}(h) \subseteq \bigcup_{n\in\mathbb{Z}} (\beta g^{2n}, \beta g^{2n+1})$ and $\tau g^{2n} h = \sigma g^{2n+1}$ $(n \in \mathbb{Z})$. Again $h \wedge g^{-1}hg = e$, so $h_1 = h \vee g^{-1}hg \in P$. We now show $g \leq (f_1h_1)^2 \in P$ (whence $g \in P$). Let $\delta \in \text{supp}(g)$. Then $\delta \in (\sigma g^m, \sigma g^{m+1})$ for some $m \in \mathbb{Z}$. Thus $\delta(f_1h_1)^2 \geq \sigma g^m(f_1h_1)^2 = \sigma g^{m+2} > \delta g$. Hence $(f_1h_1)^2 \geq g$ as desired.

If g has more than one bump, do the same as above for each bump simultaneously to get $e \leq |g| \leq (f_1h_1)^2 \in P$. But $|g|^{-1} \leq g \leq |g|$, so $g \in P$ and the theorem is proved.

Observe that the same proof works for any doubly transitive $A(\Omega)$.

We now prove Theorem 11D. Suppose that $G \notin \mathcal{N}$ and let $\mathcal{V}$ be the variety of ℓ-groups generated by G. We must prove $\mathcal{V} = \mathcal{L}$, the variety of all ℓ-groups. By Holland's theorem (Appendix I), (G,Ω) is an ℓ-permutation group for some chain Ω. By Theorem

11A, there is a primitive component (G_K,Ω_K) that is transitively derived from a periodic primitive or doubly transitive ℓ-permutation group. Now $G_K = G_{(\alpha\mathcal{C}^K)}/L$ for some $\alpha \in \mathrm{dom}(K)$. Since $G_{(\alpha\mathcal{C}^K)}$ is an ℓ-subgroup of G, it belongs to $\mathcal{U}$ by Birkhoff's theorem. Hence $G_K \in \mathcal{U}$, again by Birkhoff's theorem. But if (G_K,Ω_K) is periodic primitive with period z, then $((G_K)_{(\Delta)}/L',\Delta)$ is doubly transitively derived by Theorem 4.3.1, where $\Delta = (\alpha\mathcal{C}_K,\alpha\mathcal{C}_K z)$ and L' is the lazy subgroup. Since $(G_K)_{(\Delta)}$ is an ℓ-subgroup of G_K, it belongs to $\mathcal{U}$. Hence $\mathcal{U}$ contains a doubly transitive ℓ-permutation group (H,T)--strictly there is $H \in \mathcal{U}$ such that (H,T) is doubly transitive for some chain T.

Let $w = \bigvee_{i=1}^{m} \bigwedge_{j=1}^{n} \prod_{k=1}^{p} x_{ijk}$ be a word in the alphabet $X \cup X^{-1} \cup \{e\}$; i.e., each x_{ijk} belongs to $X \cup X^{-1} \cup \{e\}$. Assume that $w = e$ fails in some ℓ-group. Let F_ω be the free ℓ-group on a countably infinite set of generators containing $X' = \{x \in X\colon x_{ijk} = x \text{ or } x^{-1} \text{ for some } i,j,k\}$. So (F_ω,Λ) is an ℓ-permutation group for some chain Λ by Holland's theorem. Since $w \neq e$ in F_ω, there exists $\sigma \in \Lambda$ such that $\sigma w \neq \sigma$. For each (i,j) $(1 \leq i \leq m;\ 1 \leq j \leq n)$, let $\sigma(i,j,0) = \sigma$ and $\sigma(i,j,k) = \sigma(i,j,k-1)x_{ijk}$ $(1 \leq k \leq p)$. So $\sigma(i,j,k) = \sigma x_{ij1}\ldots x_{ijk}$. For each $x \in X'$ and each (i,j), let $M_{i,j}(x) = \{k\colon x = x_{ijk}\}$ and $N_{i,j}(x) = \{k\colon x^{-1} = x_{ijk}\}$. Let $\Lambda_0 = \{\sigma(i,j,k)\colon 1 \leq i \leq m,\ 1 \leq j \leq n,\ 0 \leq k \leq p\}$, a finite subset of Λ. Choose $T_0 \subseteq T$ ordermorphic to Λ_0 and let $\tau(i,j,k)$ be the image of $\sigma(i,j,k)$ under this ordermorphism. Now the action of x maps $\sigma(i,j,k-1)$ to $\sigma(i,j,k)$ if $k \in M_{i,j}(x)$, and $\sigma(i,j,k)$ to $\sigma(i,j,k-1)$ if $k \in N_{i,j}(x)$. So $\tau(i,j,k-1) \mapsto \tau(i,j,k)$ $(k \in M_{i,j}(x))$ and $\tau(i,j,k) \mapsto \tau(i,j,k-1)(k \in N_{i,j}(x))$ yields a one-to-one order preserving correspondence. Since (H,T) is a doubly transitive ℓ-permutation group (and hence q-transitive for all $q \in \mathbb{Z}^+$), there exist $h(x) \in H$ such that $\tau(i,j,k-1)h(x) = \tau(i,j,k)$ whenever $k \in M_{i,j}(x)$, and $\tau(i,j,k)h(x) = \tau(i,j,k-1)$ whenever

$k \in N_{i,j}(x)$. Note that $\tau(i,j,0) = \tau(i',j',0)$ since both are images of $\sigma = \sigma(i,j,0)$. So $\tau = \tau(i,j,0) \in T$ is independent of (i,j). Now $\tau(\prod_{k=1}^{p} h(x_{ijk})) = \tau h(x_{ij1})\ldots h(x_{ijp}) = \tau(i,j,p)$. Hence $\tau(\bigvee_{i=1}^{m} \bigwedge_{j=1}^{n} \prod_{k=1}^{p} h(x_{ijk})) = \max_i \min_j \tau(i,j,p) \neq \tau$ because $\sigma \neq \max_i \min_j \sigma(i,j,p)$. Consequently, $w = e$ fails in H (under the substitution $h(x)$ for x). So H satisfies no laws that are not true in all ℓ-groups. Thus $\mathscr{A} = \mathscr{L}$.

COROLLARY 11.4. *If G is an ℓ-group and $G \notin \mathscr{N}$, then G satisfies only those identities which are true in all ℓ-groups.*

An analysis of the last part of the proof of Theorem 11D shows:

THEOREM 11.5. *The free ℓ-group on a finite number of generators has soluble word problem.*

For the details, see Holland and McCleary [79]. Contrast Theorem 11.5 with Theorem 13A. Note that Theorem 11.5 is not trivial (as it is in the group case) since there is no uniqueness of normal form $\bigvee_{i=1}^{m} \bigwedge_{j=1}^{n} \prod_{k=1}^{p} x_{ijk}$ of ℓ-group words.

We use a similar technique, together with earlier theorems, to prove Theorem 11E.

We first note that $Wr\{(\mathbb{R},\mathbb{R}): 1 \leq K \leq n\} \in A^n$ $(n \in \mathbb{Z}^+)$ by induction on n. Since $\mathscr{N}\cdot\mathscr{N} = \mathscr{N}$ and $A \subseteq \mathscr{N}$, $A^n \subseteq \mathscr{N}$ for all $n \in \mathbb{Z}^+$. So $\mathscr{N}$ contains the variety of ℓ-groups generated by $\bigcup_{n\in\mathbb{Z}^+} A^n$. To prove equality, we must show that any law which fails in $\mathscr{N}$ fails in some A^n $(n \in \mathbb{Z}^+)$. So let $w = \bigvee_{i=1}^{m} \bigwedge_{j=1}^{n} \prod_{k=1}^{p} x_{ijk}$ and suppose $w = e$ fails in some element of $\mathscr{N}$. Then it fails in some subdirectly irreducible element of $\mathscr{N}$. But any subdirectly irreducible ℓ-group has a transitive representation (Appendix I Corollary 6) and so can be ℓ-embedded in $(W,\Omega) = \mathrm{Wr}\{(\mathbb{R},\mathbb{R}): K \in k\}$ for some set k by Corollary 11B. Hence $w = e$ fails in (W,Ω). Let $g_{ijk} \in W$ and $\sigma \in \Omega$ be such that $\sigma(\bigvee_{i=1}^{m} \bigwedge_{j=1}^{n} \prod_{k=1}^{p} g_{ijk}) \neq \sigma$. Let $\sigma(i,j,k)$ be defined as in the proof of Theorem 11D with g_{ijk} in place of x_{ijk}, and let $\Omega_0 = \{\sigma(i,j,k): 1 \leq i \leq m, 1 \leq j \leq n, 0 \leq k \leq p\}$. Then Ω_0 is a finite subset of Ω. Let $k_0 = \{K \in k: (\exists\alpha_1 \in \Omega_0)(\exists\alpha_2 \in \Omega_0)(\alpha_1 \neq \alpha_2 \;\&\; K = \mathrm{Val}(\alpha_1,\alpha_2))\}$, a finite subset of k. Let $(W',\Omega') = \mathrm{Wr}\{(\mathbb{R},\mathbb{R}): K \in k_0\}$. So $W' \in A^{|k_0|}$. We now show that $w = e$ fails in W' whence the proof is complete.

Let $\alpha \mapsto \alpha'$ be the projection of Ω onto Ω'. Note that the restriction of this map to Ω_0 is one-to-one and order preserving. For each $g \in \{g_{ijk}\}$, let $g' = (g'_{K,\alpha'}) \in W'$ be defined by:

$$g'_{K,\alpha'} = \begin{cases} g_{K,\beta} & \text{if } (\exists\beta,\gamma \in \Omega_0)(\beta g \equiv^K \gamma \text{ (in } \Omega) \text{ and} \\ & \qquad \beta' \equiv^K \alpha' \text{ (in } \Omega')) \\ e & \text{otherwise} \end{cases}$$

If $\beta_1,\beta_2 \in \Omega_0$ and $\beta_1' \equiv^K \beta_2'$ in Ω', then $\beta_1 \equiv^K \beta_2$ in Ω and so $\beta_1 g \equiv^K \beta_2 g$ if $g \in W$. Hence $g'_{K,\alpha'}$ is well defined. Moreover, if $x_{ijk}, x_{abc} = x$, then $g'_{ijk} = g'_{abc}$; and similarly $g'_{ijk} = (g'_{abc})^{-1}$ if $x_{ijk} = x \doteq x^{-1}_{abc}$. Thus replacing x_{ijk} by g'_{ijk} (i,j,k) is a bona fide substitution. Pick (i,j,k) and let $g = g_{ijk}$, $\beta = \sigma(i,j,k-1) \in \Omega_0$ and $K \in k_0$. Then $(\beta'g')_K = \beta'_K g'_{K,\beta'} = \beta_K g_{K,\beta} = (\beta g)_K = (\beta g)'_K$ (since $\beta,\beta g = \sigma(i,j,k) \in \Omega_0$). Thus $\beta'g' = (\beta g)'$. Hence for each i,j, $\sigma'(\prod_{k=1}^{p} g'_{ijk}) = (\sigma\prod_{k=1}^{p} g_{ijk})'$. It follows that $\sigma'(\bigvee_{i=1}^{m} \bigwedge_{j=1}^{n} \prod_{k=1}^{p} g'_{ijk}) = \max_i \min_j \sigma(i,j,p)' \neq \sigma'$ since $\alpha \mapsto \alpha'$ is one-to-one order preserving on Ω_0. Thus $w = e$ fails in W' as desired. //

We now prove Theorem 11F using Theorems 11A and 11D. The crux is the following lemma (If G is an ℓ-group and $X \subseteq G$, we say that *X generates G as an ℓ-group* if the convex ℓ-subgroup of G generated by X is G. If $X = \{g\}$, we say that *g generates G as an ℓ-group*).

LEMMA 11.6. *Let G be a normal valued subdirectly irreducible ℓ-group generated (as an ℓ-group) by $\{g_1, \ldots, g_n\}$. Then G is generated as an ℓ-group by some g_i $(1 \leq i \leq n)$.*

Proof: By Appendix I Corollary 6, (G,Ω) is a transitive ℓ-permutation group for some chain Ω. Let $\alpha \in \Omega$. By rechristening, $\Delta(\alpha,g_n) \subseteq \ldots \subseteq \Delta(\alpha,g_1)$ since (G,Ω) is non-overlapping by Theorem 11A. Now $\Delta(\alpha,g_1) = \Omega$ (If not, say $\bar\beta = \sup(\Delta(\alpha,g_1)) \in \bar\Omega$. By the non-overlapping of supporting intervals, $\bar\beta g_i = \bar\beta$ for $1 \leq i \leq n$. Hence if $\sigma \in \Omega$ with $\sigma > \bar\beta$, there is no $f \in G$ with $\alpha f = \sigma$. This contradicts (G,Ω) is transitive). Let $K = \mathrm{Val}(\alpha,\alpha g_1)$. Then K is regular so (G_K,Ω_K) is Archimedean. Since $\Omega_K = \Omega/\mathcal{C}_K$, g_1 generates G as an ℓ-group.

Proof of Theorem 11F: Let $\{G_i : i \in I\}$ be a family of convex ℓ-subgroups of an ℓ-group H and let G be the convex ℓ-subgroup of H generated by $\bigcup_{i \in I} G_i$. If $\mathcal{V}$ is a variety and $G_i \in \mathcal{V}$ for each $i \in I$, we must show that $G \in \mathcal{V}$. If $\mathcal{V} = \mathcal{L}$ there is nothing to prove. So assume $\mathcal{V} \neq \mathcal{L}$; i.e., $\mathcal{V} \subseteq \mathcal{N}$ by Theorem 11D. Then $G_i \in \mathcal{N}$ for each $i \in I$, and hence $G \in \mathcal{N}$. Let $w(x_1,\ldots,x_m) = e$ hold in $\mathcal{V}$ and let $h_1,\ldots,h_m \in G$. Thus there are $g_{jk} \in \bigcup_{i \in I} G_i$ such that $h_j = \prod_{k=1}^{n} g_{jk}$. Let H_0 be the ℓ-subgroup of G generated by $\{g_{jk} : 1 \leq j \leq m,\ 1 \leq k \leq n\}$. If $w(h_1,\ldots,h_m) \neq e$, there is a subdirectly irreducible factor $\bar H_0$ of H_0 such that $w(\bar h_1,\ldots,\bar h_m) \neq e$ where $h \mapsto \bar h$ denotes the natural map of H_0 onto $\bar H_0$. Since $H_0 \subseteq G \in \mathcal{N}$, $H_0 \in \mathcal{N}$. Hence $\bar H_0 \in \mathcal{N}$ and as it is generated by $\{\bar g_{jk} : 1 \leq j \leq m,\ 1 \leq k \leq n\}$, there is a $\bar g_{j_0 k_0}$ which generates $\bar H_0$ (by Lemma 11.6). But

$\bar{g}_{j_0k_0} \in G_i$ for some $i \in I$ and $\overline{H_0 \cap G_i}$ is a convex ℓ-subgroup of $\bar{H}_0$. Thus $\bar{H}_0 = \overline{H_0 \cap G_i} \in V$. Therefore $w(\bar{h}_1,\ldots,\bar{h}_m) = e$, the desired contradiction. Consequently $G \in V$.

Finally we prove Theorem 11G. We first establish a lemma due to Glass and Holland (unpublished).

<u>LEMMA</u> <u>11.7</u>. *Let (G,Ω) be a doubly transitive ℓ-permutation group with $G \subseteq B(\Omega)$. Assume that each element of G has only finitely many bumps, and that no point of $\bar{\Omega}$ is both the infimum of a supporting interval of an element of G and the supremum of a supporting interval of an element of G. Then each proper prime subgroup of G contains a unique minimal prime subgroup of G.*

<u>Proof:</u> For each $\bar{\alpha} \in \bar{\Omega}$, let $M_{\bar{\alpha}} = \{g \in G: g \in G_\Delta$ for some open interval Δ with $\bar{\alpha} \in \bar{\Delta}\}$. We will prove that $\{M_{\bar{\alpha}}: \bar{\alpha} \in \bar{\Omega}\}$ is the set of minimal prime subgroups of G. Clearly, each $M_{\bar{\alpha}}$ is a convex ℓ-subgroup of G. Let $f,g \in G$ with $f \wedge g = e$. If $f \notin G_{\bar{\alpha}}$, then $g \in G_\Delta$ where $\bar{\Delta} = (\bar{\alpha}f^{-1},\bar{\alpha}f)$, so $g \in M_{\bar{\alpha}}$. Thus if $M_{\bar{\alpha}}$ is not prime, there are $f,g \in G_{\bar{\alpha}}\backslash M_{\bar{\alpha}}$ with $f \wedge g = e$. Since f and g have only finitely many supporting intervals and move points arbitrarily close to $\bar{\alpha}$, $\bar{\alpha}$ is the endpoint of a supporting interval of f and of a supporting interval of g. But as $f \wedge g = e$, this contradicts the hypothesis on $\bar{\alpha}$. Hence $M_{\bar{\alpha}}$ is prime.

If P is prime and $P \subseteq G_{\bar{\alpha}}$, let $e < g \in M_{\bar{\alpha}}$; say $\bar{\alpha} \in \bar{\Delta}$ with $g \in G_\Delta$. By Corollary 1.10.6, there is $e < h \in G$ with $\bar{\alpha} \in \mathrm{supp}(h) \subseteq \bar{\Delta}$. Then $h \notin G_{\bar{\alpha}}$ so $h \notin P$. Since $g \wedge h = e$, $g \in P$; i.e., $M_{\bar{\alpha}} \subseteq P$.

If P is prime and $P \nsubseteq G_{\bar{\alpha}}$ for all $\bar{\alpha} \in \bar{\Omega}$, then $P = G$ by Lemma 7.1.3. Hence $\{M_{\bar{\alpha}}: \bar{\alpha} \in \bar{\Omega}\}$ is the set of minimal prime subgroups of G.

Note that $M_{\bar{\beta}} \subseteq G_{\bar{\alpha}}$ only if $\bar{\beta} = \bar{\alpha}$ since (G,Ω) has the interval support property. By the previous two paragraphs, each proper prime subgroup of G contains a unique minimal prime.

Proof of Theorem 11G: Let $\Omega = {}^{\omega_1}2$, the set of all functions from ω_1 into $\{0,1\}$. Ω is a chain under the lexicographic order: $\alpha < \beta$ if $(\exists\, \mu \in \omega_1)[\alpha(\mu) < \beta(\mu)\ \&\ \alpha(\nu) = \beta(\nu)$ if $\nu < \mu]$. Ω has largest element $\underline{1}$ and smallest element $\underline{0}$. If $\mu \in \omega_1$, $p \in {}^{\mu}2$ and $\alpha \in {}^{\omega_1}2$, $p\alpha$ will denote the element $\beta \in \Omega$ defined by: $\beta(\nu) = \begin{cases} p(\nu) & \text{if } \nu < \mu \\ \alpha(\lambda) & \text{if } \nu = \mu + \lambda\ (\lambda \in \omega_1) \end{cases}$. Let $\lambda,\mu \in \omega_1$, $p \in {}^{\lambda}\omega_1$ and $q \in {}^{\mu}\omega_1$. A partial map f from $[p\underline{0},p\underline{1}]$ onto $[q\underline{0},q\underline{1}]$ is said to be linear if for each $\alpha \in \Omega$, $(p\alpha)f = q\alpha$. Note the ordinal map $[\lambda,\omega_1) \to [\mu,\omega_1)$ given by $\lambda + \nu \mapsto \mu + \nu$ is the identity for all $\nu \geq \max\{\lambda^2,\mu^2\}$. We now show that if f_1,f_2 are linear maps, $\alpha,\beta \in \mathrm{dom}(f_1) \cap \mathrm{dom}(f_2)$, and $\alpha f_1 \leq \alpha f_2\ \&\ \beta f_1 \geq \beta f_2$, then there is a non-trivial interval Δ between α and β such that f_1 and f_2 agree on Δ.

Let $f_i\colon [p_i\underline{0},p_i\underline{1}] \to [q_i\underline{0},q_i\underline{1}]$ with $p_i \in {}^{\lambda_i}2$, $q_i \in {}^{\mu_i}2$ $(i = 1,2)$. If $\alpha,\beta \in \mathrm{dom}(f_1) \cap \mathrm{dom}(f_2)$, then $p_1 \subseteq p_2$ or $p_2 \subseteq p_1$. Without loss of generality, $p_1 \subseteq p_2$, and we write $p_2 = p_1 r$. So $[p_2\underline{0},p_2\underline{1}] \subseteq [p_1\underline{0},p_1\underline{1}]$. Let $\alpha = p_2\alpha_1$ and $\beta = p_2\beta_1$. If $\alpha f_1 \leq \alpha f_2$ and $\beta f_2 \leq \beta f_1$, $q_1 r\alpha_1 \leq q_2\alpha_1$ and $q_2\beta_1 \leq q_1 r\beta_1$. Again $q_1 r \subseteq q_2$ or $q_2 \subseteq q_1 r$. By the observation about ordinal maps in the previous paragraph, in either case we get the desired conclusion as is easily checked.

Also observe that any non-trivial interval of Ω has size $\aleph_1$.

Now let $G = \{g \in B(\Omega)\colon$ for some $m \in \mathbb{Z}^+$, $\Omega = \bigcup_{j=1}^{m} [p_j\underline{0},q_j\underline{1}]$ and $g|[p_j\underline{0},q_j\underline{0}]$ is the restriction of a linear map $(1 \leq j \leq m)\}$, the bounded "piecewise linear" order-preserving permutations of Ω. Then G is an ℓ-subgroup of $B(\Omega)$ and by Lemma 1.10.2, $(G,\alpha_0 G)$ is a doubly transitive ℓ-permutation group if $\alpha_0 \in \Omega$ with $\alpha_0 \neq \underline{0},\underline{1}$. If g_1,g_2 are restrictions of linear maps on $[p_i\underline{0},q_i\underline{1}] \subseteq [r_i\underline{0},r_i\underline{1}]$ $(i = 1,2)$ and for some $\alpha,\beta \in [p_1\underline{0},q_1\underline{1}] \cap [p_2\underline{0},q_2\underline{1}]$, $\alpha g_1 \leq \alpha g_2$ and $\beta g_2 \leq \beta g_1$, then g_1

and g_2 agree on an interval between α and β (by the previous paragraph). Hence $(G,\alpha_0 G)$ satisfies the hypotheses of Lemma 11.7. Consequently, G satisfies Theorem 11G.

PART V

THE AUTHOR'S PREROGATIVE

CHAPTER 12
ALGEBRAICALLY CLOSED LATTICE-ORDERED GROUPS

A field F is algebraically closed if every polynomial with coefficients in F has a root (zero) in F. Any algebraically closed field F satisfies Hilbert's Nullstellensatz: If $p_1(\underset{\sim}{x}),\ldots,p_m(\underset{\sim}{x})$ are polynomials with coefficients in F that have a common root in some field $H \supseteq F$, then $p_i(\underset{\sim}{x})$ $(1 \leq i \leq m)$ have a common root in F (It is assumed here that $m \in \mathbb{Z}^+$).

If G is an ℓ-subgroup of an ℓ-group H, we will write $\underline{G \leq H}$. An ℓ-group G is said to be *algebraically closed* if whenever $u_1(\underset{\sim}{x},\underset{\sim}{g}),\ldots,u_m(\underset{\sim}{x},\underset{\sim}{g})$ are ℓ-group words (with parameters $\underset{\sim}{g} \in G$), and for some $H \geq G$ and $\underset{\sim}{h} \in H$, $u_i(\underset{\sim}{h},\underset{\sim}{g}) = e$ $(1 \leq i \leq m)$, then there are $\underset{\sim}{f} \in G$ such that $u_i(\underset{\sim}{f},\underset{\sim}{g}) = e$ $(1 \leq i \leq m)$.

The following theorem is easy (and nothing whatever to do with ℓ-groups!).

THEOREM 12A. *If G_0 is any ℓ-group, there is an algebraically closed ℓ-group $G \geq G_0$.*

A field F is *existentially closed* if for any finite sets of polynomials $p_i(\underset{\sim}{x})$, $q_j(\underset{\sim}{x})$ $(i \in I,\ j \in J)$ with coefficients in F, if there is a field $H \supseteq F$ and $\underset{\sim}{h} \in H$ such that $p_i(\underset{\sim}{h}) = 0 \neq q_j(\underset{\sim}{h})$ $(i \in I,\ j \in J)$, then there are $\underset{\sim}{f} \in F$ such that $p_i(\underset{\sim}{f}) = 0 \neq q_j(\underset{\sim}{f})$ $(i \in I,\ j \in J)$. Since $y \neq 0$ is equivalent (for fields) to $yz = 1$ has a solution, any algebraically closed field is existentially closed.

An ℓ-group G is said to be *existentially closed* if for any finite sets $u_i(\underset{\sim}{x},\underset{\sim}{g})$, $v_j(\underset{\sim}{x},\underset{\sim}{g})$ $(i \in I,\ j \in J)$ of ℓ-group words with parameters $\underset{\sim}{g} \in G$, if there is an ℓ-group $H \geq G$ and $\underset{\sim}{h} \in H$ such that $u_i(\underset{\sim}{h},\underset{\sim}{g}) = e \neq v_j(\underset{\sim}{h},\underset{\sim}{g})$ $(i \in I,\ j \in J)$, then there are

$\underset{\sim}{f} \in G$ such that $u_i(\underset{\sim}{f},\underset{\sim}{g}) = e \neq v_j(\underset{\sim}{f},\underset{\sim}{g})$ $(i \in I,\ j \in J)$.

THEOREM 12B. *If* $G \neq \{e\}$ *is an algebraically closed* ℓ*-group, then it is existentially closed.*

Our main aim (besides establishing Theorem 12B) is to use our previous material (especially Chapters 2, 5, 6 and 10) to establish some properties of algebraically closed ℓ-groups. An ℓ-group G is said to be *laterally complete* if $\bigvee_{i \in I} g_i \in G$ whenever $\{g_i : i \in I\} \subseteq G$ with $g_i \wedge g_j = e$ if $i \neq j$.

THEOREM 12C. *Let* $G \neq \{e\}$ *be an algebraically closed* ℓ*-group. Then* G *is simple and divisible. Every element of* G *is a commutator--indeed, every element is conjugate to its square. Any two strictly positive elements of* G *are conjugate. If* $e < g \in G$, *there exist* $f_1, f_2 \in G$ *with* $f_1 \wedge f_2 = e$ *and* $e < f_i < g$ $(i = 1,2)$. *The* ℓ*-group variety generated by* G *is* $\mathcal{L}$. G *is neither finitely generated nor finitely related and contains no finite maximal pairwise disjoint set of elements. Moreover,* G *is not laterally complete.*

THEOREM 12D. *There is an algebraically closed* ℓ*-group* G *and* $e < g \in G$ *such that if* $X \subseteq \mathbb{Z}^+$, *there is* $e < g_X \in G$ *with* $g^{-n} g_X g^n \wedge g_X = e$ *if and only if* $n \in X$.

Finally in this chapter we prove a result about representations of algebraically closed ℓ-groups. By Theorem 12C and Appendix I Corollary 8, any algebraically closed ℓ-group has a transitive representation.

THEOREM 12E. *If* (G,Ω) *is a transitive primitive* ℓ*-permutation group with* $G \neq \{e\}$ *algebraically closed, then* (G,Ω) *is pathological. Moreover, there are algebraically closed* ℓ*-groups having pathological representations.*

All the results in this chapter (with the exception of the last part of Theorem 12E which is due to O. H. Kegel) are due to Glass and Pierce [80c].

We first prove Theorem 12A: The proof is cheap universal algebra. List all finite conjunctions of $u(\underset{\sim}{x},\underset{\sim}{g}) = e$, u an ℓ-group word and $\underset{\sim}{g} \in G_0$: $\theta_0(\underset{\sim}{x}),\theta_1(\underset{\sim}{x}),\ldots,\theta_\lambda(\underset{\sim}{x}),\ldots$ $(\lambda \in \kappa = |G_0|)$. We define G_λ $(\lambda \in \kappa)$ by induction: If $\theta_\mu(\underset{\sim}{x})$ has a solution in some ℓ-group $H \geq G_\mu$, let $G_{\mu+1}$ be the ℓ-subgroup of H generated by G_μ and a solution. For $\lambda \in \kappa$ a limit ordinal, let $G_\lambda = \bigcup_{\mu<\lambda} G_\mu$. Now let $G_0^\flat = \bigcup_{\lambda<\kappa} G_\lambda$. Then for any finite set of ℓ-group words $u_i(\underset{\sim}{x},\underset{\sim}{g})$ $(i \in I)$ with parameters $\underset{\sim}{g} \in G_0$, if $u_i(\underset{\sim}{x},\underset{\sim}{g}) = e$ $(i \in I)$ has a solution in some ℓ-group $H \geq G_0^\flat$, it has a solution in $G_0^\flat$. Let $G(0) = G_0$ and $G(n+1) = G_n^\flat$. If $G = \bigcup_{n\in\omega} G(n)$, then $G \geq G_0$ and G is algebraically closed (If $u_i(\underset{\sim}{x},\underset{\sim}{g})$ $(i \in I)$ is a finite set of ℓ-group words with $\underset{\sim}{g} \in G$, then $\underset{\sim}{g} \in G(m)$ for some $m \in \omega$. Hence if $u_i(\underset{\sim}{x},\underset{\sim}{g}) = e$ $(i \in I)$ has a solution in some ℓ-group $H \geq G$ $(\geq G(m))$, then it has a solution in $G(m)^\dagger = G(m+1) \leq G$). This proves Theorem 12A.

We next prove all but the last four sentences of Theorem 12C. Let $G \neq \{e\}$ be algebraically closed. By Corollary 2K, every ℓ-group can be ℓ-embedded in a divisible ℓ-group. So if $g \in G$ and $m \in \mathbb{Z}^+$, $x^m = g$ has a solution in some extension of G. Hence it has a solution in G ($x^m = g$ is the same as $x^m g^{-1} = e$). Thus G is divisible. Moreover, by Corollary 2L, $G \leq B(\Omega)$ for some doubly homogeneous Ω. Let $g \in G$. By Corollary 2.2.6, $g^2 = x^{-1}gx$ has a solution in $A(\Omega)$. Hence it has one in G; i.e., every element of G is conjugate to its square. Thus every element of G is a commutator. Therefore, by Lemma 6.4, if $g,h \in G$ with $g \neq e$, there exist $f_i \in G$ $(1 \leq i \leq 4)$ such that $h = f_1^{-1}g^{-1}f_1\cdot f_2^{-1}gf_2\cdot f_3^{-1}g^{-1}f_3\cdot f_4^{-1}gf_4$. Hence h belongs to the normal subgroup of G generated by g. It follows that G is simple.

By Theorem 10B, G can be ℓ-embedded in an ℓ-group H in which any two strictly positive elements of H are conjugate.

Hence if $e < g_1, g_2 \in G$, $x^{-1}g_1x = g_2$ has a solution in an extension of G and so in G.

To complete the proof of Theorem 12C we must first establish Theorem 12B. We need two lemmas.

LEMMA 12.1. *Let G be an ℓ-group, $e < g_0 \in G$ and $\underset{\sim}{g} \in G$. Suppose $u_i(\underset{\sim}{x},\underset{\sim}{g})$ and $v_j(\underset{\sim}{x},\underset{\sim}{g})$ $(i \in I,\ j \in J)$ are ℓ-group words, where I and J are finite. Assume there is an ℓ-group $H \geq G$ satisfying for some $\underset{\sim}{h} \in H$*

(1) $u_i(\underset{\sim}{h},\underset{\sim}{g}) = e$ $(i \in I)$, and

(2) $v_j(\underset{\sim}{h},\underset{\sim}{g}) \neq e$ $(j \in J)$.

Then there is an ℓ-group $\hat{G} \geq H$ in which (1)-(5) are satisfied where

(3) $z_j^2 = v_j(\underset{\sim}{h},\underset{\sim}{g})^{-1} z_j v_j(\underset{\sim}{h},\underset{\sim}{g})$ $(j \in J)$

(4) $y_j^{-1} z_j y_j = z$ $(j \in J)$

(5) $y^{-1} z g_0 y = z$.

Proof: Assume the hypotheses and write v_j for $v_j(\underset{\sim}{h},\underset{\sim}{g})$ $(j \in J)$. By Corollary 2L, we can ℓ-embed G in $B(\Omega)$ for some doubly homogeneous Ω. Define $e < f_j \in B(\Omega)$ by: $\alpha f_j = \alpha$ if $\alpha v_j = \alpha$. If $\alpha < \alpha v_j$, choose $\gamma \in (\alpha, \alpha v_j)$. There exists $a_j \in A(\Omega)^+$ such that $\gamma \in \operatorname{supp}(a_j) \subseteq (\alpha, \alpha v_j)$. Let $f_{j,0} = a_j$ and define $f_{j,n}$ inductively $(n \in \mathbb{Z}^+)$ to satisfy $f_{j,n+1}^2 = v_j^{-1} f_{j,n} v_j$ (By Theorem 2E, $B(\Omega)$ is divisible so $f_{j,n+1}$ exists in $B(\Omega)$). For $n \leq 0$, we let $f_{j,n+1} = v_j f_{j,n}^2 v_j^{-1}$. Now for $\beta \in \Delta(\alpha, v_j)$, say $\beta \in [\alpha v_j^n, \alpha v_j^{n+1})$, let $\beta f_j = \beta f_{j,n}$. Similarly, if $\alpha v_j < \alpha$. This gives $e < f_j \in B(\Omega)$ such that $f_j^2 = v_j^{-1} f_j v_j$ $(j \in J)$ so (3) holds. By Theorem 10B, we can ℓ-embed $B(\Omega)$ in $\hat{G}$ an ℓ-group in which any two strictly positive elements are conjugate. So if $e < f \in \hat{G}$, then $fg_0 > e$ and (4) and (5) have a solution in $\hat{G}$.

The following lemma is obvious.

LEMMA 12.2. *Let* $e < g_0 \in G$ *and let the* ℓ*-group* $L \geq G$ *satisfy (3), (4) and (5); then any solution satisfies (2).*

We now prove Theorem 12B: Assume $G \neq \{e\}$ is algebraically closed and that (1) and (2) are satisfied in some ℓ-group $H \geq G$. By Lemma 12.1, there is an ℓ-group $\hat{G} \geq H$ such that (1)-(5) are satisfied in $\hat{G}$. Since G is algebraically closed, it satisfies (1), (3), (4) and (5). By Lemma 12.2, G also satisfies (2). Hence G satisfies (1) and (2). Consequently, G is existentially closed.

We now return to the proof of Theorem 12C. We may assume that G is existentially closed. Since G is simple, it has a transitive representation (G,Ω) as an ℓ-permutation group (Appendix I Corollary 8). Let $(H_0,T_0) = (\mathbb{Z},\mathbb{Z})$ Wr (G,Ω) and ℓ-embed G in H_0 via $g \mapsto (\{0\},g)$. Let $e < g \in G$ and $\alpha \in \text{supp}(g)$. Let $a = (\{\hat{a}_\beta\},e)$, $b = (\{\hat{b}_\beta\},e) \in H_0$, where

$$\hat{a}_\beta = \begin{cases} 1 & \text{if } \beta = \alpha \\ 0 & \text{if } \beta \neq \alpha \end{cases}, \quad \hat{b}_\beta = \begin{cases} 1 & \text{if } \beta = \alpha g \\ 0 & \text{if } \beta \neq \alpha g \end{cases}.$$

Then $a \wedge b = e$, $a,b \neq e$, $a,b \leq g$ (i.e., $a \vee g = g = b \vee g$). Actually, $g^{-1}ag = b$. Thus, in G, there are f_1,f_2 with $f_1 \wedge f_2 = e$ and $e < f_1,f_2 < g$ (G is existentially closed); let $(H_1,T_1) = (G,\Omega)$ Wr $(\mathbb{Z},\mathbb{Z})$ and ℓ-embed G in H_1 via: $g \mapsto g^*$ where

$$(\alpha,n)g^* = \begin{cases} (\alpha g,0) & \text{if } n = 0 \\ (\alpha,n) & \text{if } n \neq 0 \end{cases}.$$

If $\{g_i\colon 1 \leq i \leq m\}$ is a pairwise disjoint set of positive elements of G, let $h \in H_1$ be defined by: $\bar{h} = 0$,

$$\hat{h}_n = \begin{cases} 1 & \text{if } n = 1 \\ 0 & \text{if } n \neq 1 \end{cases}.$$

Then $h = (\{\hat{h}_n\},\bar{h}) \neq e$ and $h \wedge g_i = e$ $(1 \leq i \leq m)$. Thus, in G, we can find $f \neq e$ with $f \wedge g_i = e$ $(1 \leq i \leq m)$. Hence $\{g_i\colon 1 \leq i \leq m\}$ is not a maximal pairwise disjoint subset of G.

Let $w(\underset{\sim}{x}) = e$ fail in some ℓ-group H_2. Then for some $\underset{\sim}{f} \in G \oplus H_2$, $w(\underset{\sim}{f}) \neq e$. Since $G \leq G \oplus H_2$, $w(\underset{\sim}{f}') \neq e$ for some $\underset{\sim}{f}' \in G$. Thus the ℓ-group variety generated by G is $\mathcal{L}$.

Let $H_3 = G \oplus \mathbb{Z}$. If $g_1,\ldots,g_m \in G$, then there is $e \neq h \in H_3$ which commutes with each $(g_i,0)$ $(1 \leq i \leq m)$. Thus $C(\{g_1,\ldots,g_n\}) \neq \{e\}$, where $C(X) = \{f \in G: (\forall x \in X)(xf = fx)\}$. But if $g_1,\ldots,g_n$ generate G, $C(\{g_1,\ldots,g_n\})$ is the centre of G. Since G is non-Abelian (the ℓ-group variety G generates is $\mathcal{L}$) and simple, its centre is $\{e\}$. Consequently, G is not finitely generated. Therefore, if G were finitely related, infinitely many generators of G would not occur in the finite number of defining relations. Hence they would generate a free ℓ-group F which is a free factor of G; say G is the ℓ-group free product of F and H_4. Let $g \neq e$ be one of the generators of F. Since g is conjugate to g^2, there is $x \in G$ such that $g^2 = x^{-1}gx$. Let $L = H_4 \oplus \mathbb{Z}$ and map G onto L via ψ, where $\psi|H_4$ is the identity function, $g\psi = (e,1)$ and $f\psi = (e,0)$ for all other generators f of F. This is an ℓ-homomorphism so $(g\psi)^2 = (x\psi)^{-1}(g\psi)(x\psi)$. But $g\psi$ is in the centre of L. Hence $(g\psi)^2 = g\psi$; i.e., $(e,2) = (e,1)$, a contradiction.

To complete the proof of Theorem 12C, we need a lemma.

<u>LEMMA</u> <u>12.3</u>. *If G is an ℓ-group and $\{g_i: i \in I\}$ is an infinite pairwise disjoint set of elements with supremum f existing in G, then there is an ℓ-group $H \geq G$ in which $\{g_i: i \in I\}$ has a supremum $x < f$.*

<u>Proof:</u> Let C be the convex ℓ-subgroup of G generated by $\{g_i: i \in I\}$. Since $f \notin C$, there is a convex ℓ-subgroup $P \supseteq C$ maximal with respect to not containing f (by Zorn's lemma). By Appendix I Lemmas 2 and 3, $R(P)$ is a chain. Now (G,Ω) is an ℓ-permutation group for some chain Ω. We can extend the action of G to $\Omega \overset{\leftarrow}{\cup} R(P)$ via: $(Ph)g = Phg$ $(h,g \in G)$. Hence $(G,\Omega \overset{\leftarrow}{\cup} R(P))$ is an ℓ-permutation group. Each g_i fixes $P \in \Omega \overset{\leftarrow}{\cup} R(P)$ but f does not. By Theorem 8B, $A(\Omega \overset{\leftarrow}{\cup} R(P))_P$ is closed. Therefore f is not the supremum of $\{g_i: i \in I\}$ in $A(\Omega \overset{\leftarrow}{\cup} R(P))$. Since

$\{g_i: i \in I\}$ is a pairwise disjoint subset of $A(\Omega \overset{\leftarrow}{\cup} R(P))$, it has a supremum $x \in A(\Omega \overset{\leftarrow}{\cup} R(P))$.

We now complete the proof of Theorem 12C. Note that lateral completeness is not something we would expect to capture. In $A(\mathbb{R})$, let $e < f_0$ with $\mathrm{supp}(f_0) = (0,1)$ and f_1 be translation by $+1$. Then $\{f_1^{-n} f_0 f_1^n : n \in \mathbb{Z}^+\}$ is a pairwise disjoint set. Let f_2 be its supremum in $A(\mathbb{R})$. Then $f_2 \wedge f_0 = e$, $f_1^{-1} f_0 f_1 \le f_2$ and $f_1^{-1} f_2 f_1 \le f_2$. Hence $(\exists g_0 > e)(\exists g_1 > e)(\exists g_2)(g_2 \wedge g_0 = e \ \& \ g_1^{-1} g_0 g_1 \le g_2 \ \& \ g_1^{-1} g_2 g_1 \le g_2)$ holds in $G \oplus A(\mathbb{R})$. Thus, if G is algebraically closed, there are $g_0, g_1, g_2 > e$ in G with $g_2 \wedge g_0 = e$, $g_1^{-1} g_0 g_1 \le g_2$ and $g_1^{-1} g_2 g_1 \le g_2$. It follows by induction on $n \in \mathbb{Z}^+$ that $g_1^{-n} g_0 g_1^n \le g_2$. Hence $\{g_1^{-m} g_0 g_1^m : m \in \mathbb{Z}\}$ is an infinite pairwise disjoint set. If G were laterally complete, it would have a supremum, f say. By Lemma 12.3, there is $H \ge G$ satisfying: $\{g_1^{-m} g_0 g_1^m : m \in \mathbb{Z}\}$ has a supremum $x < f$. Clearly $(*)$ $x < f$, $g_0 \le x$ and $x g_1 = g_1 x$. Since G is existentially closed, there is $y \in G$ satisfying $(*)$. Now $g_1^{-n} g_0 g_1^n \le g_1^{-n} y g_1^n = y < f$ for all $n \in \mathbb{Z}$, contradicting $f = \bigvee_{n \in \mathbb{Z}} g_1^{-n} g_0 g_1^n$ in G.

We next prove a lemma reminiscent of Lemma 2.1.10. Note that $f \in \langle g \rangle$ is not a first order concept but $f \in C(C(g))$ is: $(\forall x)(xg = gx \rightarrow xf = fx)$.

<u>LEMMA</u> <u>12.4</u>. *If G is algebraically closed and $e < g \in G$, then $\langle g \rangle^+ = C(C(g))^+$.*

<u>Proof:</u> The proof of Lemma 2.1.10 shows that if $e < f \in C(C(g))$, then $\mathrm{supp}(f) \subseteq \mathrm{supp}(g)$ and for each $\alpha \in \mathrm{supp}(g)$, there is $n(\alpha) \in \mathbb{Z}^+$ such that $f|\Delta(\alpha,g) = g^{n_\alpha}|\Delta(\alpha,g)$. Suppose $\alpha, \beta \in \mathrm{supp}(g)$ and $n_\alpha \ne n_\beta$. By Theorem 10.8, there is $H \ge G$ and $x \in H$ such that $x^{-1} g x = g$ and $\alpha x = \beta$. Hence $\alpha f x = \alpha g^{n_\alpha} x = \alpha x g^{n_\alpha} = \beta g^{n_\alpha} \ne \beta g^{n_\beta} = \beta f = \alpha x f$ so $xf \ne fx$. Since

G is existentially closed, there is $x \in C(g) \cap G$ such that $xf \neq fx$. This contradicts $f \in C(C(g))$. Hence $n_\alpha = n_\beta$ for all $\alpha,\beta \in \mathrm{supp}(g)$; i.e., $f = g^n$ for some $n \in \mathbb{Z}^+$.

We now prove Theorem 12D. We first establish:

<u>LEMMA</u> <u>12.5</u>. *Let* $X \subseteq \mathbb{Z}^+$. *Then there exist* $e < g_0, g_1 \in A(\mathbb{R})$ *such that if* $n \in \mathbb{Z}^+$, $g_1^{-n} g_0 g_1^n \wedge g_0 = e$ *if and only if* $n \in X$.

<u>Proof</u>:

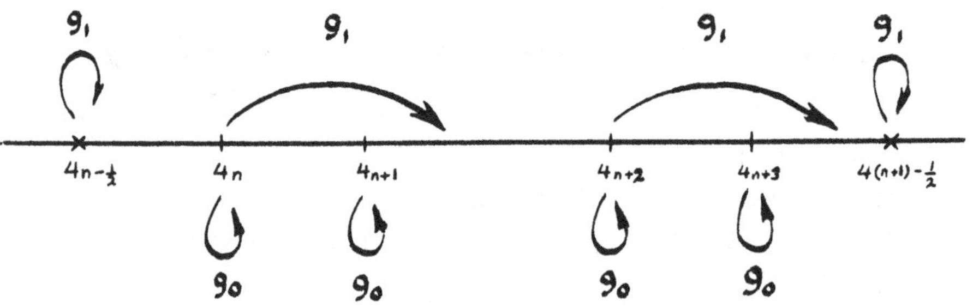

Enumerate $\mathbb{Z}^+\backslash X$ in increasing order, say $m_0 < m_1 < m_2 < \ldots$. Let $e < g_0 \in A(\mathbb{R})$ with $\mathrm{supp}(g_0) = \bigcup_{n=0}^{\infty}(2n, 2n+1)$. Define $e < g_1 \in A(\mathbb{R})$ so that, for all $n \in \omega$, $(4n - \frac{1}{2})g_1 = 4n - \frac{1}{2}$, $(4n)g_1 > 4n + 1$, $(4n+2)g_1 > 4n + 3$, $(4n)g_1^{m_n} = 4n + 2$ and $(4n+1)g_1^{m_n} = 4n + 3$; if $\mathbb{Z}^+\backslash X$ is finite so m_n doesn't exist, change the last part to $(4n)g_1 \geq 4n + 3$. Note that if $k \neq m_n$, then $g_1^{-k} g_0 g_1^k \wedge g_0 | (4n, 4(n+1)) = e|(4n, 4(n+1))$; but $g_1^{-k} g_0 g_1^k \wedge g_0 | (4n, 4(n+1)) \neq e|(4n, 4(n+1))$ if $k = m_n$. Hence $g_1^{-k} g_0 g_1^k \wedge g_0 \neq e$ if and only if $k \in \mathbb{Z}^+\backslash X$.

By Theorem 12A, there is an algebraically closed ℓ-group $G \geq A(\mathbb{R})$. Let $e < g_1 \in G$. Given $X \subseteq \mathbb{Z}^+$, let $e < f_0, f_1 \in A(\mathbb{R})$ be such that $f_1^{-n} f_0 f_1^n \wedge f_0 = e$ if and only if $n \in X$. But

$e < f_1, g_1 \in G$ so for some $h \in G$, $h^{-1}f_1h = g_1$. Then, for $g_X = h^{-1}f_0h$, $g_1^{-n}g_Xg_1^n \wedge g_X = e$ if and only if $n \in X$. Thus Theorem 12D is proved.

We next prove Theorem 12E. Let (G,Ω) be a transitive primitive ℓ-permutation group with $G \neq \{e\}$ algebraically closed. Let $e < g \in G$. ℓ-embed G in $\prod_{n \in \mathbb{Z}} G$ diagonally and thereby ℓ-embed G in H, where $(H,T) = (G,\Omega)\ \mathrm{Wr}\ (\mathbb{Z},\mathbb{Z})$. Let $g_1 \in G$. Now, in H, we can find $x > e$ such that $x^{-1}g_1^{-1}gg_1x \wedge g \neq e$; so we can find such an x in G. By Lemma 6.1, G does not have bounded support. Since any two strictly positive elements of G are conjugate, (G,Ω) is doubly transitive (If (G,Ω) is periodic with period z, and $\alpha \in \Omega$, f and h are not conjugate where $\alpha f = \alpha$ and $\alpha h \geq \alpha z$ (so H fixes no point of $\bar{\Omega}$)). Thus (G,Ω) is pathological.

We now prove the converse. The concept of existentially closed applies equally to ℓ-permutation groups. The proof of Theorem 12A can be easily adapted to show that existentially closed ℓ-permutation groups exist. Now if (G,Ω) is an existentially closed ℓ-permutation group, it is clearly doubly transitive (use Corollary 2L). Let H be an ℓ-group with $H \geq G$ such that H satisfies a particular finite set of ℓ-group equations and inequations (with parameters from G). By Appendix I Theorem 11, there is a chain $T \supseteq \Omega$ such that (H,T) is an ℓ-permutation group and (G,Ω) can be ℓ-embedded in (H,T). Hence $(G,\Omega) \leq (H,T)$. Thus the finite set of ℓ-group equations and inequations hold in G (since they hold in (H,T) and so in (G,Ω)). Thus, for any existentially closed ℓ-permutation group (G,Ω), G is an existentially closed ℓ-group. Thus (G,Ω) is pathological and the proof of Theorem 12E is complete.

Finally we prove the following easy consequence of Lemma 12.4 which in model-theoretic language says that the theory of ℓ-groups has no model companion.

THEOREM 12.6. *If $G \neq \{e\}$ is an algebraically closed ℓ-group, there is an ℓ-group $H \equiv G$ such that H is not algebraically closed.*

Proof: Let $\mathcal{U}$ be any non-principal ultrafilter on $\mathbb{Z}^+$ and $H = \prod_{n \in \mathbb{Z}^+} G/\mathcal{U}$, an ultrapower of G; i.e., H is the quotient of the ℓ-group $\prod_{n \in \mathbb{Z}^+} G$ by the ℓ-ideal $N = \{f \in \prod_{n \in \mathbb{Z}^+} G : \{n \in \mathbb{Z}^+ : f(n) = e\} \in \mathcal{U}\}$. By Łos' theorem (see Chang and Keisler [73]), $H \equiv G$. Let $e < g \in G$ and let $h = Nh'$ where $h'(n) = g^n$ $(n \in \mathbb{Z}^+)$. Then h does not belong to the subgroup of H generated by $\hat{g}$, where $\hat{g} = N\bar{g}$ and $\bar{g}(n) = g$ $(n \in \mathbb{Z}^+)$. However, $h'x = xh'$ whenever $\hat{g}x = x\hat{g}$ (in H) since this holds at each coordinate in $\prod_{n \in \mathbb{Z}^+} G$. By Lemma 12.4, H is not algebraically closed.

For further results about algebraically closed ℓ-groups, see Glass and Pierce [80a] and [80c]; the Abelian case is considered in Glass and Pierce [80b].

CHAPTER 13
THE WORD PROBLEM FOR LATTICE-ORDERED GROUPS

In this last chapter, we prove a result due to Glass and Gurevich [81]:

THEOREM 13A. *There exists a finitely presented lattice-ordered group with insoluble word problem.*

A *presentation* of an ℓ-group with generators x_i $(i \in I)$, and relations $r_j(\underset{\sim}{x}) = e$ $(j \in J)$ is the quotient of the free ℓ-group F on $\{x_i: i \in I\}$ by the ℓ-ideal (normal convex ℓ-subgroup) of F generated by the elements $\{r_j(\underset{\sim}{x}): j \in J\}$. If I and J are both finite we will say that the ℓ-group is *finitely presented* and write it $(x_i$ $(i \in I); r_j(\underset{\sim}{x}) = e$ $(j \in J))$.

We will obtain a certain primitive recursive set of relations R, involving only a finite set of letters correlated with elements of $A(\mathbb{R})$, such that the subset S of R comprising those members of R which hold in $A(\mathbb{R})$ is non-recursive. We will then show that there is a finite set of relations S_0 which hold in $A(\mathbb{R})$ and imply S. S_0 will be constructed from a finite set of letters X_0, consisting of the letters which appear in R and new elements correlated with additional elements of $A(\mathbb{R})$. A member of R will then be implied by S_0 if and only if it belongs to S. Hence the finitely presented ℓ-group $(X_0;S_0)$ will have insoluble word problem; i.e., there is no algorithm to decide whether or not an arbitrary ℓ-group word in the generators X_0 is equal to e in $(X_0;S_0)$.

We will use $\underline{^\omega\omega}$ for the set of functions from ω into ω. In contradistinction to our notation for permutation groups, we will write the elements of $^\omega\omega$ on the left; so if $f,g \in {}^\omega\omega$, fg will be that member of $^\omega\omega$ defined by $fg(n) = f(g(n))$ $(n \in \omega)$.

A function $f \in {}^{\omega}\omega$ is obtained from $g,h,u,v \in {}^{\omega}\omega$ by *general recursion* if (i) $fg = u$, (ii) $fh = vf$, and (iii) each $m \in \omega$ belongs to the range of one of the functions $h^k g$ $(k \in \omega)$.

We will need two results from recursion theory:

(I) Julia Robinson [68]. *The class of recursive functions is the smallest class of numerical functions which is closed under composition and general recursion, and contains the zero function* θ *(*$\theta(n) \equiv 0$*) and the successor function* s *(*$s(n) = n + 1$*).*

For those unfamiliar with recursion theory, (I) can be taken as the definition of a recursive function. A set $A \subseteq \omega$ is *recursive* if there is an algorithm such that given $n \in \omega$, we can decide, on the basis of the algorithm, whether $n \in A$ or $n \notin A$.

(II). *There is a recursive function whose range is not a recursive set.*

See, for example, Hartley Rogers [67] Theorem VI page 62.

In the ensuing, we will write down several relations involving simultaneously conjugating one set of elements to another by a single conjugator. That this is indeed possible (in the absence of the amalgamation property) without all the elements in the sets reducing to e hinges on the techniques of Section 2.2--in especial, see Lemma 2.2.9.

We now discuss the interpretation in $A(\mathbb{R})$.

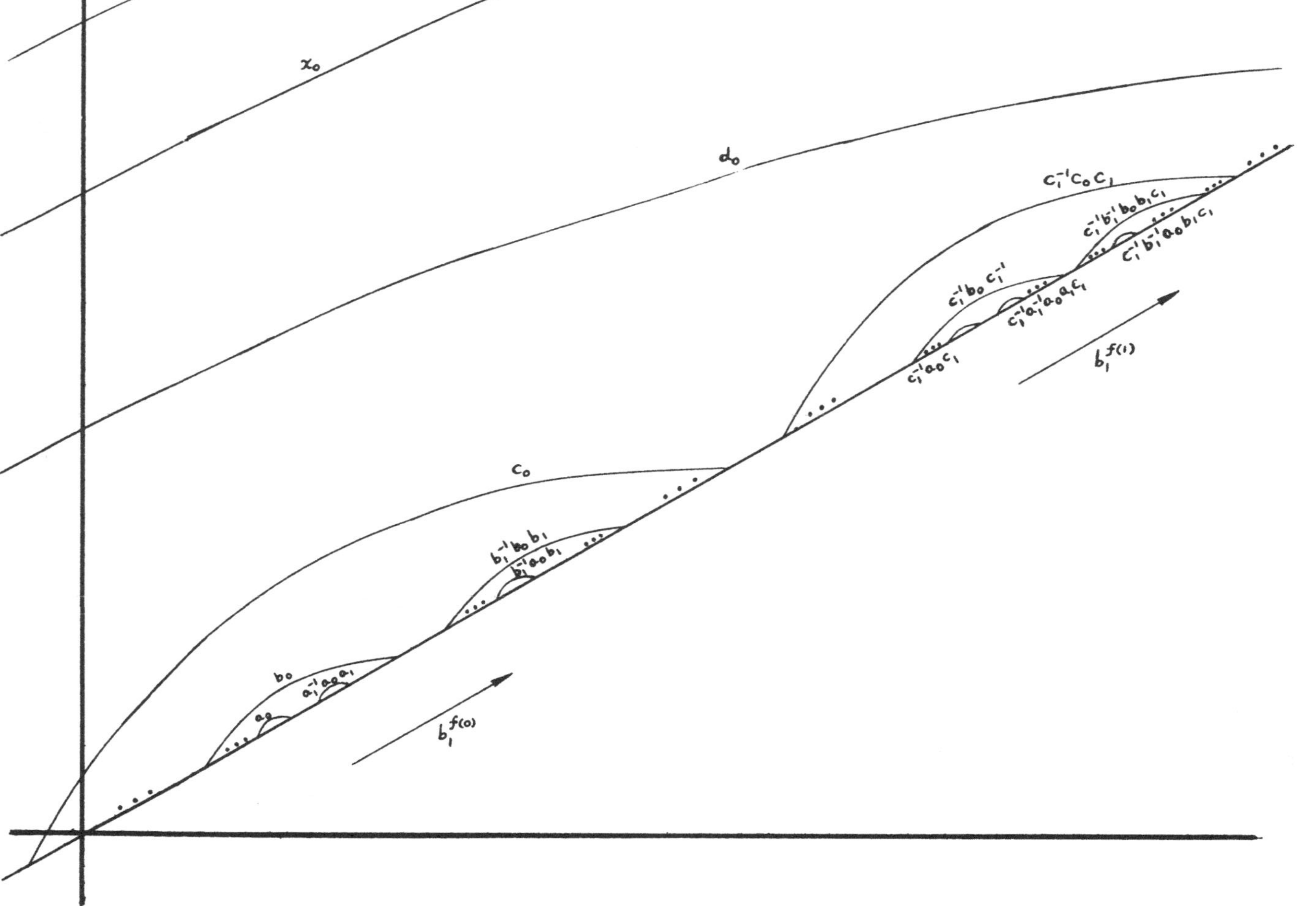
x_0
d_0
$c_1^{-1}c_0c_1$
$c_1^{-1}b_1^{-1}b_0b_1c_1$
$c_1^{-1}b_1^{-1}a_0b_1c_1$
$c_1^{-1}b_0c_1^{-1}$
$c_1^{-1}a_1^{-1}a_0a_1c_1$
$c_1^{-1}a_0c_1$
$b_1^{f(1)}$
c_0
$b_1^{-1}b_0b_1$
$b_1^{-1}a_0b_1$
b_0
$a_1^{-1}a_0a_1$
a_0
$b_1^{f(0)}$

In $A(\mathbb{R})$, there are elements $a_0, b_0, c_0 > e$ each bounded and having one bump, such that:

(a) $a_0 \wedge b_0^{-n} a_0 b_0^n = e = b_0 \wedge c_0^{-n} b_0 c_0^n$ for all $n \in \mathbb{Z}\backslash\{0\}$,

(b) $b_0^{-m} a_0 b_0^m < b_0$ and $c_0^{-m} b_0 c_0^m < c_0$ for all $m \in \mathbb{Z}$,

(c) between $\mathrm{supp}(b_0^{-m} a_0 b_0^m)$ and $\mathrm{supp}(b_0^{-n} a_0 b_0^n)$ there is a non-degenerate interval of $\mathbb{R}$ whenever m and n are distinct integers, and

(d) similarly for $\mathrm{supp}(c_0^{-m} b_0 c_0^m)$ and $\mathrm{supp}(c_0^{-n} b_0 c_0^n)$.

Because of the bump structure, there is $a_3 \in A(\mathbb{R})$ such that $a_3^{-1} a_0 a_3 = b_0$ and $a_3^{-1} b_0 a_3 = c_0$ (by the techniques of Section 2.2). Let $d_0 = a_3^{-1} c_0 a_3$, $x_0 = a_3^{-1} d_0 a_3$ and $y_n = a_3^{-n} x_0 a_3^n$ $(n \in \mathbb{Z}^+)$. Let $W = \{y_1^{n_1} \ldots y_k^{n_k}: k \in \mathbb{Z}^+, n_1, \ldots, n_k \in \mathbb{Z}\}$ and write *x∗y* for $y^{-1}xy$ ($x{*}yz$ is shorthand for $x{*}(yz)$). By (a), (b) and the definition of a_3, $\{x_0{*}w: w \in W\}$ forms a pairwise disjoint set. Let $d_1 \in A(\mathbb{R})$ be the pointwise supremum of it. (Note that $x_0 \wedge d_1 x_0^{-1} = e$--i.e., $a_3^{-1} d_0 a_3 \wedge d_1 (a_3^{-1} d_0 a_3)^{-1} = e$--so $x_0 \vee d_1 x_0^{-1} = d_1$.) Let $c_1 \in A(\mathbb{R})$ be the pointwise supremum of the pairwise disjoint set of elements $\{d_0{*}d_1^n w: n \in \mathbb{Z}, w \in W\}$, $b_1 \in A(\mathbb{R})$ be the pointwise supremum of the pairwise disjoint set of elements $\{c_0{*}c_1^m d_1^n w: m,n \in \mathbb{Z}, w \in W\}$, and $a_1 \in A(\mathbb{R})$ be the pointwise supremum of the pairwise disjoint set $\{b_0{*}b_1^m c_1^n d_1^k w: m,n,k \in \mathbb{Z}, w \in W\}$. By construction, a_1, b_1, c_1 and d_1 each commute with each other, $a_3^{-1} c_1 a_3 = d_1$, $a_3^{-1} b_1 a_3 = c_1$ and $a_3^{-1} a_1 a_3 = b_1$. Also, since $a_1 | \mathrm{supp}(b_0) = b_0 | \mathrm{supp}(b_0)$, $a_1 b_0 = b_0 a_1$. Let a_2 be the pointwise supremum of the pairwise disjoint set $\{a_0{*}a_1^n: n \in \mathbb{Z}^+\}$. So $a_0 \wedge a_2 = e$ and $a_1^{-1} a_2 a_1 \leq a_2$.

Let a_6 and a_7 be the pointwise suprema of the pairwise disjoint sets $\{a_0{*}c_1^m: m \in \omega\}$ and $\{a_0{*}c_1^{-m}: m \in \mathbb{Z}^+\}$. Note that $a_6 \wedge a_7 = e$, $c_1^{-1}(a_6 \vee a_7) c_1 = a_6 \vee a_7$, $a_0 \wedge a_6 a_0^{-1} = e$, $c_1^{-1} a_6 c_1 \leq a_6$, $c_1 a_0 c_1^{-1} \leq a_7$ and $c_1 a_7 c_1^{-1} \leq a_7$.

Because of the bump structure, we can find $a_4, a_5 \in A(\mathbb{R})$ such that $a_4^{-1}a_0a_4 = a_0$, $a_4^{-1}a_1a_4 = b_1$, $a_4^{-1}b_1a_4 = c_1$ and $a_4^{-1}c_1a_4 = d_1$; $a_5^{-1}a_0a_5 = a_0$, $a_5^{-1}a_1a_5 = b_1$, and $a_5^{-1}c_1a_5 = c_1$ (by the techniques of Section 2.2; indeed, the existence of a_5 is specifically Lemma 2.2.9 applied to each bump of c_1).

We now code $f \in {}^{\omega}\omega$ into $A(\mathbb{R})$ via a_f, where $\text{supp}(a_f) \subseteq \bigcup\{\text{supp}(c_0 * c_1^m d_1^n): m \in \omega, n \in \mathbb{Z}\}$ and $a_f|\text{supp}(c_0*c_1^m d_1^n) = b_1^{f(m)}|\text{supp}(c_0*c_1^m d_1^n)$ for all $m \in \omega$ and $n \in \mathbb{Z}$. Now that a_f commutes with a_1, b_1 and d_1, and $a_0*c_1^m a_f = a_0*b_1^{f(m)}c_1^m$ for all $m \in \omega$.

Finally, for technical reasons, we need one extra element $\hat{a} \in A(\mathbb{R})$ defined by: $\text{supp}(\hat{a}) \subseteq \bigcup\{\text{supp}(c_0*c_1^m d_1^n w): m,n \in \mathbb{Z}, w \in W\}$, and $\hat{a}|\text{supp}(c_0*c_1^m d_1^n w) = b_1^{m+1}|\text{supp}(c_0*c_1^m d_1^n w)$ for all $m,n \in \mathbb{Z}$, $w \in W$. Note that $a_s = (\hat{a} \vee e) \wedge a_3^{-1}d_0a_3$, s being the successor function.

We are now ready to prove Theorem 13A.

In $A(\mathbb{R})$ we constructed five "levels" of $\mathbb{Z}$. We now write down a finite presentation which will capture this information.

Let G be the finitely presented ℓ-group with generators a_0, a_1, a_2, a_3, a_4, a_5, a_6 and a_7 and relations: $a_0 \wedge a_2 = e$, $a_1 \geq e$, $a_1^{-1}a_0a_1 \leq a_2$, $a_1^{-1}a_2a_1 \leq a_2$, $a_0 \leq b_0$, $a_1b_0 = b_0a_1$, $a_1b_1 = b_1a_1$, $a_1c_1 = c_1a_1$, $a_1d_1 = d_1a_1$, $a_3^{-1}d_0a_3 \wedge d_1(a_3^{-1}d_0a_3)^{-1} = e$, $a_0a_4 = a_4a_0$, $a_4^{-1}a_1a_4 = b_1$, $a_4^{-1}b_1a_4 = c_1$, $a_4^{-1}c_1a_4 = d_1$, $a_0a_5 = a_5a_0$, $a_5^{-1}a_1a_5 = b_1$, $c_1a_5 = a_5c_1$, $a_6 \wedge a_7 = e$, $c_1^{-1}(a_6 \vee a_7)c_1 = a_6 \vee a_7$, $a_0 \wedge a_6a_0^{-1} = e$, $c_1^{-1}a_6c_1 \leq a_6$, $c_1a_0c_1^{-1} \leq a_7$, $c_1a_7c_1^{-1} \leq a_7$, where $b_j \equiv a_3^{-1}a_ja_3$, $c_j \equiv a_3^{-2}a_ja_3^2$, $d_j \equiv a_3^{-3}a_ja_3^3$ $(j = 0,1)$.

Observe that all of the above relations hold in $A(\mathbb{R})$ under the natural interpretation.

Note that we write $x \leq y$ as a shorthand for $x \wedge y = x$.

We will continue to use $x*y$ for $y^{-1}xy$ and $x*yz$ for $x*(yz)$. Clearly $(x*y)*z = x*yz$.

LEMMA 13.1. *In G, the following facts hold:*

(i) $b_1c_0 = c_0b_1$; $c_1d_0 = d_0c_1$.

(ii) $b_1c_1 = c_1b_1$; $c_1d_1 = d_1c_1$; $b_1d_1 = d_1b_1$.

(iii) $a_0 * a_1^n \leq a_2$ *for all* $n \in \mathbb{Z}^+$.

(iv) $(a_0 * a_1^n) \wedge (a_0 * a_1^m) = e$ *unless* $n = m$.

(v) $(b_0 * b_1^n) \wedge (b_0 * b_1^m) = e$ *unless* $n = m$.

(vi) $(c_0 * c_1^n) \wedge (c_0 * c_1^m) = e$ *unless* $n = m$.

(vii) $(d_0 * d_1^n) \wedge (d_0 * d_1^m) = e$ *unless* $n = m$.

(viii) $(a_0 * a_1^m b_1^n c_1^k d_1^p) \wedge (a_0 * a_1^{m'} b_1^{n'} c_1^{k'} d_1^{p'}) = e$ *unless* $(m = m'$, $n = n'$, $k = k'$ *and* $p = p')$.

(ix) $a_0 * c_1^m \leq a_6$ *if and only if* $m \in \omega$.

Proof: (i) and (ii) follow immediately by conjugating the appropriate relation of G by a_3 or a_3^2.

(iii) follows from the relations $a_1^{-1}a_0a_1 \leq a_2$ and $a_1^{-1}a_2a_1 \leq a_2$ by induction on n.

(iv) follows from (iii) and the relation $a_0 \wedge a_2 = e$.

(v), (vi) and (vii) now follow by conjugating (iv) by a_3, a_3^2 and a_3^3 respectively.

(viii) Since a_1 commutes with b_0, $a_0 \leq b_0$ implies $a_0 * a_1^n \leq b_0$ for all $n \in \mathbb{Z}$. Conjugating by a_3 and a_3^2 gives $b_0 * b_1^n \leq c_0$ and $c_0 * c_1^n \leq d_0$, respectively, for all $n \in \mathbb{Z}$. Hence $a_0 * a_1^m b_1^n c_1^k \leq d_0$ and $a_0 * a_1^{m'} b_1^{n'} c_1^{k'} \leq d_0$. (viii) now follows by successively using (vii), (vi), (v) and (iv).

(ix) Similar to (iii) using the relations involving a_6 and a_7. //

We write $(\underset{\sim}{a};\underset{\sim}{r}(\underset{\sim}{a}) = e)$ for the G described above.

We observe that G contains the five "levels" of $\mathbb{Z}$, as desired.

$f \in {}^{\omega}\omega$ is said to be *presentable* if there are a finite number of generators $\underset{\sim}{x}(f)$--including $\underset{\sim}{a}$ and a_f--and a finite number of words $\underset{\sim}{s}(f)$ in these generators such that the relations $\underset{\sim}{s}(f) = e$ hold in $A(\mathbb{R})$ in the natural interpretation and, in $G(f) = (\underset{\sim}{x}(f);\underset{\sim}{r}(\underset{\sim}{a}) = e, \underset{\sim}{s}(f) = e)$, a_f commutes with a_1, b_1 and

d_1, $a_f(a_f \wedge d_0)^{-1} \wedge d_0 = e$ and for all $m \in \omega$

$$(*)_m \qquad a_0 * c_1^m a_f = a_0 * b_1^{f(m)} c_1^m .$$

Since $a_f(a_f \wedge d_0)^{-1} \wedge d_0 = e$, $\alpha a_f \leq \alpha d_0$ for all $\alpha \in \mathrm{supp}(d_0)$ for any ℓ-permutation group $(G(f),\Omega)$. Hence, if $m \in \omega$, $a_0 * c_1^m(a_f \wedge d_0) = a_0 * c_1^m a_f$.

Note that in $A(\mathbb{R})$, a_f satisfies these relations. Indeed, all the relations that we will write down hold in $A(\mathbb{R})$ in the natural interpretation as can easily be checked.

LEMMA 13.2. *Every recursive functions is presentable.*

Proof: Firstly, $\theta(n) \equiv 0$ is clearly presentable--adjoin to G the generator a_θ and the relation $a_\theta = e$.

Secondly, for s the successor function, adjoin to G the generators $\hat{a}$ and a_s, and the relations: $\hat{a}$ commutes with a_1, b_1 and d_1; $c_1\hat{a} = \hat{a}b_1c_1$; $a_0 * \hat{a} = a_0 * b_1$; $a_s(a_s \wedge d_0)^{-1} \wedge d_0 = e$, $a_s = (\hat{a} \vee e) \wedge a_3^{-1}d_0a_3$. Then a_s commutes with a_1, b_1 and d_1. We now prove $a_0 * c_1^m\hat{a} = a_0 * b_1^{s(m)}c_1^m$ for all $m \in \omega$ by induction.

For $m = 0$, $a_0 * c_1^0\hat{a} = a_0 * \hat{a} = a_0 * b_1 = a_0 * b_1^{s(0)}c_1^0$. Assume $a_0 * c_1^m\hat{a} = a_0 * b_1^{s(m)}c_1^m$. Then $a_0 * c_1^{m+1}\hat{a} = a_0 * c_1^m c_1\hat{a} = a_0 * c_1^m\hat{a}b_1c_1 = a_0 * b_1^{s(m)}c_1^m b_1 c_1 = a_0 * b_1^{s(m+1)}c_1^{m+1}$, as required.

Now $e = a_3^{-1}d_0a_3 \wedge d_1(a_3^{-1}d_0a_3)^{-1}$, so $d_1 = a_3^{-1}d_0a_3 \vee d_1(a_3^{-1}d_0a_3)^{-1}$. Hence, by the above and the definition of a_s $(= (\hat{a} \vee e) \wedge a_3^{-1}d_0a_3)$, $a_0 * c_1^m a_s = a_0 * b_1^{s(m)}c_1^m$ for all $m \in \omega$. Thus s is presentable.

We next show that if f and g are presentable, then so is $h = gf$. Let $G(h)$ have generators $\underset{\sim}{x}(f) \cup \underset{\sim}{x}(g) \cup \{a_h\}$, and relations: $\underset{\sim}{s}(f) = e$, $\underset{\sim}{s}(g) = e$, a_h commutes with a_1, b_1 and d_1, $a_h(a_h \wedge d_0)^{-1} \wedge d_0 = e$, $a_h \wedge d_0 = (a_g \dagger a_f) \wedge d_0$ and $(a_g \dagger a_f)(a_g \dagger a_f \wedge d_0)^{-1} \wedge d_0 = e$, where $\underline{x \dagger y}$ is shorthand for $x * a_4^{-1}y^{-1}a_5$. Now, in $G(h)$,

$$\begin{aligned} a_0 * c_1^m a_h &= a_0 * c_1^m(a_5^{-1}a_fa_4a_ga_4^{-1}a_f^{-1}a_5) = a_0 * b_1^{f(m)}c_1^m a_4a_ga_4^{-1}a_f^{-1}a_5 \\ &= a_0 * c_1^{f(m)}d_1^m a_ga_4^{-1}a_f^{-1}a_5 = a_0 * b_1^{gf(m)}c_1^{f(m)}d_1^m a_4^{-1}a_f^{-1}a_5 \end{aligned}$$

$$= a_0*a_1^{h(m)}b_1^{f(m)}c_1^m a_f^{-1}a_5 = a_0*a_1^{h(m)}c_1^m a_5 = a_0*b_1^{h(m)}c_1^m.$$

Hence the composition of presentable functions is presentable.

Lastly, we must show that if g, h, u and v are presentable and f is obtained from them by general recursion, then f is presentable. So assume the hypothesis and let $G(f)$ have as generators a_f together with those used to show the presentability of g, h, u and v, and relations those required for g, h, u and v together with: a_f commutes with a_1, b_1 and d_1; $(a_f \dagger a_g) \wedge d_0 = a_u \wedge d_0$, $(a_f \dagger a_h) \wedge d_0 = (a_v \dagger a_f) \wedge d_0$, $a_f(a_f \wedge d_0)^{-1} \wedge d_0 = e$, $(a_f \dagger a_g)(a_f \dagger a_g \wedge d_0)^{-1} \wedge d_0 = e$, $(a_f \dagger a_h)(a_f \dagger a_h \wedge d_0)^{-1} \wedge d_0 = e$ and $(a_v \dagger a_f)(a_v \dagger a_f \wedge d_0)^{-1} \wedge d_0 = e$. Now each $m \in \omega$ belongs to the range of some $h^k g$. We prove, by induction on k, that for each $m \in rg(h^k g)$, $a_0*c_1^m a_f = a_0*b_1^{f(m)}c_1^m$.

For $k = 0$, let $m = g(n)$, say.

Thus $a_0*c_1^n a_u = a_0*b_1^{u(n)}c_1^n = a_0*b_1^{f(m)}c_1^n$ since u is presentable and $u = fg$. Since $a_u \wedge d_0 = (a_f \dagger a_g) \wedge d_0$, it follows that

$$a_0*c_1^n a_u = a_0*c_1^n(a_f*a_4^{-1}a_g^{-1}a_5).$$

Hence $a_0*b_1^{f(m)}c_1^n a_5^{-1}a_g a_4 = a_0*c_1^n a_5^{-1}a_g a_4 a_f$. So $a_0*a_1^{f(m)}c_1^n a_g a_4 = a_0*b_1^{g(n)}c_1^n a_4 a_f$. Therefore $a_0*b_1^{f(m)}c_1^m d_1^n = a_0*c_1^m d_1^n a_f$. Since $d_1 a_f = a_f d_1$, $a_0*b_1^{f(m)}c_1^m = a_0*c_1^m a_f$.

Now assume that if $n \in rg(h^k g)$, then $a_0*c_1^n a_f = a_0*b_1^{f(n)}c_1^n$, and suppose $m \in rg(h^{k+1}g)$. So $m = h(n)$ for some $n \in rg(h^k g)$.

But $a_0*c_1^n(a_f \dagger a_h) = a_0*c_1^n(a_v \dagger a_f)$, so

$$a_0*c_1^n a_5^{-1}a_h a_4 a_f a_4^{-1}a_h^{-1}a_5 = a_0*c_1^n a_5^{-1}a_f a_4 a_v a_4^{-1}a_f^{-1}a_5.$$

By the hypotheses, $a_0*c_1^{h(n)}d_1^n a_f a_4^{-1}a_h^{-1} = a_0*c_1^{f(n)}d_1^n a_v a_4^{-1}a_f^{-1}$. Thus $a_0*c_1^m a_f d_1^n a_4^{-1}a_h^{-1} = a_0*b_1^{vf(n)}c_1^{f(n)}d_1^n a_4^{-1}a_f^{-1}$. Hence $a_0*c_1^m a_f d_1^n = a_0*b_1^{vf(n)}c_1^{f(n)}d_1^n a_4^{-1}a_f^{-1}a_h a_4$. Since $vf(n) = fh(n) = f(m)$ and a_1 commutes with b_1, c_1 and a_f,

$$a_0*c_1^m a_f = a_0*b_1^{f(n)} c_1^n a_f^{-1} a_1^{f(m)} a_h a_4 d_1^{-n}.$$

Therefore $a_0*c_1^m a_f = a_0*c_1^n a_h a_1^{f(m)} a_4 d_1^{-n}$. So $a_0*c_1^m a_f = a_0*b_1^{h(n)} c_1^n a_1^{f(m)} a_4 d_1^{-n}$. As $h(n) = m$ and b_1, c_1 and d_1 commute, $a_0*c_1^m a_f = a_0*b_1^{f(m)} c_1^m$.

By (I), every recursive function is presentable, and the proof of the lemma is complete.

We now prove the theorem.

By (II), there is a recursive function f whose range, X, is not a recursive set. Let h be the characteristic function of $\omega\backslash X$; i.e., $h(m) = \begin{cases} 0 & \text{if } m \in X \\ 1 & \text{if } m \notin X \end{cases}$. By Lemma 13.2, f is presentable. Adjoin to $G(f)$ a new generator a_h and relations: a_h commutes with a_1, b_1 and d_1, $a_h(a_h \wedge d_0)^{-1} \wedge d_0 = e$ and $(a_h + a_f) \wedge a_6 = e$. Let H be the resulting finitely presented lattice-ordered group. Now since $a_0*c_1^n \leq a_6$ for all $n \in \omega$, $a_0*c_1^n(a_h + a_f) = a_0*c_1^n$, so $a_0*c_1^n a_5^{-1} a_f a_4 a_h = a_0*c_1^n a_5^{-1} a_f a_4$. Thus $a_0*c_1^{f(n)} a_h = a_0*c_1^{f(n)} = a_0*b_1^{hf(n)} c_1^{f(n)}$ $(hf(n) = 0)$. However, in $A(\mathbb{R})$, if $m \notin X$, $a_0*c_1^m a_h = a_0*b_1 c_1^m \neq a_0*c_1^m$. Therefore, in H, $a_0*c_1^m a_h = a_0*c_1^m$ if and only if $m \in X$. Consequently, the finitely presented ℓ-group H has insoluble word problem. Q.E.D.

APPENDIX I

The purpose of this appendix is to establish Holland's theorem and some immediate consequences of it.

THEOREM (Holland). *Every ℓ-group can be ℓ-embedded in* $A(\Omega)$ *for some chain* Ω. *Indeed, if* G *is an ℓ-group, then* (G,Ω) *is an ℓ-permutation group for some chain* Ω *with* $|\Omega| \leq |G|$.

This is the natural analogue of Cayley's theorem for groups. In order to prove the theorem, we first need to establish some lemmas. If C is a subgroup of G, write $R(C)$ for the set of right cosets of C in G.

LEMMA 1. *Let* C *be a convex subgroup of a p.o. group* G. *Then* $R(C)$ *is a p.o. set if we define* $Cg \leq Cf$ *if and only if* $cg \leq f$ *for some* $c \in C$. *If* G *is an ℓ-group and* C *is a convex ℓ-subgroup of* G, *then* $R(C)$ *becomes a lattice with* $Cg \vee Ch = C(g \vee h)$ *and* $Cg \wedge Ch = C(g \wedge h)$ $(g,h \in G)$.

Proof: By routine verification.

A convex ℓ-subgroup C of an ℓ-group G is said to be *prime* if $f,g \in G$ and $f \wedge g = e$ imply $f \in C$ or $g \in C$.

LEMMA 2. *Let* C *be a convex ℓ-subgroup of an ℓ-group* G. *Then the following are equivalent:*

1. *C is prime.*
2. *$R(C)$ is a chain.*
3. *The set of convex ℓ-subgroups of G that contain C forms a chain under inclusion.*

4. *For* $f, g \in G$, $f \wedge g \in C$ *implies* $f \in C$ *or* $g \in C$.

Proof: We prove that 1 is equivalent to each of 2, 3 and 4.

1 → 2. Let $g, h \in G$ and suppose that C is prime. Since $g(g \wedge h)^{-1} \wedge h(g \wedge h)^{-1} = (g \wedge h)(g \wedge h)^{-1} = e$, $g(g \wedge h)^{-1} \in C$ or $h(g \wedge h)^{-1} \in C$; i.e., $Cg = C(g \wedge h)$ or $Ch = C(g \wedge h)$. Assume the former, without loss of generality. Now, by Lemma 1, $Cg \wedge Ch = C(g \wedge h) = Cg$ so $Cg \leq Ch$.

2 → 1. Assume $R(C)$ is a chain and let $g \wedge h = e$. Then $Cg \wedge Ch = C(g \wedge h) = C$ by Lemma 1, and hence $Cg = C$ or $Ch = C$; i.e., $g \in C$ or $h \in C$.

1 → 3. Suppose that C is prime and A, B are convex ℓ-subgroups of G that contain C but $A \not\subseteq B$ and $B \not\subseteq A$. Let $e < a \in A \setminus B$ and $e < b \in B \setminus A$. Now $e \leq a \wedge b \leq a, b$ so $a \wedge b \in A \cap B$, since A and B are convex. But $a(a \wedge b)^{-1} \wedge b(a \wedge b)^{-1} = e$ so, without loss of generality, $a(a \wedge b)^{-1} \in C \subseteq B$. Hence $a \in B$, a contradiction.

3 → 1. Assume 3. For each $f \in G$, $p(f, C) = \{g \in G: |g| \wedge |f| \in C\} \supseteq C$ and is a convex ℓ-subgroup of G, where $|f| = f \vee f^{-1} \geq e$. Now if $e < g, h \in G$ and $g \wedge h = e$, then without loss of generality, $p(g, C) \subseteq p(h, C)$. But $h \in p(g, C)$. Hence $h \in p(h, C)$. Thus $h \wedge h \in C$; i.e., $h \in C$.

1 → 4. For all $f, g \in G$, $f(f \wedge g)^{-1} \wedge g(f \wedge g)^{-1} = e$. So if C is prime and $f \wedge g \in C$, $f \in C$ or $g \in C$.

4 → 1. A fortiori.

LEMMA 3. *Let* G *be an* ℓ*-group and* $e \neq g \in G$. *Let* C *be a convex* ℓ*-subgroup of* G *maximal with respect to* $g \notin C$ *(existence by Zorn's Lemma). Then* C *is a prime subgroup of* G.

Proof: Let $f \wedge h = e$. If $f, h \notin C$, then $p(f, C), p(h, C) \supsetneq C$ since $h \in p(f, C)$ and $f \in p(h, C)$. Let H be the intersection of all convex ℓ-subgroups of G that properly contain C. Then H is a convex ℓ-subgroup of G and since $g \in H$, $C \subsetneq H$. Now $H \subseteq p(f, C) \cap p(h, C)$. $C \subseteq p(g, C)$ and if $C \neq p(g, C)$, then $g \in H \subseteq p(g, C)$. Thus $|g| = |g| \wedge |g| \in C$ so $g \in C$, a contradiction. Hence $C = p(g, C)$. But $g \in H$ so

$|g| \wedge |f| \in C$ and $|g| \wedge |h| \in C$. Therefore, $f,h \in p(g,C) = C$, a contradiction. Consequently, C is prime.

LEMMA 4. *Let C be a prime subgroup of an ℓ-group G. The map $\theta: G \to A(R(C))$ defined by $(Cf)(g\theta) = Cfg$ $(f,g \in G)$ is an ℓ-homomorphism of G onto the transitive ℓ-permutation group $(G\theta, R(C))$.*

Proof: By routine verification.

LEMMA 5. *If G is an ℓ-group, then G is ℓ-isomorphic to a subcartesian product of ℓ-groups $\{B_g : e \neq g \in G\}$, where (B_g, Ω_g) is a transitive locally primitive ℓ-permutation group (and Ω_g is a chain) for each $e \neq g \in G$.*

Proof: Let $e \neq g \in G$. Choose C_g a convex ℓ-subgroup of G maximal with respect to $g \notin C_g$. By Lemma 3, C_g is a prime subgroup of G, so $R(C_g) = \Omega_g$ is a chain by Lemma 2. By Lemma 4, the map $\theta_g : G \to A(R(C_g))$ is an ℓ-homomorphism and $B_g = G\theta_g$ is an ℓ-group such that (B_g, Ω_g) is a transitive ℓ-permutation group. Let H_g be the intersection of all convex ℓ-subgroups of G that properly contain C_g. Then $C_g \subsetneqq H_g$. By Theorem 1.6.2, the H_g-orbits of Ω_g are the classes of the smallest non-trivial congruence of (B_g, Ω_g). Hence (B_g, Ω_g) is locally primitive. Now $\ker(\theta_g) = \bigcap_{h \in G} h^{-1}C_g h$. Order $\prod_{e \neq g \in G} B_g$ by: $f \geq e$ if and only if $f_g \geq e$ for all $g \in G$, $g \neq e$. Then $\theta: G \to \prod_{e \neq g \in G} B_g$ is an ℓ-homomorphism where $(h\theta)_g = h\theta_g$ $(h \in G;$ $e \neq g \in G)$. $\ker(\theta) = \bigcap_{e \neq g \in G} \ker(\theta_g) \subseteq \bigcap_{e \neq g \in G} C_g$. Since $g \notin C_g$ if $g \neq e$, $\bigcap_{e \neq g \in G} C_g = \{e\}$; i.e., θ is an ℓ-embedding.

COROLLARY 6. *If G is a subdirectly irreducible ℓ-group, then (G,Ω) is a transitive ℓ-permutation group for some chain Ω.*

We now prove Holland's Theorem: By Lemma 5, there exists an ℓ-embedding $\theta: G \to \prod_{e \neq g \in G} A(\Omega_g)$. Totally order $G \setminus \{e\}$ by $\prec$, say, and let $\Omega = \overleftarrow{\bigcup_{e \neq g \in G}} \Omega_g$; i.e. $\alpha < \beta$ (in Ω) if $\alpha \in \Omega_g$, $\beta \in \Omega_h$ and $g \prec h$ or ($g = h$ & $\alpha < \beta$ (in Ω_g)). We can ℓ-embed $\prod_{e \neq g \in G} A(\Omega_g)$ in $A(\Omega)$ via ψ where: $\alpha(f\psi) = \alpha(f\theta_g)$ if $\alpha \in \Omega_g$, as can easily be checked. Each Ω_g is a set of right cosets in G so $|\Omega| \leq |G|^2 = |G|$.

Note that we have actually proved that any ℓ-group G can be represented on a totally ordered set of chains, and the action on any one of these chains is transitive and locally primitive.

A prime subgroup C of an ℓ-group G with the property that C contains no ℓ-ideal (normal convex ℓ-subgroup) of G other than $\{e\}$ is called a *representing subgroup* of G.

COROLLARY 7. *Let G be an ℓ-group. (G,Ω) is a transitive ℓ-permutation group for some chain Ω if and only if G contains a representing subgroup.*

Proof: If (G,Ω) is transitive, let $\alpha \in \Omega$. Then G_α is a prime subgroup of G and, by the fundamental triviality, $\bigcap_{g \in G} g^{-1}G_\alpha g = \bigcap_{g \in G} G_{\alpha g} = \bigcap_{\beta \in \Omega} G_\beta = \{e\}$. Hence G_α is a representing subgroup of G.

Conversely, if C is a representing subgroup, let $\Omega = R(C)$ (a chain). As in Lemma 4, the map $\theta: G \to A(\Omega)$ has kernel $= \bigcap_{g \in G} g^{-1}Cg = \{e\}$. Hence θ is an ℓ-embedding of G onto a "transitive" ℓ-subgroup of $A(\Omega)$; i.e., (G,Ω,θ) is a transitive ℓ-permutation group.

COROLLARY 8. *If G is an ℓ-simple ℓ-group, then (G,Ω) is a transitive ℓ-permutation group for some chain Ω.*

Proof: Let G be an ℓ-simple ℓ-group and let $e \neq g \in G$. Since G is ℓ-simple, C_g is a representing subgroup, where C_g is defined as in the proof of Lemma 5. Alternatively, apply Corollary 6 directly.

<u>COROLLARY</u> <u>9</u>. *If* (G,Ω) *is a transitive* ℓ*-permutation group and* G *is Abelian, then* G *is an o-group and* (G,Ω) *is its right regular representation.*

<u>Proof</u>: By Lemma 4.2.4, it is enough to show that (G,Ω) is uniquely transitive. If $\gamma \in \Omega$ and $e \neq f \in G_\gamma$, let $\delta \in \mathrm{supp}(f)$. Let $g \in G$ be such that $\gamma g = \delta$. Now $\delta g^{-1} f g = \delta$ so $g^{-1} f g \neq f$. This contradicts the fact that G is Abelian. Hence $G_\gamma = \{e\}$ for all $\gamma \in \Omega$; i.e., (G,Ω) is uniquely transitive. //

(Cf., Theorem 4.5.1.)

A group G can be <u>*right*</u> <u>*totally*</u> <u>*ordered*</u> if we can endow G with a total order $\preccurlyeq$ so that $fh \preccurlyeq gh$ whenever $f \preccurlyeq g$.

<u>COROLLARY</u> <u>10</u>. *Every* ℓ*-group can be right totally ordered.*

<u>Proof</u>: We first show that $A(\Omega)$ can be right totally ordered for any chain Ω. Well-order Ω by $\preccurlyeq$, say, and define $f \prec g$ if $\alpha f < \alpha g$, where α is the least element (with respect to $\preccurlyeq$) of $\{\beta \in \Omega: \beta f \neq \beta g\}$. Then $\preccurlyeq$ is a right total order on $A(\Omega)$. Moreover, if $f \leq g$ (in the pointwise ordering), then $f \preccurlyeq g$.

Now, by Holland's Theorem, every ℓ-group G is a subgroup of some $A(\Omega)$. The restriction of $\preccurlyeq$ to G makes G right totally ordered.

We conclude this appendix with one more application of Holland's Theorem.

<u>THEOREM</u> <u>11</u>. *Let* (G,Ω) *be a doubly transitive* ℓ*-permutation group and* H *be an* ℓ*-group containing* G *as an* ℓ*-subgroup. Then there is a chain* $T \supseteq \Omega$ *such that* (H,T) *is an* ℓ*-permutation group and* (G,Ω) *can be* ℓ*-embedded in it.*

<u>Proof</u>: Let $\alpha \in \Omega$ and $e < g_0 \in G$ with $\alpha g_0 \neq \alpha$. Let C be a convex ℓ-subgroup of H containing G_α maximal with respect to not containing g_0 (existence by Zorn's lemma). Since

$C \not\supseteq G$ ($g_0 \in G \setminus C$) and G_α is a maximal subgroup of G (Theorem 4.1.5), $C \cap G = G_\alpha$. For each $h \in H \setminus G$, let C_h be a convex ℓ-subgroup of H maximal with respect to not containing h. Let $C_{g_0} = C$ and let $I = (H \setminus G) \cup \{g_0\}$. Totally order I. Let $T = \overleftarrow{\bigcup_{i \in I}} R(C_i)$. Let $\beta \in \Omega$. Then $\beta = \alpha f$ for some $f \in G$. Define $\phi\colon \Omega \to T$ by $\beta\phi = Cf$. Since $C \cap G = G_\alpha$, ϕ is a well-defined order-embedding. Define $\psi\colon H \to A(T)$ by $(C_i h_1)(h\psi) = C_i h_1 h$ ($i \in I$; $h_1, h \in H$). As in the proof of Holland's Theorem, ψ is an ℓ-embedding (If $g \in G$ and $(Cf_1)g\psi = Cf_1$ for all $f_1 \in H$, then $g \in \bigcap_{f_1 \in G} f_1^{-1} C f_1 \cap G = \bigcap_{f_1 \in G} G_{\alpha f_1} = \{e\}$). If $g \in G$ and $\beta = \alpha f \in \Omega$, then $(\beta g)\phi = (\alpha f g)\phi = Cfg = (Cf)(g\psi) = (\beta\phi)(g\psi)$. Thus (ψ, ϕ) is an ℓ-embedding, as desired.

APPENDIX II

THEOREM 1. *Let G be an ℓ-group and P a closed prime subgroup of G. If $Q \supseteq P$ is a convex ℓ-subgroup of G, then Q is a closed prime subgroup of G.*

Proof: We first show that if $g = \bigvee_{i \in I} g_i$, then $Pg = \sup\{Pg_i : i \in I\}$. Clearly, $Pg_i \leq Pg$ for all $i \in I$. If $Pg_i \leq Ph$ for all $i \in I$, then $g_i h^{-1} \vee e \in P$ $(i \in I)$. Since P is closed, $gh^{-1} \vee e = \bigvee_{i \in I} (g_i h^{-1} \vee e) \in P$. Thus $Pgh^{-1} \leq P$, so $Pg \leq Ph$. Hence $Pg = \sup\{Pg_i : i \in I\}$.

Now assume $Q \supsetneqq P$. Clearly Q is prime. Suppose $\{g_i : i \in I\} \subseteq Q^+$ and $g = \bigvee_{i \in I} g_i$. Let $e < f \in Q \backslash P$. If $g \wedge f \in P$, then $g \in P \subseteq Q$ since P is prime (Appendix I Lemma 2). If $g \wedge f \notin P$, then $P(g \wedge f)^{-1} g < Pg = \sup\{Pg_i : i \in I\}$. Hence $P(g \wedge f)^{-1} g < Pg_j$ for some $j \in I$. Thus $e \leq (g \wedge f)^{-1} g < hg_j$ for some $h \in P \subseteq Q$. Since Q is convex, $(g \wedge f)^{-1} g \in Q$. But $e \leq g \wedge f \leq f \in Q$. Consequently, $g \in Q$ and Q is closed.

THEOREM 2. *Let G be an ℓ-group. Then G is completely distributive if and only if $D(G) = \{e\}$.*

Proof: First observe that we may assume Theorem 8.2.5 (iv) $\rightarrow$ (i) since its proof did not depend on the characterisation of complete distributivity given by $D(G)$.

Assume that G is completely distributive. Let $\{M_i : i \in I\}$ be the collection of *all* prime subgroups of G (so $\bigcap_{i \in I} M_i = \{e\}$ by Appendix I Lemma 3). Let $e \leq g \in D(G)$. Then for each $i \in I$, $g \in M_i^*$; so $g = \bigvee_{i \in I} g_{ij}$ where $\{g_{ij} : j \in J\}$ is the set of all elements of M_i^+ which are less than or equal to g. Therefore $g = \bigwedge_{i \in I} \bigvee_{j \in J} g_{ij} = \bigvee_{f \in {}^I J} \bigwedge_{i \in I} g_{if(i)}$. But $g_{if(i)} \in M_i^+$ so $\bigwedge_{i \in I} g_{if(i)} \in \bigcap_{i \in I} M_i = \{e\}$. Thus $g = \bigvee_{f \in {}^I J} e = e$; i.e., $D(G) = \{e\}$.

Conversely, suppose that $D(G) = \{e\}$ and let $\{G_\lambda : \lambda \in \Lambda\}$ be the family of all *closed* prime subgroups of G. Totally order Λ and let $\Omega_\lambda = R(G_\lambda)$ $(\lambda \in \Lambda)$. We may assume that $\Omega_\lambda \cap \Omega_\mu = \emptyset$ if λ and μ are distinct elements of Λ. Let $\Omega = \overleftarrow{\bigcup_{\lambda \in \Lambda}} \Omega_\lambda$. Define $\psi: G \to A(\Omega)$ by: $(G_\lambda f)(g\psi) = G_\lambda fg$ $(f,g \in G;\ \lambda \in \Lambda)$. Since $\bigcap_{\lambda \in \Lambda} G_\lambda = D(G) = \{e\}$, ψ is an ℓ-embedding. Moreover, $(G\psi)_{G_\lambda f} = (f^{-1}G_\lambda f)$ a closed prime subgroup of G. Thus $(G\psi, \Omega)$ has closed stabilisers. By Theorem 8.2.5 (iv) $\to$ (i), $(G\psi, \Omega/\mathcal{K})$ has pointwise suprema (and infima). But chains are completely distributive. Hence for all $\alpha \in \Omega/\mathcal{K}$, $\alpha(\bigwedge_{i \in I} \bigvee_{j \in J} g_{ij}) = \inf\{\sup\{\alpha g_{ij} : j \in J\} : i \in I\} = \sup\{\inf\{\alpha g_{if(i)} : i \in I\} : f \in {}^I J\} = \alpha(\bigvee_{f \in {}^I J} \bigwedge_{i \in I} g_{if(i)})$, whenever the indicated suprema and infima in G exist. It follows that $\bigwedge_{i \in I} \bigvee_{j \in J} g_{ij} = \bigvee_{f \in {}^I J} \bigwedge_{i \in I} g_{if(i)}$ whenever the indicated suprema and infima exist. Consequently, G is completely distributive.

SOME UNSOLVED PROBLEMS

THE ABELIAN GROUPABILITY PROBLEM.

Let $(\Omega,\leq)$ be a chain. Assume that for some binary operation $\circ$ on Ω, $(\Omega,\leq,\circ)$ is an o-group. Does there exist a binary operation $*$ on Ω such that $(\Omega,\leq,*)$ is an Abelian o-group?

This question was asked by Reinhold Baer in 1961. If $(\Omega,\leq)$ satisfies the hypothesis and is countable, then it has the form of a countable ordinal number of copies of $\mathbb{Z}$ with possibly a copy of $\mathbb{Q}$ on top. Hence the answer is yes in this special case.

ORDERABILITY PROBLEMS.

1. Classify those p.o. sets Ω for which $A(\Omega)$ is a (transitive) ℓ-permutation group. This has been done by M. A. Bardwell [80a] if it is further assumed that $f \wedge g = e$ if and only if $\mathrm{supp}(f) \cap \mathrm{supp}(g) = \emptyset$. This is a definite restriction as is shown by letting $\Omega = \mathbb{R} \oplus \mathbb{Z}$ ordered by: $(r,m) \leq (s,n)$ if $r \leq s$ & $m \leq n$.

2. If G is a group of permutations on a set Ω, find necessary and sufficient conditions (on G and Ω) so that Ω can be totally ordered by a G-invariant $\leq$; i.e., $\alpha \leq \beta$ implies $\alpha g \leq \beta g$ ($\alpha,\beta \in \Omega$; $g \in G$)--also see P. J. Cameron [76], M. Droste [80], G. Higman [77] and H. Putt [77].

MULTIPLE TRANSITIVITY PROBLEMS.

1. Can $A(\mathbb{Q})$ be embedded as a group (as an ℓ-group) in every doubly transitive $A(\Omega)$?

2. If (G,Ω) is triply (four, five, etc.) transitive, is it m-transitive for all $m \in \mathbb{Z}^+$? (Assume Ω is infinite.)

If (G,Ω) is an ℓ-permutation group, the answer is yes by Lemma 1.10.1. The proof of Lemma 1.10.4 shows that the answer is also yes provided G contains an element $\neq e$ whose support is bounded above or below. However, if G contains no such elements, the problem remains open.

3. Find two doubly homogeneous chains Ω_1, Ω_2 such that $A(\Omega_1)$ and $A(\Omega_2)$ satisfy the same ℓ-group sentences but are not ℓ-isomorphic.

Since there are only $2^{\aleph_0}$ complete theories of ℓ-groups, such chains must exist. The problem is to prove specific $A(\Omega_1)$ and $A(\Omega_2)$ satisfy the same ℓ-group sentences.

4. Give further examples of pathological ℓ-permutation groups. Our current list is limited by Glass and McCleary [76] and Chapter 12. Surely there are many others.

THE PRIMITIVITY PROBLEM.

If (G,Ω) is a transitive primitive ordered permutation group, must it be weakly doubly transitive, the right regular representation of a subgroup of $\mathbb{R}$, or periodic?

Theorem 4.3.7 shows that the answer is yes if (G,Ω) is coherent but leaves open this problem if (G,Ω) is incoherent. If we can completely classify transitive primitive ordered permutation groups, we can completely classify all primitive ordered permutation groups by Theorem 4C.

PROBLEMS ON SIMPLICITY AND ℓ-SIMPLICITY.

1. Does there exist an ℓ-simple ℓ-group that has no primitive faithful representation and is not an O-group?

By Appendix I Corollary 8, every ℓ-simple ℓ-group has a transitive faithful representation (G,Ω) for some chain Ω. The only examples we know of that have no primitive such representation are o-groups (see the proof of Theorem 6F, Dlab [68] and Feil [80]).

2. If (G,Ω) is a transitive ℓ-permutation group with G ℓ-simple, then are all the primitive components of (G,Ω) also ℓ-simple?

3. Are Example 6.6 and those ℓ-simple pathological examples given in Glass and McCleary [76] actually simple as groups?

Theorem 6.5 suggests that they might be.

4. Does there exist a (primitive) ℓ-permutation group (G,Ω) with G both finitely generated and simple as a group?

5. Can every ℓ-group be ℓ-embedded in an ℓ-group with exactly four conjugacy classes? That is, can we strengthen Theorem 10B so that, in addition, any two elements unrelated (by the order) to e are conjugate? If so, algebraically closed ℓ-groups will enjoy this property.

EXISTENTIALLY CLOSED ℓ-PERMUTATION GROUPS.

Determine further properties of existentially closed ℓ-permutation groups (see Chapter 12 and Glass and Pierce [80c]).

We conclude with two classes of problems from ℓ-groups that might well lend themselves to an ℓ-permutation group approach.

THE LATERAL COMPLETION PROBLEM.

If G is an ℓ-group, then an ℓ-group H containing G as an ℓ-subgroup is said to be a *lateral completion* of G if (i) for each $e < h \in H$, there is $g \in G$ with $e < g \leq h$, (ii) H is laterally complete, and (iii) no proper ℓ-subgroup of H that contains G is laterally complete. For example, for any chain Ω, $A(\Omega)$ is a lateral completion of $B(\Omega)$. S. J. Bernau [75] proved that every ℓ-group has a lateral completion and that it is essentially unique. The technique is a formal construction that is simple in concept but messy in detail (This is still true of McCleary's improvement [81a] though to a lesser extent). An alternative proof was provided by R. N. Ball [80a] which uses a more sophisticated concept (a generalisation of uniform spaces) but has a simpler proof. However, in both constructions, it is not easy to describe the lateral completion of given ℓ-groups. It would be nice to give an ℓ-permutation group construction of the lateral completion of an ℓ-group. However, a simple computation shows that $G = \{(\{g_n\},m) \in \mathbb{Z} \text{ Wr } \mathbb{Z}: (\exists k \in \mathbb{Z}^+)(g_{n_1} = g_{n_2} \text{ if } n_1 \equiv n_2 (\text{mod } k))\}$ has no lateral completion inside $\mathbb{Z} \text{ Wr } \mathbb{Z} = A(\mathbb{Z} \overleftarrow{\times} \mathbb{Z})$ even though $(G,\mathbb{Z} \overleftarrow{\times} \mathbb{Z})$ is transitive and has only the obvious non-trivial congruence. This means that there is no facile answer to an ℓ-permutation group construction of the lateral completion in the general setting; this is a shame but, in the immortal words, "sic biscuitus disintegrat." For a nice ℓ-permutation answer in the completely distributive case, see Davis and McCleary [81b].

WORD PROBLEM TYPE PROBLEMS.

Since the word problem for ℓ-groups was solved (negatively) using ℓ-permutation groups (Chapter 13), it may be possible to use ℓ-permutation groups to answer some of the following sample problems (cf. McKenzie and Thompson [73] and Thompson [80]):

1. Does there exist a finitely presented ℓ-group (whose relations only use group words) having insoluble group word problem?

   Without the parenthetic part, we saw in Chapter 13 that the answer is yes.

2. Does there exist a finitely presented group with insoluble word problem which can be ordered so as to be an ℓ-group or an o-group?

3. Can every finitely generated ℓ-group defined by a recursively enumerable set of relations be ℓ-embedded in a finitely presented ℓ-group?

   This is the ℓ-group analogue of Graham Higman's famous embedding theorem for groups [61]. All proofs of this phenomenon use either free products with amalgamated subgroup or the Higman-Neumann-Neumann construction. Unfortunately, these two constructions are not possible for ℓ-groups (Theorem 10C and Glass [75b]).

4. R. J. Thompson [80] proved: A finitely generated group has a soluble word problem if and only if there is an embedding of

the group (such that two elements not conjugate in the original group will not be conjugate in the embedding group) into a finitely generated simple group which is a subgroup of a finitely presented group. This is a strong form of a theorem due originally to Boone and Higman [74]. Is the ℓ-group analogue true? Note that Thompson's proof uses permutation groups.

BIBLIOGRAPHY

Note that 81 means: will appear in 1981 at the earliest. The *Mathematical Reviews* (MR) reference is given where possible.

R. D. Anderson:

[58] The algebraic simplicity of certain groups of homeomorphisms, *Amer. J. Math.*, 80(1958), 955-963; MR 20(1959), #4607.

R. Arens:

[45] On the construction of linear homogeneous continua, *Bol. Soc. Matematica Mexicana*, 2(1945), 33-36; MR 7(1946), p. 277.

[51] Ordered sequence spaces, *Portugaliae Math.*, vol10 (1951), 25-28; MR 13(1952), p. 543.

A. K. Arora:

[81] Quasi-varieties of lattice-ordered groups, to appear.

A. K. Arora and S. H. McCleary:

[81] Work in progress on free lattice-ordered groups.

W. W. Babcock:

[64] On linearly ordered topological spaces, Ph.D. Thesis, Tulane University, New Orleans, La., U.S.A., 1964.

K. A. Baker:

[74] Primitive satisfaction and equational problems for lattices and other algebras, *Trans. Amer. Math. Soc.*, 190(1974), 125-150; MR 50(1975), #2025.

R. N. Ball:

[74] Full convex ℓ-subgroups of a lattice-ordered group, Ph.D. Thesis, University of Wisconsin, Madison, Wis., U.S.A., 1974.

[75] Full convex ℓ-subgroups and the existence of a*-closures of lattice-ordered groups, *Pacific J. Math.*, 61(1975), 7-16; MR 53(1977), #5407.

[79] Topological lattice-ordered groups, *Pacific J. Math.*, 83(1979), 1-26.

R. N. Ball (cont.)

[80a] Convergence and Cauchy structures on lattice-ordered groups, *Trans. Amer. Math. Soc.*, 259(1980), 357-392.

[80b] Cut completions of lattice-ordered groups by Cauchy constructions, in *Ordered Groups: Proc. Boise State Conference 1978*, ed. J. E. Smith, G. O. Kenny & R. N. Ball, Lecture Notes in Pure and Applied Math., No. 62, Marcel Dekker 1980, 81-92.

[81] The distinguished completion of a lattice-ordered group, in *Algebra, Carbondale 1980*, ed. R. K. Amayo, Springer Lecture Notes No. 848, 208-217.

R. N. Ball and G. Davis:

[81] The α-completion of a lattice-ordered group, *J. Australian Math. Soc.* (to appear).

M. A. Bardwell:

[79] The o-primitive components of a regular ordered permutation group, *Pacific J. Math.*, 84(1979), 261-274.

[80a] Lattice-ordered groups of order automorphisms of partially ordered sets, *Houston J. Math.*, 6(1980), 191-225.

[80b] A class of partially ordered sets whose group of order automorphisms are lattice-ordered, in *Ordered Groups: Proc. Boise State Conference 1978*, ed. J. E. Smith, G. O. Kenny & R. N. Ball, 53-69.

J. S. Batty:

[47] Sets of non-integral functional powers, *Quarterly J. Math.*, 18(1947), 85-96; MR 9(1948), p. 81.

J. S. Batty and A. G. Walker:

[46] Non-integral functional powers, *Quarterly J. Math.*, 17 (1946), 145-152; MR 8(1947), p. 199.

G. M. Bergman:

[81] Ordering free groups and coproducts of ordered groups, unpublished preprint.

S. J. Bernau:

[75] The lateral completion of an arbitrary lattice group, *J. Australian Math. Soc.*, 19(1975), 263-289; MR 52(1976), #5513.

[77] Varieties of lattice groups are closed under $\mathcal{L}$-completion, *Symposia Math.*, 21(1977), 349-355; MR 57(1979), #12327.

A. Bigard, K. Keimel and S. Wolfenstein:

[77] *Groupes et Anneaux Réticulés*, Springer Lecture Notes No. 608, 1977; MR 58(1979), #27688.

G. Birkhoff:

[46] Sobre los grupos de automorphismos, *Revista Unión Mat. Argentina*, 11(1946), 155-157; MR 7(1946), p. 411.
[57] *Lattice Theory*, 3rd ed., Amer. Math. Soc. Colloquium Publications, vol. 25; MR 37(1969), #2638.

W. W. Boone and G. Higman:

[74] An algebraic characterization of groups with solvable word problem, *J. Australian Math. Soc.*, 18(1974), 41-53; MR 50 (1975), #10093.

W. Buszkowski:

[79] Undecidability of the theory of lattice-orderable groups, *Functiones et Approximatio Comment. Math.*, 7(1979), 23-28; MR 80m(1980), #03078.

R. D. Byrd and J. T. Lloyd:

[67] Closed subgroups and complete distributivity in lattice-ordered groups, *Math. Zeit.*, 101(1967), 123-130; MR 36 (1968), #1371.

P. J. Cameron:

[76] Transitivity of permutation groups of unordered sets, *Math. Zeit.*, 148(1976), 127-139; MR 53(1977), #5711.

C. C. Chang and A. Ehrenfeucht:

[62] A characterization of abelian groups of automorphisms of a simple ordering relation, *Fund. Math.*, 51(1962), 141-147; MR 26(1963), #198.

C. C. Chang and H. J. Keisler:

[73] *Model Theory*, North Holland, 1973; MR 53(1977), #12927.

C. G. Chehata:

[52] An algebraically simple ordered group, *Proc. London Math. Soc.*, 2(1952), 183-197; MR 13(1952), p. 817.

P. M. Cohn:

[57] Groups of order automorphisms of ordered sets, *Mathematika*, 4(1957), 41-50; MR 19(1958), p. 940.
[65] *Universal Algebra*, Harper & Row, 1965; MR 31(1966), #224; erratum 32(1966), p. 1754.

P. F. Conrad:

[70] Free lattice-ordered groups, *J. Algebra*, 16(1970), 191-203; MR 42(1971), #5875.

G. Davis:

[75] Automorphic compactifications and the fixed point lattice of a totally ordered set, *Bull. Australian Math. Soc.*, 12 (1975), 101-109; MR 51(1976), #1762.

[81a] Certain transitive groups of homeomorphisms with zero homology, to appear.

[81b] Permutation representations of groups with Boolean orthogonalities, to appear.

G. Davis and C. D. Fox:

[76] Compatible tight Riesz orders on the group of automorphisms of an o-2 homogeneous set, *Canad. J. Math.*, 28(1976), 1076-1081; MR 54(1977), #7350. Addendum: *Canad. J. Math.*, 29 (1977), 664-665; MR 55(1978) #7879.

G. Davis and E. Loci:

[76] Compatible tight Riesz orders on ordered permutation groups, *J. Australian Math. Soc.*, 21(1976), 317-333; MR 53(1977), #13068.

G. Davis and S. H. McCleary:

[81a] The π-full tight Riesz orders on $A(\Omega)$, *Canad. Math. Bull.*, 24(1981), 137-151.

[81b] The lateral completion of a completely distributive lattice-ordered group (revisited), *J. Australian Math. Soc.* (to appear).

V. Dlab:

[68] On a family of simple ordered groups, *J. Australian Math. Soc.*, 8(1968), 591-608; MR 37(1969), #3978.

M. Droste:

[80] Über k-homogene unendliche Permutationsgruppen, *Archiv der Math.*, 34(1980), 494-495.

[81] Personal Communication.

B. Dushnik and E. W. Miller:

[40] Concerning similarity transformations of linearly ordered sets, *Bull. Amer. Math. Soc.*, 46(1940), 322-326; MR 1 (1940), p. 318.

D. Eisenbud:

[69] Groups of order automorphisms of certain homogeneous ordered sets, *Michigan Math. J.*, 16(1969), 59-63; MR 39(1970), #1373.

D. B. A. Epstein:

[70] The simplicity of certain groups of homeomorphisms, *Compositio Math.*, 22(1970), 165-173; MR 42(1971), #2491.

A. G. Fadell and K. D. Magill Jr.:

[69] Automorphisms of semigroups of polynomials, *Comp. Math.*, 21(1969), 223-239; MR 40(1970), #1505.

T. Feil:

[80] A comparison of Chehata's and Clifford's ordinally simple ordered groups, *Proc. Amer. Math. Soc.*, 79(1980), 512-514.

[81] An uncountable tower of ℓ-group varieties, *Algebra Universalis* (to appear).

N. J. Fine and G. E. Schweigert:

[55] On the group of homeomorphisms of an arc, *Annals of Math.*, 62(1955), 237-253; MR 17(1956), p. 288.

I. Fleischer:

[77] The automorphism group of a scattered set can be non-commutative, *Bull. Australian Math. Soc.*, 16(1977), p. 306; MR 57(1979), #6157.

C. D. Fox:

[75] Commutators in orderable groups, *Comm. in Algebra*, 3 (1975), 213-217; MR 51(1976), #7984.

J. Franchello:

[77] On the structure of free products of lattice-ordered groups, Ph.D. Thesis, Bowling Green State Univ., Bowling Green, Ohio, U.S.A., 1977.

[78] Sublattices of free products of lattice-ordered groups, *Alg. Universalis*, 8(1978), 101-110; MR 56(1978), #8461.

E. Fried:

[65] Representations of partially ordered groups, *Acta Sci. Math.* (Szeged), 26(1965), 15-18; MR 31(1966), #1307.

[67] Note on the representation of partially ordered groups, *Ann. Univ. Sci. Budapest Eötvös Sect. Math.* 10(1967), 85-87; MR 39(1970), #1375.

L. Fuchs:

[63] *Partially Ordered Algebraic Systems*, Addison-Wesley, 1963; MR 30(1965), #2090.

L. Gillman and M. Jerison:

[60] *Rings of Continuous Functions*, Van Nostrand, 1960; MR 22(1961), #6994.

A. M. W. Glass:

[72] Polars and their applications in directed interpolation groups, *Trans. Amer. Math. Soc.*, 166(1972), 1-25; MR 45(1973), #5052.

[74] ℓ-simple lattice-ordered groups, *Proc. Edinburgh Math. Soc.*, 19(1974), 133-138; MR 53(1977), #13069.

[75a] The word problem for lattice-ordered groups, *Proc. Edinburgh Math. Soc.*, 19(1975), 217-219; MR 51(1976), #3313.

[75b] Results in partially ordered groups, *Comm. in Algebra*, 3(1975), 749-761; MR 52(1976), #13558.

[76] Compatible tight Riesz orders, *Canad. J. Math.*, 28 (1976), 186-200; MR 53(1977), #13070.

[79] Compatible tight Riesz orders II, *Canad. J. Math.*, 31 (1979), 304-307; MR 80h(1980), #06014.

[81] Elementary types of automorphisms of linearly ordered sets--a survey, in *Algebra, Carbondale 1980*, ed. R. K. Amayo, Springer Lecture Notes No. 848, 218-229.

A. M. W. Glass and Y. Gurevich:

[81] The word problem for lattice-ordered groups, to appear.

A. M. W. Glass, Y. Gurevich, W. C. Holland and M. Jambu-Giraudet:

[81] Elementary theory of automorphisms of doubly homogeneous chains, in *Logic Year 1979-80, The University of Connecticut*, ed. M. Lerman, J. H. Schmerl & R. I. Soare, Springer Lecture Notes No. 859, 67-82.

A. M. W. Glass, Y. Gurevich, W. C. Holland and S. Shelah:

[81] Rigid homogeneous chains, *Math. Proc. Cambridge Philos. Soc.*, 89(1981), 7-17.

A. M. W. Glass and W. C. Holland:

[73] A characterization of normal valued lattice-ordered groups, *Notices Amer. Math. Soc.*, 20(1973), Oct., #73T-A237, p. A563.

A. M. W. Glass, W. C. Holland and S. H. McCleary:

[75] a*-closures of completely distributive lattice-ordered groups, *Pacific J. Math.*, 59(1975), 43-67; Correction: *ibid.*, 61(1975),606; MR 52(1976), #7994.

[80] The structure of ℓ-group varieties, *Alg. Universalis*, 10(1980), 1-20.

A. M. W. Glass and S. H. McCleary:

[76] Some ℓ-simple pathological lattice-ordered groups, *Proc. Amer. Math. Soc.*, 57(1976), 221-226; MR 53(1977), #7894.

A. M. W. Glass and K. R. Pierce:

[80a] Equations and inequations in lattice-ordered groups, in *Ordered Groups: Proc. Boise State Conference 1978*, ed. J. E. Smith, G. O. Kenny & R. N. Ball, Lecture Notes in Pure and Applied Math., No. 62, Marcel Dekker, 1980, 141-171.

[80b] Existentially complete abelian lattice-ordered groups, *Trans. Amer. Math. Soc.*,,261(1980), 255-270.

[80c] Existentially complete lattice-ordered groups, *Israel J. Math.*, 36(1980), 257-272.

Y. Gurevich:

[67] Hereditary undecidability of the theory of lattice-ordered abelian groups, *Algebra i Logika*, 6(1967), 45-62 (In Russian); MR 36(1968), #92.

Y. Gurevich and W. C. Holland:

[81] Recognizing the real line, *Trans. Amer. Math. Soc.*, 265(1981), 527-534.

E. Harzheim:

[64] Beiträge zur Theorie der Ordnungstypen, insbesondere der η_α-Mengen, *Math. Ann.*, 154(1964), 116-134; MR 28(1964), #5011.

F. Hausdorff:

[08] Grundzüge einer Theorie der geordneten Mengen, *Math. Ann.*, 65(1908), 435-505.

[14] *Grundzüge der Mengenlehre*, Leipzig, 1914.

J. L. Hickman:

[76] Groups of automorphisms of linearly ordered sets, *Bull. Australian Math. Soc.*, 15(1976), 13-32; MR 55(1978), #204.

[77] Groups of automorphisms of linearly ordered sets: corrigenda, *Bull Australian Math. Soc.*, 16(1977), 317-318; MR 57 (1979), #6156.

G. Higman:

[54] On infinite simple groups, *Publ. Math. Debrecen, 3* (1954), 221-226; MR 17(1956), p. 234.

[61] Subgroups of finitely presented groups, *Proc. Royal Soc. London*, Series A, 262(1961), 455-475; MR 24(1962), #A152.

[77] Homogeneous relations, *Quart. J. Math. Oxford*, 28 (1977), 31-39; MR 55(1978), #3090.

W. C. Holland:

[63] The lattice-ordered group of automorphisms of an ordered set, *Michigan Math. J.*, 10(1963), 399-408; MR 28(1964), #1237.

[65a] Transitive lattice-ordered permutation groups, *Math. Zeit.*, 87(1965), 420-433; MR 31(1966), #2310.

W. C. Holland (cont.):

[65b] A class of simple lattice-ordered permutation groups, *Proc. Amer. Math. Soc.*, 16(1965), 326-329; MR 30(1965), #3927.

[65c] The interval topology of a certain lattice-ordered group, *Czech. Math. J.*, 15(1965), 311-314; MR 31(1966), #1308.

[69] The characterization of generalized wreath products, *J. Algebra*, 13(1969), 155-172; MR 41(1971), #1884.

[72] Ordered permutation groups, in *Permutations: Actes du colloque, Gautier-Villars, Paris 1972*, 57-64; MR 49(1975), #8914.

[75a] Outer automorphisms of ordered permutation groups, *Proc. Edinburgh Math. Soc.*, 19(1975), 331-344; MR 52(1976), #7995.

[75b] MR 49(1975), #4876.

[76a] Equitable partitions of the continuum, *Fund. Math.*, 92(1976), 131-133; MR 54(1977), #5076.

[76b] The largest proper variety of lattice-ordered groups, *Proc. Amer. Math. Soc.*, 57(1976), 25-28; MR 53(1977), #10688.

[77] Group equations which hold in lattice-ordered groups, *Symposia Math.*, 21(1977), 365-378.

[79] Varieties of ℓ-groups are torsion classes, *Czech. Math. J.*, 29(1979), 11-12; MR 80b(1980), #06017.

[80] Trying to recognize the real line, in *Ordered Groups: Proc. Boise State Conference 1978*, ed. J. E. Smith, G. O. Kenny & R. N. Ball, Lecture Notes in Pure and Applied Math., No. 62, Marcel Dekker, 1980, 131-134.

W. C. Holland and S. H. McCleary:

[69] Wreath products of ordered permutation groups, *Pacific J. Math.*, 31(1969), 703-716; MR 41(1971), #3350.

[79] Solvability of the word problem in free lattice-ordered groups, *Houston J. Math.*, 5(1979), 99-105; MR 80f(1980), #06018.

W. C. Holland and E. B. Scrimger:

[72] Free products of lattice-ordered groups, *Alg. Universalis*, 2(1972), 247-254; MR 47(1974), #8384.

M. Jambu-Giraudet:

[80] Théorie des modèles de groupes d'automorphismes d'ensembles totalement ordonnés 2-homogènes, *C. R. Acad. Sci. Paris*, Série A, 290(1980), 1037-1039.

[81a] Bi-interpretable groups and lattices, *Trans. Amer. Math. Soc.* (to appear).

[81b] Interprétations d'arithmétiques dans des groupes et des treillis, *ATP Logique et Arith. 1979/80*, l.n. ed. K. McAloon, J. P. Ressayre & C. Berline (to appear).

[81c] Interpretation of arithmetic in groups and lattices, to appear.

N. G. Khisamiev:

[66] Universal theory of lattice-ordered abelian groups, *Algebra i Logika*, 5(1966), 71-76 (In Russian); MR 34(1967), #2727.

D. Khuon:

[70a] Groupes réticulés doublement transitifs, *C. R. Acad. Sci. Paris*, Série A, 270(1970), A708-709; MR 41(1971), #6746.
[70b] Cardinal des groupes réticulés: complété archimédien d'un groupe réticulé, *C. R. Acad. Sci. Paris*, Série A, 270(1970), 1150-1153; MR 42(1971), #1735.

A. I. Kokorin and V. M. Kopytov:

[74] *Fully Ordered Groups*, Halstead Press, 1974; MR 50 (1975), #12852.

V. M. Kopytov:

[79] Free lattice-ordered groups, *Algebra and Logic*, 18 (1979), 259-270 (English translation).

V. M. Kopytov and N. I. Medvedev:

[77] Varieties of lattice-ordered groups, *Algebra and Logic*, 16(1977), 281-285 (English translation); MR 58(1979), #27686.

J. T. Lloyd:

[64] Lattice-ordered groups and o-permutation groups, Ph.D. Thesis, Tulane University, New Orleans, La., U.S.A., 1964.
[65] Representations of lattice-ordered groups having a basis, *Pacific J. Math.*, 15(1965), 1313-1317; MR 32(1966), #2489.
[67] Complete distributivity in certain infinite permutation groups, *Michigan Math. J.*, 14(1967), 393-400; MR 36 (1968), #2544.

J. Q. Longyear:

[74] The structure of linear homogeneous sets, *J. Reine Agnew. Math.*, 266(1974), 132-135; MR 49(1975), #4876.

A. Macintyre:

[72] Algebraically closed groups, *Annals of Math.*, 96(1972), 53-97; MR 47(1974), #6477.

K. D. Magill Jr.:

[64] Semigroups of continuous functions, *Amer. Math. Monthly*, 71(1964), 984-988; MR 33(1967), #689.
[67] Automorphisms of the semigroup of all differentiable functions, *Glasgow Math. J.*, 8(1967), 63-66; MR 34(1967), #7688.

A. I. Mal'cev:

[71] *The Metamathematics of Algebraic Systems*, North Holland, 1971; MR 50(1975), #1877.

J. Martinez:

[74] Varieties of lattice-ordered groups, *Math. Zeit.*, 137 (1974), 265-284; MR 50(1975), #6961.

S. H. McCleary:

[69] The closed prime subgroups of certain ordered permutation groups, *Pacific J. Math.*, 31(1969), 745-753; MR 42 (1971), #1736.

[70] Generalized wreath products viewed as sets with valuation, *J. Algebra*, 16(1970), 163-182; MR 42(1971), #380.

[72a] o-primitive ordered permutation groups, *Pacific J. Math.*, 40(1972), 349-372; MR 47(1974), #1710.

[72b] Pointwise suprema of order-preserving permutations, *Illinois J. Math.*, 16(1972), 69-75; MR 45(1973), #3275.

[72c] Closed subgroups of lattice-ordered permutation groups, *Trans. Amer. Math. Soc.*, 173(1972), 303-314; MR 47(1974), #97.

[73a] The lattice-ordered group of automorphisms of an α-set, *Pacific J. Math.*, 49(1973), 417-424; MR 50(1975), #6962.

[73b] o-2 transitive ordered permutation groups, *Pacific J. Math.*, 49(1973), 425-429; MR 50(1975), #2018.

[73c] o-primitive ordered permutation groups II, *Pacific J. Math.*, 49(1973), 431-443; MR 50(1975), #2019.

[74] The structure of ordered permutation groups applied to lattice-ordered groups, *Notices Amer. Math. Soc.*, 21(1974), Feb. #712-A14, p. A336.

[75] Which lattice-ordered groups are isomorphic to some $\boldsymbol{A}$(S)?, *Notices Amer. Math. Soc.*, 22(1975), Aug. #726-06-13, p. A540.

[76] The structure of intransitive ordered permutation groups, *Algebra Universalis*, 6(1976), 229-255; MR 54(1977), #12597.

[78a] Some simple homeomorphism groups having nonsolvable outer automorphism groups, *Comm. in Algebra*, 6(1978), 483-496.

[78b] Groups of homeomorphisms with manageable automorphism groups, *Comm. in Algebra*, 6(1978), 497-528.

[80] A solution of the word problem in free normal valued lattice-ordered groups, in *Ordered Groups: Proc. Boise State Conference 1978,*, ed. J. E. Smith, G. O. Kenny & R. N. Ball, Lecture Notes in Pure and Applied Math., No. 62, Marcel Dekker, 1980, 107-129.

[81a] The lateral completion of an arbitrary lattice-ordered group (Bernau's proof revisited), *Algebra Universalis* (to appear).

[81b] The word problem in free normal valued lattice-ordered groups: a solution and practical shortcuts, to appear.

[81c] Work in progress on free lattice-ordered groups.

G. F. McNulty:

[80] Classes which generate the variety of all lattice-ordered groups, in *Ordered Groups: Proc. Boise State Conference 1978*, ed. J. E. Smith, G. O. Kenny & R. N. Ball, Lecture Notes in Pure and Applied Math., No. 62, Marcel Dekker, 1980, 135-140.

R. N. McKenzie and R. J. Thompson:

[73] An elementary construction of unsolvable word problems in group theory, in *Word Problems*, ed. W. W. Boone, F. B. Cannonito and R. C. Lyndon, North Holland, 1973, 457-478; MR 53 (1977), #629.

N. Ja. Medvedev:

[77] The lattices of varieties of lattice-ordered groups and Lie algebras, *Algebra and Logic*, 16(1977), 27-30 (English translation); MR 58(1979), #16456.

R. B. Mura and A. H. Rhemtulla:

[77] *Orderable Groups*, Lecture Notes in Pure and Applied Math., No. 27, Marcel Dekker; MR 58(1979), #10652.

B. H. Neumann:

[60] Embedding theorems for ordered groups, *J. London Math. Soc.*, 35(1960), 503-512; MR 24(1962), #A176.

B. H. Neumann and H. Neumann:

[59] Embedding theorems for groups, *J. London Math. Soc.*, 34(1959), 465-479.

T. Ohkuma:

[55] Sur quelques ensembles ordonnés linéairement, *Fund. Math.*, 43(1955), 326-337; MR 18(1957), p. 868.

D. Passman:

[68] *Permutation Groups*, W. A. Benjamin, 1968; MR 38(1969), #5908.

K. R. Pierce:

[72] Amalgamations of lattice-ordered groups, *Trans. Amer. Math. Soc.*, 172(1972), 249-260; MR 48(1974), #3835.

H. L. Putt:

[77] Orderable permutation groups, Ph.D. Thesis, Bowling Green State Univ., Ohio, U.S.A., 1977.

E. B. Rabinovich:

[75] On linearly ordered sets with 2-transitive groups of automorphisms, *Vesti Akad. Navuk Belaruskai SSR* (Seria Fizika-Mate. Navuk), 6(1975), 10-17 (In Russian); MR 55(1978), #8157.

E. B. Rabinovich and V. E. Feinberg:

[74] Normal divisors of a 2-transitive group of automorphisms of a linearly ordered set, *Mat. USSR Sbornik* (English translation), 22(1974), 187-200; MR 49(1975), #394.

J. A. Read:

[71] ℓ-sous-groupes compressibles du groupe-réticulé $A(S)$, *Séminaire Dubreil* (Algèbre), 6(1971-2); Secrétariat mathématique, 11 rue Pierre et Marie Curie, 75231 Paris Cedex 05, France; MR 52 (1976), #10538.

[75] Wreath products of nonoverlapping lattice-ordered groups, *Canad. Math. Bull.*, 17(1975), 713-722; MR 52(1976), #5515.

N. R. Reilly:

[69] Some applications of wreath products and ultraproducts in the theory of lattice-ordered groups, *Duke Math. J.*, 36(1969), 825-834; MR 40(1970), #4184.

[72] Permutational products of lattice-ordered groups, *J. Australian Math. Soc.*, 13(1972), 25-34; MR 45(1973), #3278.

[81] A semilattice of the lattice of varieties of lattice-ordered groups, *Canad. J. Math.* (to appear).

N. R. Reilly and R. Wroblewski:

[81] Suprema of classes of generalized Scrimger varieties of lattice-ordered groups, *Math. Zeit.*, 176(1981), 293-309.

A. Robinson and E. Zakon:

[60] Elementary properties of ordered abelian groups, *Trans. Amer. Math. Soc.*, 96(1960), 222-236; MR 22(1961), #5673.

J. Robinson:

[68] Recursive functions of one variable, *Proc. Amer. Math. Soc.*, 19(1968), 815-820; MR 37(1969), #6178.

H. Rogers Jr.:

[67] *Theory of Recursive Functions and Effective Computability*, McGraw-Hill, 1967; MR 37(1969), #61.

J. Roitman:

[81] The measure of rigid homogeneous chains, to appear.

J. G. Rosenstein:

[81] *Linear Orderings*, Academic Press (to appear).

J. Schreier and S. Ulam:

[33] Über die Permutationsgruppe der natürlichen Zahlenfolge, *Studia Math.*, 4(1933), 134-141.

[35] Ein Bemerkung über die Gruppe der Topologische Abbildungen der Kreislinie auf sich selbst, *Studia Math.*, 5(1935), 155-159.

[37] Über die Automorphismen der Permutationsgruppe der natürlichen Zahlenfolge, *Fund. Math.*, 28(1937), 258-260.

T. J. Scott:

[74] Monotonic permutations of chains, *Pacific J. Math.*, 55 (1974), 583-594; MR 52(1976), #5516.

E. B. Scrimger:

[70] Intransitive lattice-ordered groups of order-preserving permutations of chains, Ph.D. Thesis, University of Wisconsin, Madison, Wis., U.S.A., 1970.

[75] A large class of small varieties of lattice-ordered groups, *Proc. Amer. Math. Soc.*, 51(1975), 301-306; MR 52(1976), #5517.

W. Sierpinski:

[49] Sur une propriéte des ensembles ordonnés, *Fund. Math.*, 36(1949), 56-67; MR 11(1950), p. 165.

F. Šik:

[58] Automorphismen geordneter Mengen, *Časopis Pěst. Mat.*, 83(1958), 1-22; MR 20(1959), #2290.

J. E. Smith:

[80a] The lattice of ℓ-group varieties, *Trans. Amer. Math. Soc.*, 257(1980), 347-357.

[80b] ℓ-group varieties, in *Ordered Groups: Proc. Boise State Conference 1978*, ed. J. E. Smith, G. O. Kenny & R. N. Ball, Lecture Notes in Pure and Applied Math., No. 62, 99-105.

[81] A new family of ℓ-group varieties, *Houston J. Math.* (to appear).

A. Tarski (in collaboration with A. Mostowski and R. M. Robinson):

[53] *Undecidable Theories*, North Holland, 1953; MR 15(1954), p. 384.

R. J. Thompson:

[80] Embeddings into finitely generated simple groups which preserve the word problem, in *Word Problems II*, ed. S. I. Adian, W. W. Boone & G. Higman, North Holland, 1980, 401-441.

L. B. Treybig:

[63] Concerning homogeneity in totally ordered, connected topological space, *Pacific J. Math.*, 13(1963), 1417-1421; MR 28 (1964), #2526.

A. A. Vinogradov:

[71] Nonaxiomatizability of lattice-orderable groups, *Siberian Math. J.*, 13(1971), 331-332 (English translation); MR 44(1972), #132.

A. A. Vinogradov and A. A. Vinogradov:

[69] Nonlocality of lattice-orderable groups, *Algebra and Logic*, 8(1969), 359-361 (English translation); MR 43(1972), #1909.

E. C. Weinberg:

[63] Free lattice-ordered abelian groups, *Math. Ann.*, 151 (1963), 187-199; MR 27(1964), #3720.

[65] Free lattice-ordered abelian groups II, *Math. Ann.*, 159(1965), 217-222; MR 31(1966), #5895.

[67] Embedding in a divisible lattice-ordered group, *J. London Math. Soc.*, 42(1967), 504-506; MR 36(1968), #91.

[80] Automorphism groups of minimal η_α-sets, in *Ordered Groups: Proc. Boise State Conference 1978*, ed. J. E. Smith, G. O. Kenny & R. N. Ball, Lecture Notes in Pure and Applied Math., No. 62, Marcel Dekker, 1980, 71-79.

[81a] Partitioned chains and amalgamations of lattice-ordered groups, *Houston J. Math.* (to appear).

[81b] Real order-automorphism groups, *J. Australian Math. Soc.* (to appear).

J. V. Whittaker:

[63] On isomorphic groups and homeomorphic spaces, *Annals Math.*, 78(1963), 74-91; MR 27(1964), #737.

H. Wielandt:

[64] *Finite Permutation Groups*, Academic Press, 1964; MR 32(1966), #1252.

[67] *Unendliche Permutationsgruppen*, 2nd ed., York University, Toronto, 1967.

[69] *Permutation Groups through Invariant Relations and Invariant Functions*, Ohio State University, 1969.

S. Wolfenstein:

[68] Valeurs normales dans un groupe réticulé, *Acad. Nazionale dei Lincei*, 44(1968), 337-342; MR 38(1969), #3202.

[70] Extensions archimédiennes de groupes réticulés transitifs, *Bull. Soc. Math. France*, 98(1970), 193-200; MR 42 (1971), #5879.

ANNOTATIONS

Unless otherwise stated, each numbered theorem in the annotations refers to a theorem in our book.

A. K. Arora:

[81] A quasi-variety of ℓ-groups is a class of ℓ-groups closed under ℓ-isomorphisms, ℓ-subgroups, cardinal Cartesian products and ultraproducts. Equivalently, it is defined by a set of implications of the form $\forall \underset{\sim}{x}(w_1(\underset{\sim}{x}) = e \rightarrow w_2(\underset{\sim}{x}) = e)$, where $w_1(\underset{\sim}{x}), w_2(\underset{\sim}{x})$ are ℓ-group words. Canonical examples are varieties, the class of all totally orderable ℓ-groups, and the class of ℓ-groups having unique extraction of roots. There are at most $2^{\aleph_0}$ distinct quasi-varieties. Whereas $\mathcal{N}$ is the maximal proper variety of ℓ-groups (Theorem 11D--Holland [76b]), $2^{\aleph_0}$ pairwise incomparable quasi-varieties are constructed containing $\mathcal{N}$. Moreover, it is shown that there is no maximal proper quasi-variety. The ℓ-semigroup of quasi-varieties is also examined.

W. W. Babcock:

[64] The main results in this thesis are (i) Any homogeneous Dedekind complete chain other than $\mathbb{Z}$ has cardinality $2^{\aleph_0}$, and (ii) There are at least $\aleph_1$ pairwise non-ordermorphic such chains. McCleary [72a] notes that each of them supports a full periodic primitive Config(1) ℓ-permutation group, and those that come from non-ordermorphic chains are not ℓ-isomorphic.

R. N. Ball:

[74] The only part of this thesis not included in Ball [75] concerns the normal subgroups of $A(\mathbb{L})$--see Theorem 2.3.7. Indeed, if N denotes the normal subgroup of $A(\mathbb{L})$ defined on page 70, it is shown that N covers $L(\mathbb{L})$ and the set of normal subgroups of $A(\mathbb{L})$ that contain N is lattice-isomorphic to the lattice of all filters on the set $\mathcal{P}(\omega_1)/\sim$, where $\mathcal{P}(\omega_1)$ is the set of all subsets of ω_1 and $A \sim B$ if $A \cap T = B \cap T$ for some closed unbounded subset T of ω_1. Moreover, $L(\mathbb{L})$ covers $B(\mathbb{L})$, $A(\mathbb{L})$ covers $R(\mathbb{L})$, and the lattice of normal subgroups between $B(\mathbb{L})$ and $R(\mathbb{L})$ is isomorphic to the lattice of normal subgroups between $L(\mathbb{L})$ and $A(\mathbb{L})$. The example given at the beginning of Chapter 7 is also taken from this thesis, where it is shown that

P has a cover which can be taken to be $A(\mathbb{R})_0$ for a suitable choice of ultrafilter $\mathcal{F}$ (assuming the Continuum Hypothesis).

[75] An ℓ-group H is an a*-extension of an ℓ-subgroup G if the map $C \mapsto C \cap G$ of the closed convex ℓ-subgroups of H onto those of G is one-to-one. If G is an o-group, this is the same definition as a-extension (see Khuon [70a]). It is shown that an arbitrary ℓ-group has an a*-closure (a maximal a*-extension)--cf., Glass, Holland and McCleary [75]. The proof has the same spirit as Khuon [70b] but is complicated by the fact that if G is an ℓ-group with (G,Ω,θ) an ℓ-permutation group, the ℓ-embedding θ of G into $A(\Omega)$ may not preserve infinite suprema (whence G and Gθ may have different closed convex ℓ-subgroups). The difficulty is overcome by considering full convex ℓ-subgroups (A convex ℓ-subgroup C of G is said to be full if whenever $c \in C$ and g and c generate the same closed convex ℓ-subgroup of G, then $g \in C$). Any full convex ℓ-subgroup of G is a union of closed convex ℓ-subgroups, so it suffices to bound $|G|$ in terms of the cardinality of the set of full convex ℓ-subgroups of G. This is done by Khuon's lemma and a lemma of McCleary.

[79] A topology $\mathcal{J}$ on an ℓ-group G in which the group and lattice operations are continuous is called an ℓ-topology, and $(G,\mathcal{J})$ is said to be a topological ℓ-group. Properties of such structures are examined. In particular, a continuous version of Holland's Theorem is established. If (G,Ω) is a transitive ℓ-permutation group and $\mathcal{J}$ is any Hausdorff ℓ-topology on $A(\bar{\Omega})$ or $A(\Omega^0)$, then $G^0 = \{f \in A(\Omega^0)$: f respects the congruences of $(G,\Omega)\}$ is shown to be closed in $\mathcal{J}$. The topological closures (in $A(\bar{\Omega})$) are found for (G,Ω) a transitive primitive ℓ-permutation group.

[80a] This is a generalisation of both Ball [79] and uniformities. It adapts the techniques of convergence and Cauchy structures to an ℓ-group setting to obtain the existence of various completions of ℓ-groups, notably the existence of a lateral completion of an arbitrary ℓ-group (cf. Bernau [75]).

[81] Let $a < b$ and $c < d$ in an ℓ-group G. Write $(a,b) \sim (c,d)$ if $\uparrow(a,b) = \uparrow(c,d)$ and $\downarrow(a,b) = \downarrow(c,d)$, where $\uparrow(a,b) = \{g \in G: g \vee a \geq b\}$ and $\downarrow(a,b) = \{g \in G: g \wedge b \leq a\}$. If (G,Ω) is an ℓ-permutation group, $(a,b) \sim (c,d)$ can be described pictorially by the "bubble" where a and b differ is the same as the "bubble" where c and d differ; i.e., $\alpha a = \alpha b$ implies $\alpha c = \alpha d$ and $\alpha a < \alpha b$ implies $\alpha c = \alpha a < \alpha b = \alpha d$. When ba^{-1} is a bump, "bubble" accurately describes the picture. We can order these equivalence classes by: $(a,b)^\sim \leq (c,d)^\sim$ if the "bubble" of some element of $(a,b)^\sim$ is inside the "bubble" of (c,d). This leads to a completion of G known as the distinguished completion. The shadow cast by a bubble also plays an important rôle.

M. A. Bardwell:

[79] Let G be a right totally ordered group. The right regular representation (G,G) is examined. It is shown that any transitive primitive ordered permutation group can occur as a primitive component of (G,G) for suitably chosen such G. Moreover, an example is given of a right totally ordered group G where the right regular representation is primitive and uniquely transitive but not isomorphic to a subgroup of $\mathbb{R}$ (cf. Lemma 4.2.4 and Theorem 4.2.1). Some results are obtained about (G,G) when it is a "large" subaction of a Wreath product and some unsolved problems are listed. The main tool is the proof of Appendix I Corollary 10.

[80a] The p.o. sets Ω for which $A(\Omega)$ is an ℓ-group having the disjoint support property ($f \wedge g = e$ if and only if $\mathrm{supp}(f) \cap \mathrm{supp}(g) = \emptyset$) are classified. This last property is equivalent to each of (i) every orbit of $A(\Omega)$ is a chain; (ii) $\alpha(f \vee g) = \max\{\alpha f, \alpha g\}$ for all $\alpha \in \Omega$, $f,g \in A(\Omega)$. It follows that such $A(\Omega)$ are laterally complete. The classification relies heavily on the congruence structure. There are six main cases when $A(\Omega)$ has precisely two orbits, and all are shown to be non-vacuous. An example is given of a p.o. set for which $A(\Omega)$ is a transitive ℓ-group without the disjoint support property.

[80b] Most of this article is an outline of Bardwell [80a], but there are a few new results. If Ω is a p.o. set such that $A(\Omega)$ is an ℓ-group, then $A(\Omega)$ has the disjoint support property if and only if $A(\Omega)_\alpha$ is prime for all $\alpha \in \Omega$. By using the classification given in Bardwell [80a], it is noted that any such $A(\Omega)$ having two orbits has closed stabilisers and is completely distributive--cf., Theorems 8B and 8.2.1.

S. J. Bernau:

[75] Let G be an ℓ-subgroup of an ℓ-group H. Each element of the ℓ-subgroup $G^{\#}$ of H generated by the suprema of all pairwise disjoint non-empty subsets of G is easily seen to be of the form $\bigvee_I \bigwedge_J h_{ij}$, where I,J are finite and each $h_{ij} = \prod_K (\bigvee M_{ijk})^{\varepsilon(i,j,k)}$ for some finite K, $\varepsilon(i,j,k) = \pm 1$, and M_{ijk} a non-empty pairwise disjoint subset of G^+ having a supremum in H. If $G_0 = G$, $G_{\nu+1} = G^{\#}_\nu$, and $G_\lambda = \bigcup_{\nu<\lambda} G_\nu$ if λ is a limit ordinal, then $G_{\mu+1} = G_\mu$ for some $\mu \leq |H|$. If G is "dense" in H, G_μ is a lateral completion of G with respect to H. In the absence of H, $G^{\#}$ has to be defined via formal symbols and factoring out a complicated equivalence relation (roughly, when such elements should be equal) on the resulting set

to get an ℓ-group. At limit stages, take direct limits. The cardinality bound is provided by κ, the next cardinal after $|G|$, since G must be "dense" in each G_ν. G_κ is a lateral completion of G. The construction gives uniqueness. The last part of the paper is devoted to showing that certain properties of G transfer to the lateral completion; e.g., complete distributivity (cf. Davis and McCleary [81b]). A simplification of the construction is provided in McCleary [81a]. For a less elementary construction of the lateral completion, see Ball [80a]. Note that the details of Ball [80a] are much easier than those of even McCleary [81a]; however, the degree of abstraction is considerably greater.

[77] It is shown that if $\mathcal{V}$ is a variety and $G \in \mathcal{V}$, then the lateral completion of G belongs to $\mathcal{V}$. The proof proceeds by showing that $G \in \mathcal{V}$ implies $G^{\#} \in \mathcal{V}$, where $G^{\#}$ is defined in the previous annotation. Actually, the proof shows that if G is an ℓ-subgroup of an ℓ-group H and $G \in \mathcal{V}$, then the closure of G in H belongs to $\mathcal{V}$. This is used a lot in Glass, Holland and McCleary [80].

A. Bigard, K. Keimel and S. Wolfenstein:

[77] This is now the main collected work on ℓ-groups. It is strongly influenced by Conrad's Tulane University lectures on "Lattice-ordered Groups" (no longer in print) and the authors' own work. It recounts research up to 1972 (approximately). For a more detailed review, see *Zentralblatt für Math.*, 384(1979), 06022.

G. Birkhoff:

[46] For a given group G, a p.o. set $\langle\Omega,\leq\rangle$ is constructed with G isomorphic to the group $\mathrm{Aut}(\langle\Omega,\leq\rangle)$ of all order-preserving permutations of the p.o. set $\langle\Omega,\leq\rangle$; cf., Scrimger [70].

W. W. Boone and G. Higman:

[74] Using free products with amalgamation and Higman's embedding theorem (Higman [61]), the authors prove: A finitely generated group G has a soluble word problem if and only if there exist a simple group H and a finitely presented group L such that G can be embedded in H and H can be embedded in L. The semigroup analogue is also proved. R. J. Thompson [80], using permutation groups, improves on the group theory part, showing (among other things) that H can be taken to be finitely generated. The construction of such an H is difficult. Can it be taken to be finitely presented?

W. Buszkowski:

[79] The theory of lattice-orderable groups is shown to be undecidable--Theorem 2.2.8. The proof given is as follows: Let $d = (\{d_n\},1) \in \mathbb{Z} \text{ Wr } \mathbb{Z}$, where $d_n = n$ $(n \in \mathbb{Z})$. Let $C(d)$ be the centraliser of d in $\mathbb{Z} \text{ Wr } \mathbb{Z}$. Write $x \sim y$ if xy^{-1} belongs to the centre of $\mathbb{Z} \text{ Wr } \mathbb{Z}$ (i.e., $\bar{x} = \bar{y}$ and $x_n - y_n$ is constant). Then $d^n \sim d^m$ only if $n = m$, and $x \in C(d)$ if and only if $x \sim d^m$ for some $m \in \mathbb{Z}$. For $x,y,z \in C(d)$, write $D(x,y)$ if $C(x) \subseteq C(y)$, and $S(x,y,z)$ if $xy \sim z$. Then, when restricted to $C(d)$, S and D respect $\sim$ and give addition and divisibility (of $\mathbb{Z}$ via $\{d^n\colon n \in \mathbb{Z}\}$). This guarantees the undecidability--just of the group part--of lattice-ordered groups; cf., Glass and Pierce [80c], Gurevich [67], Jambu-Giraudet [81b] and Mal'cev [71].

R. D. Byrd and J. T. Lloyd:

[67] The main result of this paper is Appendix II Theorem 2. Unlike our proof, theirs does not use ℓ-permutation groups. Again the key lemma is Appendix II Theorem 1.

P. J. Cameron:

[76] A permutation group G on a set Ω is said to be n-homogeneous if whenever $\Gamma,\Delta \subseteq \Omega$ and $|\Gamma| = |\Delta| = n$, there exists $g \in G$ with $\Gamma g = \Delta$. $A(\mathbb{R})$ is n-homogeneous for all $n \in \mathbb{Z}^+$ (Lemma 1.10.1). However, it is not 2-transitive as a permutation group--there is no $g \in A(\mathbb{R})$ with $0g = 3$ & $1g = 2$. The group $\mathcal{M}(\mathbb{R})$ of all monotonic permutations of $\mathbb{R}$ is still n-homogeneous for all $n \in \mathbb{Z}^+$; it is 2-transitive but not 3-transitive as a permutation group. McDermott (unpublished) has shown that any 3-homogeneous but not 2-transitive group of permutations of an infinite set Ω is a subgroup of $A(\Omega)$ for some linear order on Ω. Using a weak form of Ramsey's Theorem for infinite sets, this is extended to prove (1) Any 4-homogeneous but not 3-transitive group of permutations of an infinite set Ω is a subgroup of either $\mathcal{M}(\Omega)$ for some linear order on Ω or the group of automorphisms of some cyclic order on Ω. (2) If G is a group of permutations of an infinite set Ω that is m-transitive but not (m+1)-transitive for some $m \in \mathbb{Z}^+$, and is n-homogeneous for all $n \in \mathbb{Z}^+$, then $m \leq 3$ and there is a linear or cyclic order on Ω that is preserved or reversed by all elements of G.

C. C. Chang and A. Ehrenfeucht:

[62] Theorem 10.7 is established, even with "proper" removed. The proof proceeds by tagging elements of distinct cosets of a proper subgroup of $\mathbb{R}$ with chains associated with distinct ordinals (as in the proof of Theorem 10.5). By Corollary 10.6, it follows that an Abelian group G is isomorphic to $A(\Omega)$ for some chain Ω if and only if G is isomorphic to a direct product of subgroups of $\mathbb{R}$ (cf. Cohn [57]--Theorem 4.5.1). This is the first use of the ingenious device of "tagging."

C. G. Chehata:

[52] The set of bounded elements of Example 1.2.3 is shown to be a simple group. The proof of simplicity is more computational than ours (see the proof of Theorem 6F) since it predates Higman [54]--Lemma 6.4. Also see Dlab [68] and Feil [80].

P. M. Cohn:

[57] Theorem 4.5.1 is established--cf. Chang and Ehrenfeucht [62]. The proof is of an elementary nature and follows from considering the case that Ω is an extensive $A(\Omega)$ block. Our proof is shorter but uses Section 3.2 (which is more recent). The paper also includes a proof that $A(\Omega)$ can be right totally ordered (cf. Appendix I Corollary 10).

P. F. Conrad:

[70] The free ℓ-group over a p.o. group G is shown to exist precisely when the partial order is the intersection of all right total orders that extend it. Moreover, the free ℓ-group over such a p.o. group G is explicitly constructed. This extends the work of E. C. Weinberg [63] for the construction of free Abelian and representable ℓ-groups. The construction is as follows: Let $\{G_\lambda : \lambda \in \Lambda\}$ denote the set of all right total orders on G extending the given partial order. Represent G on $A(G_\lambda)$ via the right regular representation (for each $\lambda \in \Lambda$). The ℓ-subgroup of $\prod_{\lambda \in \Lambda} A(G_\lambda)$ generated by this image is the free ℓ-group over G. Since any free group can be right totally ordered, this yields a construction for the free ℓ-group on any set X. Its relation to the free Abelian and free representable ℓ-groups on X is examined. For further results on free ℓ-groups, see Arora and McCleary [81], Glass [74], Kopytov [79] and McCleary [81c].

G. Davis and C. D. Fox:

[76] The main result in this paper is that if Ω is doubly homogeneous, then $A(\Omega)$ endures a CTRO (see annotation of Davis and Loci [76] for definition). This result was also proved independently by R. N. Ball (unpublished). Let $T_\rho = \{e < g \in A(\Omega): (\exists \alpha \in \Omega)\text{supp}(g) \cap [\alpha,\infty) \text{ is dense in } [\alpha,\infty)\}$. It is shown that T_ρ is a maximal CTRO on such $A(\Omega)$ if and only if $\text{cf}(\Omega) = \aleph_0$. Dually for $\text{ci}(\Omega) = \aleph_0$.

G. Davis and E. Loci:

[76] The interval topology on an ℓ-group is rarely Hausdorff. On the other hand, the interval topology on a tight Riesz group is (A p.o. group G is a tight Riesz group if the partial order is directed, and $f_1, f_2 < g_1, g_2$ implies there is $h \in G$ such that $f_1, f_2 < h < g_1, g_2$. Note, any ℓ-group that is a tight

Riesz group is an o-group). If G is an ℓ-group that can be reordered to be a tight Riesz group with set of positive elements T in such a way that the two orders are compatible, then T is said to be a compatible tight Riesz order on G, or CTRO for short. More precisely, if G is an ℓ-group and $T \subseteq G$, then T is CTRO if (i) T is a normal subsemigroup of G, (ii) $T \cdot T = T$ and (iii) $\inf T = e$, where the infimum is with respect to the lattice order. If Ω is a totally ordered field, $A(\Omega)$ contains positive elements that fix no point of Ω. Since $A(\Omega)$ is divisible, the set T_0 of such elements forms a CTRO. Other CTROs are easily obtainable. Moreover, it is shown that there are prime subgroups of $A(\Omega)$ maximal with respect to not meeting T_0 which are not stabilisers. Other more technical results are proved. Like the above, they are easy--a lot simpler than the presentation given in the paper.

G. Davis and S. H. McCleary:

[81a] In an ℓ-group G, write $f \pi g$ if $(\forall h)(f \wedge h = e \leftrightarrow g \wedge h = e)$. A subset $X \subseteq G$ is said to be π-full if $y \pi x$ and $x \in X$ imply $y \in X$. The main theorem is: If (G,Ω) is a laterally complete ℓ-permutation group with no block on which the appropriate restriction acts like $\mathbb{Z}$, then (i) the π-full prime subgroups are precisely the minimal primes, (ii) every π-full tight Riesz order is a CTRO and (iii) there is a one-to-one correspondence between the π-full CTROs on G and the π-full ℓ-ideals of G. Glass and Holland (unpublished) have shown that if Ω is doubly homogeneous, any ℓ-ideal of $A(\Omega)$ is π-full. Hence there is a one-to-one correspondence between the π-full TROs on $A(\Omega)$ and the set of normal subgroups of $A(\Omega)$. When $A(\Omega)$ is locally $\mathbb{Z}$, the π-full TROs on $A(\Omega)$ are shown to correspond to the ℓ-ideals of $A(\Omega)|\tilde{\Omega}$, where $\tilde{\Omega}$ is an appropriate subset of Ω.

[81b] Let G be a completely distributive ℓ-group. There is a chain Ω such that (G,Ω) is an ℓ-permutation group with G a complete subgroup of $A(\Omega)$. Let $H = \{h \in A(\Omega): (\forall \alpha_1,\ldots,\alpha_n \in \Omega)(\exists g \in G)\alpha_i g = \alpha_i h \ (1 \leq i \leq n)\}$. Then G is dense in H and H is laterally complete. Hence G has a lateral completion which is shown to be completely distributive (For the existence of the lateral completion of a general ℓ-group, see Ball [80a], Bernau [75] or McCleary [81a]). The lateral completion is further restricted in certain nice cases. For an earlier proof of the existence of a lateral completion of a completely distributive ℓ-group not using ordered permutation groups, see R. D. Byrd and J. T. Lloyd (*J. London Math. Soc.* (2), 1(1969), 358-362).

V. Dlab:

[68] Let $\Omega = [0,1)$ and H be a subgroup of the group of positive real numbers. Let A_H be the group of locally right H-linear elements of $A(\Omega)$; i.e., $g \in A_H$ if $(\forall\alpha \in \Omega)(\exists\, \varepsilon = \varepsilon(\alpha,g) > 0)(\exists h = h(\alpha,g) \in H)(\forall\beta \in (\alpha,\alpha + \varepsilon))$ $(\beta g = \alpha g + h(\beta - \alpha))$. Let $B_H = A_H \cap B(\Omega)$, $L_H = A_H \cap L(\Omega)$ and $R_H = A_H \cap R(\Omega)$. Then $f,g \in A_H$ are conjugate in A_H precisely when there is a one-to-one left-right preserving map from the orbitals of f onto the orbitals of g such that corresponding orbitals have the same boundity and left-most slope (cf., Theorem 2.2.4). Moreover, if $e \neq g \in A_H$ and N is the normal subgroup of A_H generated by g, then $B_H \subseteq N$; if $g \notin L_H$, then $R_H \subseteq N$; and if $f \notin R_H$, then $f \in N$ if and only if its left-most slope belongs to the cyclic subgroup of H generated by the left-most slope of g (and $f \in L_H$ if $g \in L_H$). Hence B_H is simple and is the only normal subgroup of R_H. The lattice of normal subgroups of L_H containing B_H is isomorphic to the lattice of all subgroups of H. The group A_H is divisible if and only if H is; and A_H has precisely two linear orderings if and only if H has rank 1 as an Abelian group. The collection of B_H (as H ranges through all subgroups of the group of positive real numbers) provides an uncountable family of pairwise non-isomorphic simple orderable groups--see Theorem 6F. The proofs are computational and similar to those in Section 2.2. Indeed, recently, M. Droste has obtained a uniform method to obtain the results of Dlab and Section 2.2. It too is akin to the ideas in Section 2.2.

M. Droste:

[80] Let κ be a cardinal, $\langle\Omega,\leq\rangle$ be a p.o. set and G be a subgroup of $\mathrm{Aut}(\langle\Omega,\leq\rangle)$. Assume $2 \leq \kappa < |\Omega|$. G is said to be κ-homogeneous if whenever $A,B \subseteq \Omega$ have cardinality κ, there is $g \in G$ such that $Ag = B$. It is shown that if Ω is infinite and $\leq$ is not the trivial partial ordering, then $\langle\Omega,\leq\rangle$ is a chain and κ is finite. Note that another definition (which agrees with the above if κ is finite) is: If $A,B \subseteq \Omega$ have cardinality $< |\kappa+1|$ and there is an order-preserving one-to-one map of A onto B, then there is $g \in G$ extending this map. Under this definition, $\mathrm{Aut}(\langle\Omega,\leq\rangle)$ can be $\aleph_0$-homogeneous without $\langle\Omega,\leq\rangle$ being a chain. We adopted this second definition--see page 187. Also see Cameron [76] and Higman [77].

B. Dushnik and E. W. Miller:

[40] Using transfinite induction, a non-well-ordered subset of $\mathbb{R}$ of size $2^{\aleph_0}$ is constructed which is not ordermorphic to any proper subset of itself. This paper can be thought of as a precursor of Ohkuma [55] and Glass, Gurevich, Holland and Shelah [81]--see Section 4.2.

D. Eisenbud:

[69] Let Ω be a doubly homogeneous chain of countable coterminality with all elements of $\bar{\Omega}$ having character c_{00}. The key lemma is: $f \in A(\Omega)$ can be written uniquely as the product of commuting elements one from $L(\Omega)$ and one from $R(\Omega)$ precisely when f has two supporting intervals and exactly one fixed point in $\bar{\Omega}$. Since $L(\Omega)$ and $R(\Omega)$ are the maximal normal subgroups of such $A(\Omega)$--Theorem 2.3.2--$\bar{\Omega}$ is obtained from the *group* $A(\Omega)$. It follows that $A(\Omega) \cong A(\bar{\Omega})$ as groups only if $\Omega = \bar{\Omega}$; and if Λ is also a doubly homogeneous chain of countable coterminality with $A(\Lambda) \cong A(\Omega)$ (as groups), then Λ is ordermorphic or anti-ordermorphic to a dense subset of $\bar{\Omega}$ (cf. Theorem 2D). It is also shown that $\mathrm{Aut}(A(\Omega)) \cong N_{\mathcal{S}(\bar{\Omega})}(A(\Omega))$, the normaliser of $A(\Omega)$ in the symmetric group $\mathcal{S}(\bar{\Omega})$; and if $B(\Omega) \subseteq G \subseteq A(\Omega)$, then $\mathrm{Aut}(G) = N_{\mathrm{Aut}(A(\Omega))}(G)$. This work is a generalisation of Holland [63] and a precursor of some of the ideas in Sections 2.1 and 2.3.

T. Feil:

[80] A previously known o-group C that is without proper convex normal subgroups is shown to be contained in the finitely piecewise linear elements of $A(\mathbb{R})$, as the set of such with rational "corners" and all slopes integer powers of 2. $B = C \cap B(\Omega)$ is a normal subgroup of C and $[B,B]$ is simple by Lemma 6.4. Does $B = [B,B]$? For other related work, see Chehata [52] and Dlab [68].

[81] A tower of $2^{\aleph_0}$ varieties of representable ℓ-groups is given--the maximum cardinality possible. The construction of the tower is very elementary. Unlike the prior proofs of Kopytov and Medvedev [77] and Reilly [81], this proof of the existence of $2^{\aleph_0}$ varieties of ℓ-groups does not depend on any results about varieties of groups.

I. Fleischer:

[77] This gives an example to point out an error in a proof of Hickman [76]--at least if lexicographic sum is replaced by lexicographic union; then it is just Example 4.2.6.

C. D. Fox:

[75] In an orderable group, $[f^n,g] = e$ implies $[f,g] = e$, and $[f^n,g,f] = e$ implies $[f,g] = e$. An example is given of an orderable group G such that, for any set X of positive integers, there are $f,g \in G$ with $[f^n,g,f,f] = e$ if and only if $n \in X$. The group G is a natural subgroup of $(H \text{ Wr } \mathbb{Z}) \text{ wr } \mathbb{Z}$, where H is the (orderable) group given by $(x,y\colon x^{-1}yx = y^2)$. G is finitely generated. It follows that there is a finitely generated ℓ-group having a recursively enumerable set of (group) defining relations and an insoluble group word problem; cf., Glass [75a].

J. D. Franchello:

[77] Using the ideas of Holland [77] and the aid of a computer, it is shown that there are essentially 6 group words of length 16 that hold in the ℓ-group free product of two arbitrary ℓ-groups (one of these is given by Lemma 1.11.5 and is due to Holland and Scrimger [72]). Two infinite families of group words are found that hold in all such ℓ-group free products. For other results in this thesis, see Franchello [78].

[78] Let $\mathcal{D}_e$ be the class of distributive lattices with distinguished element e. Let $\{G_i\colon i \in I\}$ be a family of ℓ-groups and G be the ℓ-group free product of $\{G_i\colon i \in I\}$. Using Corollary 2L, the sublattice generated by the set of distributive lattices $\{G_i\colon i \in I\}$ is proved to be the $\mathcal{D}_e$-free product of these lattices. This contrasts with the group theory results due to Franchello [77], Holland [77], and Holland and Scrimger [72]. This may be a bit surprising since, in Section 2.1, we showed that if $A(\Omega)$ is doubly transitive, the lattice and group languages of $A(\Omega)$ are equally powerful.

E. Fried:

[65] It is shown that if Ω is a chain, then $\mathcal{S}(\Omega)$, the symmetric group on Ω, is a left partially ordered group. Furthermore, every left partially ordered group can be o-embedded in some such $\mathcal{S}(\Omega)$. Necessary and sufficient conditions are also given for a p.o. group to be o-embeddable in some $A(\Omega)$. The footnote on page 17 is doubtful. Note "left" would be "right" if αf were used instead of $f(\alpha)$.

[67] It is proved that a p.o. group G can be o-embedded in $A(\Omega)$ for some chain Ω if and only if the partial order is the intersection of all left total orders on G that extend it. This tidies up the result in Fried [65]. Also cf. Conrad [70], Cohn [57] and Appendix I Corollary 10.

L. Fuchs:

[63] This book used to take the reader to the forefront of the many aspects of partially ordered algebraic structures. However, the subject has mushroomed in the last twenty years. Nonetheless, the book is still valuable for its readability and bibliography, and is an excellent introduction to the material.

A. M. W. Glass:

[72] A directed p.o. set in which no two incomparable elements have a supremum or infimum is called an antilattice set. So any antilattice set that is a lattice is a chain. A generalisation of Holland's Theorem is proved; viz., any directed interpolation group (this class includes ℓ-groups and tight Riesz groups) can be order embedded in the automorphisms of some antilattice set.

[74] The first examples of non-ℓ-simple pathological ℓ-permutation groups are given. They include Theorem 6.7 for certain infinite cardinals κ, including $\kappa = \aleph_0$. They answer a question of McCleary [73b]: Is a transitive primitive ℓ-permutation group ℓ-simple if $G \subseteq B(\Omega)$ or $G \cap B(\Omega) = \{e\}$?

[75a] It is shown that there exist recursively presented ℓ-groups (finitely generated with a recursively enumerable set of defining relations) having insoluble word problems. This paper is completely superceded by Chapter 13 (which is stronger and more difficult). Also cf. Fox [75].

[75b] Much combinatorial group theory is predicated on a theorem by Higman-Neumann-Neumann: If A and B are isomorphic subgroups of a group G, then it is possible to adjoin to G a new element t so that in a new group H generated by t and G, the isomorphism between A and B is realised as conjugation by t. This property is shown in groups with operators to imply the amalgamation property. Hence the property fails, in general, in ℓ-groups. Conditions are given in which it holds. Also note Lemma 2.2.9. Unrelatedly, an example is given of an Abelian subgroup of $A(\mathbb{R})$ that is contained in no Abelian divisible subgroup of $A(\mathbb{R})$.

[76] The existence of CTRO's for general transitive ℓ-permutation groups (G,Ω) is considered. Theorem 3.7 of the paper is false--see Glass [79]. For the definition of CTRO's, see Davis and Loci [76]. The work relies heavily on the structure theory given in Chapters 3-5.

[79] Theorem 3.7 is corrected: $A(\Omega)$ endures a CTRO if and only if $A(\Omega) \not\cong \mathbb{Z}$. Again the proof relies heavily on the structure theory. Also see Davis and McCleary [81a].

[81] This is a survey of the work to date on the problem: If Ω_1, Ω_2 are homogeneous chains and $A(\Omega_1)$ and $A(\Omega_2)$ satisfy the same (ℓ-)group sentences, what is the relationship between Ω_1 and Ω_2? Some of the ideas used are established in Sections 2.1 and 4.2.

A. M. W. Glass and Y. Gurevich:

[81] This is Chapter 13. It extends Glass [75a].

A. M. W. Glass, Y. Gurevich, W. C. Holland and M. Jambu-Giraudet:

[81] Some of the material in Section 2.1 is developed including the proof of Theorems 2B and 2C. In addition, it is proved that if Ω is a doubly homogeneous chain that satisfies the Souslin, Lusin or Specker property, then so does any homogeneous chain Λ such that $A(\Lambda)$ and $A(\Omega)$ satisfy the same (ℓ-)group sentences. In the same setting, other properties that transfer are: Archimedean groupable, Archimedean fieldable, and various long lines. Also see Gurevich and Holland [81].

A. M. W. Glass, Y. Gurevich, W. C. Holland and S. Shelah:

[81] The material from Section 4.2 is taken from this paper. In addition, the existence of rigidly homogeneous chains of cardinality less than $2^{\aleph_0}$ is shown to be independent of the axioms of set theory. Other results about rigidly homogeneous chains are established; e.g., they cannot be covered by a set of intervals $\{I_n : n \in \mathbb{Z}^+\}$ with $\mu(I_n) < 9^{-n}$; there are rigidly homogeneous chains whose divisible closure is $\mathbb{R}$. For earlier results about rigidly homogeneous chains, see Ohkuma [55]. Also see Roitman [81].

A. M. W. Glass and W. C. Holland:

[73] The equivalence of (i), (iii) and (iv) of Theorem 11.2 is established; cf. Wolfenstein [68].

A. M. W. Glass, W. C. Holland and S. H. McCleary:

[75] For the definition of a*-extension and a*-closure, see our annotation of Ball [75]. If G is completely distributive, it is shown to have an a*-closure (which is not necessarily unique). Moreover, using the structure theory of Chapters 3-5, the a*-closures of such transitive ℓ-groups are explicitly constructed (they have the same set k of pairs of convex congruences and if (G_K, Ω_K) is a primitive component, the K-primitive component of H is contained in $A(\bar{\Omega}_K)$). R. N. Ball [75] subsequently showed that every ℓ-group has an a*-closure. Also cf. D. Khuon [70b].

[80] This paper can be thought of as the sequel to Martinez [74]. Theorems 11E and 8.3.7 are established as well as a dimension theory for varieties of ℓ-groups, laws defining product varieties and certain distributive laws for the ℓ-semigroup of varieties.

Moreover, it is shown that every proper variety of ℓ-groups other than $\mathcal{N}$ can be written uniquely as a product of indecomposable varieties (A variety is indecomposable if it cannot be written as the product of two non-trivial varieties). The proofs rely heavily on Wreath products. Questions VI and IX of Martinez [74] are answered.

A. M. W. Glass and S. H. McCleary:

[76] Four new examples of pathological ℓ-permutation groups are given, three of which are ℓ-simple. Two of these examples contain elements which fix precisely one point. This leads to a counterexample to a conjecture of Ball [74].

A. M. W. Glass and K. R. Pierce:

[80a] This is an announcement of the results that appear in Glass and Pierce [80b] and [80c]. It also includes some informal remarks about finding an algebraic sentence that distinguishes between finitely and infinitely generic ℓ-groups as well as a self-contained appendix (A Lazy Algebraist's Guide to Model-Theoretic Forcing) which gives the necessary background from model theory.

[80b] This is the Abelian analogue of the material in Chapter 12 (Glass and Pierce [80c]). It is shown that algebraically closed Abelian ℓ-groups need not be existentially closed Abelian ℓ-groups (contrast Theorem 12B), and that given an existentially closed Abelian ℓ-group, there is an Abelian ℓ-group satisfying the same sentences that is not existentially closed (cf. Theorem 12.6). Moreover, there is an $\forall\exists\forall$ sentence which is equivalent to the Archimedean property for existentially complete Abelian ℓ-groups and holds in finitely generic Abelian ℓ-groups but not in infinitely generic ones.

[80c] This includes the results of Chapter 12 (with the exception of Theorem 12E). It also includes: (i) There are $2^{\aleph_0}$ pairwise non-elementarily equivalent countable algebraically closed ℓ-groups and (ii) There is an $\forall\exists\forall$ sentence distinguishing between finitely and infinitely generic ℓ-groups. A deeper examination of algebraically closed ℓ-groups and ℓ-permutation groups remains outstanding; see, e.g., problems on simplicity and ℓ-simplicity No. 5.

Y. Gurevich:

[67] A class of Archimedean (Abelian) ℓ-groups that are real vector lattices is shown to be hereditarily undecidable (The class includes sums of copies of $\mathbb{R}$). Actually, the proof works for the lattice theory of the class--replace $z \neq u + v$ by $z \neq u \vee v$ in the definition of $\bar{P}(x,y)$. It follows that the lattice theory of (Abelian) ℓ-groups is undecidable; cf., Theorem 2.2.8 and Jambu-Giraudet [81b].

Y. Gurevich and W. C. Holland:

[81] This is the first proof of Theorems 2B, 2B*, 2C and 2C*. Lemmas 2.1.1-2.1.9 are established. The proof is completed by heavily using the arithmetic of $\mathbb{R}$. Specifically, Lemmas 2.1.10-2.1.12 are replaced by: Let Ω be doubly homogeneous and G be a subgroup of $A(\Omega)$ all of whose $\neq e$ elements are unbounded bumps. Then each $e \neq f \in C_{A(\Omega)}(G)$ is an unbounded bump if and only if $\bar{\alpha}G$ is dense in $\bar{\Omega}$ for all $\bar{\alpha} \in \bar{\Omega}$. Moreover, if H is a non-cyclic subgroup of G and G satisfies the above equivalent conditions, then so does H. Consequently, if Ω is doubly homogeneous and there are $f_1, f_2 \in A(\Omega)$ which commute and each $e \neq g \in C_{A(\Omega)}(\{f_1, f_2\})$ is an unbounded bump, then Ω has a countable dense subset. In $A(\mathbb{R})$, f_1 and f_2 can be taken to be translation by 1 and $\sqrt{2}$ respectively. This idea is used in the Archimedean groupable and fieldable part of Glass, Gurevich, Holland and Jambu-Giraudet [81].

J. L. Hickman:

[76] Let T be any chain. Then $\mathbb{Q} \overleftarrow{\times} T$ is dense and $A(T)$ can be ℓ-embedded in $A(\mathbb{Q} \overleftarrow{\times} T)$ via: $g \mapsto \bar{g}$ where $(q,\tau)\bar{g} = (q,\tau g)$ $(q \in \mathbb{Q};\ \tau \in T)$. This proves a stronger theorem than that of this paper--and in ten pages less. What is worse, Lemma 8 is palpably false (See Example 4.2.6--which is clearly non-Abelian--or Fleischer [77]).

[77] This corrects Hickman [76] by recourse to wreath products thus making the trivial even more inscrutable.

G. Higman:

[54] Let Ω be an infinite set and G be a group of permutations of Ω satisfying: For all $f,g,h \in G$ with $h \neq e$, there is $k \in G$ such that $\Delta kh \cap \Delta k = \emptyset$, where $\Delta = \mathrm{supp}(f) \cup \mathrm{supp}(g)$. Then $[G,G]$ is proved to be simple (Lemma 6.4 and its proof are just a special case of Higman's work). As a consequence, if κ and λ are infinite cardinals with $\kappa \leq \lambda \leq 2^{\kappa}$, it is shown there is a simple group of size λ having a subgroup of index κ.

[61] It is proved that a finitely generated group can be embedded in a finitely presented group if and only if it has a recursively enumerable set of defining relations. The proof of this superb embedding theorem uses the amalgamation property (free products with amalgamation). The theorem yields several important corollaries including an easy negative solution to the word problem for groups. Also see Boone and Higman [74] and Thompson [80].

[77] An n-ary relation R on a set Ω is said to be homogeneous if whenever $\Gamma, \Delta \subseteq \Omega$ with $|\Gamma| = |\Delta| < \aleph_0$, then $\langle \Gamma, R\rangle \cong \langle \Delta, R\rangle$. For example, if G is a group of permutations of Ω that is n-homogeneous, take R to be the orbit under G of n-tuples of distinct elements of Ω (see our annotation of Cameron [76]). R is said to be derived from a binary relation S if for all $\alpha_1, \ldots, \alpha_n, \beta_1, \ldots, \beta_n \in \Omega$, $R(\alpha_1, \ldots, \alpha_n) \leftrightarrow R(\beta_1, \ldots, \beta_n)$ whenever $(\forall i,j \in \{1, \ldots, n\})(S(\alpha_i, \alpha_j) \leftrightarrow S(\beta_i, \beta_j))$. By an application of Ramsey's Theorem for infinite sets and the Compactness Theorem for first order logic, it is shown that: A relation R on an infinite set Ω is homogeneous if and only if it is derived from some linear ordering on Ω. If $(\alpha, \beta, \gamma, \delta) \in D$ when precisely one of γ and δ lies strictly between α and β, D is called the derived separation relation. The main thrust of the rest of the paper is to classify the group of permutations of a set Ω that respect a given relation derived from a linear order on Ω. In the case that Ω is infinite and the group is transitive, this reduces to one of five groups: $A(\Omega)$, the full symmetric group, the group of all permutations that preserve the derived betweenness relation, the derived cyclic order, and the derived separation relation. Hence: A transitive permutation group on an infinite set Ω which preserves a non-trivial homogeneous relation preserves the derived separation relation for some linear order on Ω. This generalises the main theorem of Cameron [76].

W. C. Holland:

[63] The main results in this paper are Holland's Theorem (Appendix I), Theorem 2H (stated in a somewhat weaker form), Theorems 2E, 2F and 2J, and the first part of Section 2.2 (including Corollary 2.2.3 with a much longer proof). A most significant beginning to the whole subject of ℓ-permutation groups! This paper lead to the study of ℓ-permutation groups as a tool to develop the theory of non-abelian ℓ-groups.

[65a] Uniqueness of representation of transitive ℓ-permutations is considered. First, Theorems 4.1.1 and 4B are proved. Next Theorems 7A and 7B are established (together with some of their consequences, Corollaries 7D and 7E). Theorems 2D and 3A are also proved. Note that this paper predates the classification of transitive primitive ℓ-permutation groups given in Theorem 4A. Indeed, the form of Theorem 4B proved in the paper (If a transitive primitive ℓ-permutation group has an element whose support is bounded above or below, then it is doubly transitive) together with Example 1.9.2 (Holland [65b]) were the clues that lead to the classification of transitive primitive ℓ-permutation groups (McCleary [72a]) and the consequent Corollary 7C (McCleary, unpublished). Caution: there are several typographical errors in the paper.

[65b] Theorem 6.2 is stated and proved. A special case (Example 1.9.2) of Theorem 6C is also shown.

[65c] Example 1.9.2 is shown to be a Hausdorff space that is both a topological group and lattice in its open interval topology. However, it is not an o-group. This answers in the negative a question of G. Birkhoff raised in the 1948 edition of [67].

[72] This is a readable exposition of some of the results about ordered permutation groups. It is mainly confined to facts about transitive ℓ-permutation groups.

[75a] Theorems 9A and 9C are established, as is Theorem 9.5. The existence of a homogeneous chain Ω with $A(\Omega)$ having an outer ℓ-automorphism had been outstanding for ten years previously.

[75b] This review gives an example of a doubly homogeneous chain Ω having a coterminal set of intervals each of which is ordermorphic to itself with the reverse ordering, yet Ω and Ω^* are not ordermorphic (This disproves the only new theorem of the paper being reviewed (Longyear [74])). This result was independently obtained by Rabinovich [75].

[76a] It is proved that no dense homogeneous subset of $\mathbb{R}$ is ordermorphic to its complement. Hence if Ω is a dense homogeneous subset of $\mathbb{R}$, $A(\Omega)$ has no outer ℓ-automorphisms, cf. Theorem 9A and Holland [75a]. The proof proceeds by examining the two cases (i) Ω is doubly homogeneous and (ii) Ω is rigidly homogeneous. The first case is dismissed as in Treybig [63]; the second is disposed of by establishing that $\mathbb{R}\backslash\Omega$ is a coset of Ω in $\mathbb{R}$.

[76b] Theorem 11D is stated and proved. Also see Holland and McCleary [79] and McNulty [80].

[77] It is shown that if G and H are ℓ-groups, then no reduced group word of length less than 14 reduces to the identity in the ℓ-group free product of G and H. The proof is an ingenious combinatorial argument using wreath products. It also shows that there are reduced group words of length 14 that are the identity in this ℓ-group, and indeed finds them all. This paper is the first in "combinatorial ℓ-group theory." See Franchello [77] for length 16; also Holland and Scrimger [72].

[79] Theorem 11F (every variety of ℓ-groups is a torsion class) is proved.

[80] This is a series of remarks about doubly homogeneous chains Ω such that $A(\Omega)$ and $A(\mathbb{R})$ satisfy the same ℓ-group sentences. The work is completed in Gurevich and Holland [81]--

see Theorem 2B (also Theorems 2B* and 2B†). Also see Glass, Gurevich, Holland and Jambu-Giraudet [81].

W. C. Holland and S. H. McCleary:

[69] This paper is basically Chapter 5 and builds on Holland [69], the construction of the generalised wreath product for groups.

[79] By analysing Holland [76b] (the proof of Theorem 11D) it is first noted that to test whether an arbitrary ℓ-group word reduces to e in all ℓ-groups, it is enough to test it in $A(\mathbb{R})$. An algorithm is then established to do this. Hence the free ℓ-group on any number of generators has soluble word problem (cf. Chapter 13). This connection was recently underlined by the completion of the proof of Theorem 6.7 in the finite case by McCleary [81c].

W. C. Holland and E. B. Scrimger:

[72] The p.o. group free product of a family of p.o. groups is constructed as is the ℓ-group free product of a family of ℓ-groups. This ℓ-group free product is shown to be the group free product only if each ℓ-group in the family is an o-group. Lemma 1.11.5 is proved to show that the ℓ-group free product of two ℓ-groups need not contain the group free product (of the ℓ-groups). For further results, see Franchello [77] and Holland [77]. Also see Bergman [81] for ordering the free product of o-groups.

M. Jambu-Giraudet:

[80] This is an announcement of the results in the author's Thèse 3ème Cycle (1979). The proofs appear in her papers [81a], [81b] and [81c].

[81a] Theorems 2A and 2A† (and consequently Theorems 2D and 2D†) are established. The proofs given in Section 2.1 are taken from this paper. Theorem 2D had previously been established in the ℓ-group case by Holland [65a], and in the group case by McCleary [78b] and Rabinovich [75], though via identifiable subgroups rather than directly using the elements of $A(\Omega)$.

[81b] Using Theorems 2A and 2A†, Theorem 2.2.8 and its lattice analogue are established. The proof is as follows: The set of bumps of f is a set of bumps of g if each bump of f is a bump of g. This latter is expressible in the ℓ-group language by Lemma 1.9.3. Hence g has a finite set of bumps if each non-empty subset of the set of bumps has a leftmost and rightmost member. We now interpret each $n \in \omega$ with the set of positive elements of $A(\Omega)$ having n bounded non-adjacent bumps. Equality between elements of ω is easily interpretable. $m + n = p$ if whenever $g,h > e$ have m and n bumps, respectively, there is $x \in A(\Omega)$ such that $g \wedge x^{-1}hx = e$ and $g \cdot x^{-1}hx$ has p bumps.

$m \times n = p$ is obtained by putting conjugates of h under each bump of g and requiring the product of these conjugates have p bumps.

[81c] It was announced in Jambu-Giraudet [81b] that second order arithmetic can be interpreted in the group (lattice) of every doubly transitive $A(\Omega)$. If, in addition, Ω has a set of $\beth_{n-1}$ pairwise disjoint non-empty open intervals, then nth order arithmetic can be interpreted in the group (lattice) of $A(\Omega)$. These are now proved. The idea is to interpret ",n," by a permutation with $\omega \cup n \cup \omega^*$ bumps, and "{" and "}" by permutations with $\omega + \omega$ and $(\omega + \omega)^*$ bumps respectively.

N. G. Khisamiev:

[66] Y. Gurevich and A. I. Kokorin showed that any two Abelian o-groups satisfy the same universal sentences [*Algebra i Logika*, 2(1963), 37-39]. Let G be an Abelian ℓ-group and n(G) be the supremum of the cardinals of pairwise disjoint subsets of G. The main result proved is that Abelian ℓ-groups G and H satisfy the same universal sentences if and only if $m(G) = m(H)$, where $m(G) = \min\{n(G), \aleph_0\}$. Hence every finitely generated Abelian ℓ-group has soluble word problem. It is also shown that any free Abelian ℓ-group is contained in a cardinal product of copies of $\mathbb{Z}$ (also see Weinberg [65]).

D. Khuon:

[70a] An ℓ-group H is an a-extension of an ℓ-group G if the map $C \to C \cap G$ of the prime subgroups of H onto those of G is one-to-one. It is shown that if G is doubly transitive, so is any a-extension of G and $A(\bar{\Omega})$ is the unique a-closure of $A(\Omega)$ if Ω is doubly homogeneous. The proofs rely on Wolfenstein [70].

[70b] This paper is devoted to proving: Theorem: Every ℓ-group has an a-closure (a maximal a-extension). If (G,Ω) is a transitive ℓ-permutation group and $\alpha \in \Omega$, then $(G_\alpha|\Omega_\alpha,\Omega_\alpha)$ is uniquely transitive, where $\Omega_\alpha = \{\beta \in \Omega\colon G_\beta = G_\alpha\}$. By a result of Conrad, this puts a cardinality bound on Ω in terms of the prime subgroups of G. The proof of Holland's Theorem (Appendix I) now yields a cardinality bound on a-extensions of an ℓ-group in terms of the cardinality of its prime subgroups as required. Note that the main theorem was first proved by McCleary (unpublished), and the proof was generalised by Ball [75] to cover a*-extensions.

A. I. Kokorin and V. M. Kopytov:

[74] This book is similar in scope to Mura and Rhemtulla [77], but is not so elegant and has several errors in statements and theorems. The books have considerable overlap but include a lot of material not found in the other.

V. M. Kopytov:

[79] The class of groups which can be made into ℓ-groups is not an axiomatisable class (Vinogradov [71]). However, if $\mathcal{V}$ is a variety of ℓ-groups, the class $\mathcal{K}(\mathcal{V})$ of all groups that can be embedded (as a group) in some member of $\mathcal{V}$ is proved to be a quasi-variety of groups. Moreover, if $F_{\mathcal{V}}$ is the $\mathcal{V}$-free ℓ-group on $\{x_n : n \in \omega\}$, the subgroup F_0 generated by $\{x_n : n \in \omega\}$ is the $\mathcal{K}(\mathcal{V})$ free group on $\{x_n : n \in \omega\}$. Conversely, let F_0 be the $\mathcal{K}(\mathcal{V})$ free group on $\{x_n : n \in \omega\}$. Let $\{F_0^\lambda : \lambda \in \Lambda\}$ be the set of all right total orders on F_0. Let H_0^λ be the minimal convex subgroup of F_0^λ for which the ℓ-subgroup A_λ of $A(R(H_0^\lambda))$ generated by the natural embedding of F_0^λ belongs to $\mathcal{V}$. If F^* is the cardinal product $\prod_{\lambda \in \Lambda} A_\lambda$ and $\bar{F}$ is the ℓ-subgroup generated by $\bar{x}_n$ (where $(H_0^\lambda f)(\bar{x}_n(\lambda)) = H_0^\lambda f x_n$), then $\bar{F}$ is the $\mathcal{V}$-free ℓ-group on $\{\bar{x}_n : n \in \omega\}$. As a consequence, if a free ℓ-group F in $\mathcal{V}$ becomes a transitive ℓ-permutation group (F,Ω) for some chain Ω, then there is a right total order F_0^λ on F_0 for which $(F,R(H_0^\lambda))$ is a (faithful) transitive ℓ-permutation group. By Theorem 6.7, if F_0 is the free group on any set of generators, then (F,F_0^λ) is a transitive ℓ-permutation group for some $\lambda \in \Lambda$, where F is the free ℓ-group on the same set of generators as F_0. Similar results are obtained for the ℓ-group varieties $\mathcal{N}$, $\mathcal{A}$, $\mathcal{R}$ and $\mathcal{W}$, where $\mathcal{R}$ is the variety defined by: $(x \vee e) \wedge y^{-1}(x^{-1} \vee e)y = e$ (or equivalently $(x \vee e)^2 \wedge (y \vee e)^2 = [(x \wedge y) \vee e]^2$) and $\mathcal{W}$ is defined by: $x^{-1}|y|x \vee |y|^2 = |y|^2$. The results in this paper extend Conrad [70] and use only the material covered in Appendix I. Other theorems are also proved.

V. M. Kopytov and N. Ja. Medvedev:

[77] Olshansky proved that there are $2^{\aleph_0}$ distinct varieties $\{\mathfrak{X}_i : i \in I\} \subseteq \mathcal{A}^5$. Let F be the free group on $\aleph_0$ generators, N_i the verbal subgroup of F corresponding to $\mathfrak{X}_i$, and M_i its second derived subgroup. It is well-known that F/M_i admits linear orderings so the class of ℓ-groups in distinct $\mathcal{A}^2\mathfrak{X}_i$ are distinct $(i \in I)$. Hence there are $2^{\aleph_0}$ distinct varieties of ℓ-groups. For more recent proofs, see Feil [81] and Reilly [81]. An example is given of a variety of ℓ-groups that is not generated by any finitely generated ℓ-group. This shows that Theorem 5.1 of

Martinez [74] is false. The lattice of varieties of ℓ-groups is shown not to be Brouwerian. The proof is the same as Smith [80a]. The last part of the paper contains an oversight. It is not at all clear that the defined $\mathcal{W}_n$ is closed under ℓ-homomorphic images. Hence Theorem 4 is not yet established.

J. T. Lloyd:

[64] First, $A(\Omega)$ is shown to be completely distributive and have closed stabilisers (Theorems 8B and 8.1.2)--see Lloyd [67] and Read [71]. Moreover, if Ω is doubly homogeneous, it is proved that every proper normal subgroup of $A(\Omega)$ is an ℓ-subgroup that contains $B(\Omega)$ (cf. Theorems 2.3.1 and 2G) and that stabiliser subgroups are maximal (cf. Theorem 4.1.5). This last fact, together with Holland [63] and [65a], lead to the central use of stabilisers in studying ℓ-automorphisms (and ℓ-isomorphisms) of $A(\Omega)$. It was employed in the original proofs of Theorems 2.3.1 and 2G, and later extended by McCleary [69], [78a] and [78b]--see also Chapters 7, 8 and 9. Indeed, Lloyd uses it here to prove that if transitive $A(\Omega)$ has the interval support property, then each ℓ-automorphism of $A(\Omega)$ is induced by conjugation by an element of $A(\bar{\Omega})$--cf. Theorem 9B. He thus obtains Corollary 2.4.4. Next it is shown that if for transitive $A(\Omega)$, $A(\Omega)_\alpha = \{e\}$ for all $\alpha \in \Omega$, then $A(\Omega)$ is an o-group (cf. Lemma 4.2.4). Corollary 2K is proved in special cases without using G.C.H.--Weinberg [67] manages this in total generality. Lloyd also gives a proof of Lemma 10.3 and hence Corollary 10.4. However, the statement at the beginning of the proof of his Lemma 1.29 is false, so the proof given fails. McCleary [73a] corrects the error.

[65] It is shown that an ℓ-group with a basis admits a minimal irreducible representation as a product of transitive ℓ-permutation groups, and that this minimal irreducible representation is essentially unique.

[67] It is proved that $A(\Omega)$ is completely distributive for any chain Ω. Using this, Theorem 8B is deduced (for the special case $\bar{\alpha} \in \Omega$). For an amplification of the method used, see the annotation of Read [71]. Also Example 1.10.3 is shown to be ℓ-simple (cf. Theorem 6.5) but not completely distributive. This last fact is generalised in Corollary 8.3.5 due to McCleary [73b].

J. Q. Longyear:

[74] The only new theorem in this paper is unfortunately false--see Holland [75b] or Rabinovich [75].

A. Macintyre:

[72] This is the group theory analogue of our Chapter 12. The results are obtained more cleanly by using free products with amalgamation (a luxury not afforded those who study ℓ-groups!) and

are stronger. For example, two algebraically closed groups are elementarily equivalent if and only if the sets of two generator groups that can be embedded in each are equal; in an algebraically closed group, $f \in \langle g_1,\ldots,g_n \rangle$ if and only if $f \in C(C(\{g_1,\ldots,g_n\}))$--cf. Lemma 12.4. There is an $\exists\forall\exists\forall$ sentence, equivalent for algebraically closed groups to a universal two generator group, which holds in some algebraically closed groups but fails in others. Indeed, there are $2^{\aleph_0}$ pairwise non-elementarily equivalent countable algebraically closed groups (In an earlier paper, the same author showed that a finitely generated group can be embedded in all algebraically closed groups $\neq \{e\}$ if and only if it has soluble word problem).

A. I. Mal'cev:

[71] Only Chapter 15 of this book is referred to. It gives an elegant interpretation of the ring of integers in the free nilpotent class 2 group on two generators ("metabelian" used by the translator means "nilpotent class 2"). Since torsion-free nilpotent groups are orderable (see Fuchs [63]), it follows that the theory of orderable groups is undecidable, cf. Buszkowski [79], Glass and Pierce [80c], Gurevich [67] and Jambu-Giraudet [81b].

J. Martinez:

[74] This paper is the first on varieties of ℓ-groups. It is more important for the questions it asks, the examples it gives, and the research of others it spawned than for the results it proves. Caution: there are many errors. 3.5 Proposition (b) needs 4.1 Lemma for its proof, 4.1 Lemma (e) should be: $\mathcal{V}$ is idempotent if and only if $[R(R(G,\mathcal{V}),\mathcal{V}) = 0 \iff G \in \mathcal{V}]$ and Section 5 is wrong ($\mathcal{V} = \bigvee_{n=1}^{\infty} \mathcal{V}_{(n)}$, not $\mathcal{V} = \bigcup_{n=1}^{\infty} \mathcal{V}_{(n)}$ as asserted). For further results on varieties of ℓ-groups, see Bernau [77], Feil [81], Glass, Holland and McCleary [80], Kopytov and Medvedev [77] (where Martinez's Theorem 5.1 is proved false), Medvedev [77], Reilly [81], Reilly and Wroblewski [81], Scrimger [75] and the works of Smith.

S. H. McCleary:

[69] Theorems 8.3.8 and 9B are proved--indeed, Theorem 9B is proved for certain depressible ℓ-subgroups of $A(\Omega)$. The results generalise some of Lloyd [64].

[72a] Theorem 4.3.7 (and hence Theorem 4A) is established. So is Theorem 4.3.10. This paper is a polished version of McCleary's Ph.D. thesis (1967). Its central importance is clear to anyone who has read this book. For prior results in this direction, see Holland [65a] and [65b].

[72b] The transitive part of Theorem 8.2.1 is proved. Note that this work was done in 1969--the journal had a big backlog.

[72c] Theorem 8.2.1 is established and extended to Theorem 8C. Theorem 8F is also proved. Indeed, much of Sections 8.2 and 8.3 come from this paper (including Theorems 8.3.1-8.3.4, 8.3.6, 8.3.9 and 8.3.10). The techniques used are considerable extensions of the ideas of McCleary [72b].

[73a] An error in Lloyd [64] is corrected, Lemma 10.3 being proved correctly. Hence, assuming G.C.H., every ℓ-group can be ℓ-embedded in an ℓ-group which has only inner ℓ-automorphisms. (Weinberg [80] removed the dependence on G.C.H.) Further, a variant form of Theorem 7A is established: If Ω is doubly homogeneous and ψ is an ℓ-embedding of $A(\Omega)$ into $A(T)$ such that $A(\Omega)\psi$ is a complete ℓ-subgroup of $A(T)$, then there is $\bar{\sigma} \in \bar{\Omega}$ and an ordermorphism $\phi: \bar{\sigma}A(\Omega) \cong T$ such that $g\psi = \phi^{-1}g\phi$ for all $g \in G$ and (ψ,ϕ) is a ℓ-permutation group isomorphism between $(A(\Omega),\bar{\sigma}A(\Omega))$ and $(A(\Omega)\psi,T)$. Truncated α-sets are also considered but Proposition 18 is false if β or γ equals $\alpha + 1$ (see Holland [75b] or Rabinovich [75]).

[73b] Theorem 8D is proved and Example 6.6 is given and shown to be ℓ-simple. Glass [74] answers a question left open in this paper.

[73c] This is a sequel to McCleary [72a] and is rather technical. The only part we have used is the last sentence of our Theorem 4.3.1. Full periodic primitive ℓ-permutation groups of each Config(n) ($n \in \mathbb{Z}^+ \cup \{\infty\}$) are constructed using α-sets. Note that there are two minor typographical errors: p. 433 line -13: $(\omega_n g - \omega g)/(\omega_n - \omega) \to 0$ or ∞; p. 434 line 4 of (4): $\bar{\Gamma} \cap \bar{\omega}G$.

[74] Some of the ideas of McCleary [76] are mentioned. Also included is the approach we have adopted in Chapter 11 (especially Theorem 11A)--rather than that of Read [75] which really predates it by three years.

[76] This paper examines the structure of intransitive ordered permutation groups and brings order to the chaos that had previously existed. The classification of primitive ℓ-permutation groups is obtained. This leads to straightforward proofs of prior results of Lloyd [64] and [67] and Read [71] and [75]. We have adopted its approach in Sections 3.2, 8.1 and Chapter 11 but have not included all the results to be found in this paper whose central importance cannot be overemphasized. For earlier work on the subject, see Scrimger [70].

[78a] Examples are given of infinite simple groups having insoluble outer automorphism groups (Schreier conjectured that this cannot occur if the simple group is finite--the conjecture is still open). The example given in Theorem 2.4.2 comes from this paper. The proof we give relies on the work of Section 2.1; McCleary's, on stabilisers.

[78b] Aut(G) is examined when (G,Ω) is triply transitive. If $\mathcal{M}(\Omega)$ denotes the group of all monotonic permutations of the chain Ω and G contains strictly positive elements of bounded support, then $\mathrm{Aut}(G) \cong N_{\mathcal{M}(\Omega)}(G)$ is shown (where N denotes the normaliser); and no two of $G \cap B(\Omega)$, $G \cap L(\Omega)$ and $G \cap R(\Omega)$ are ℓ-isomorphic (cf. Corollary 2.1.4). Using α-sets, an example is given of a doubly homogeneous chain Ω such that $|\mathrm{Outer}(A(\Omega))| = 4$ but $|\mathrm{Outer}(\mathcal{M}(\Omega))| = 2$ (cf. Holland [75a] and Theorem 9A). The main tool is Corollary 1.10.8 which is established and used to obtain Section 7.2 and the above. This lemma is also centrally used by Jambu-Giraudet [81a] who obtained it independently. Also compare with Rabinovich [75].

[80] In Glass, Holland and McCleary [80] it was shown that the variety $\mathcal{N}$ of normal-valued ℓ-groups is generated by $\{\mathrm{Wr}^n\mathbb{Z}: n \geq 1\}$, where $\mathrm{Wr}^1\mathbb{Z} = \mathbb{Z}$ and $\mathrm{Wr}^{m+1}\mathbb{Z} = (\mathrm{Wr}^m\mathbb{Z})\,\mathrm{Wr}\,\mathbb{Z}$. Hence to decide whether or not an ℓ-group identity $\forall \underset{\sim}{x}(w(\underset{\sim}{x}) = e)$ holds in every normal-valued ℓ-group ($w(\underset{\sim}{x})$ an ℓ-group word), it is enough to check it in $\{\mathrm{Wr}^n\,\mathbb{Z}: n \geq 1\}$. An algorithm is explicitly given to do this; indeed, given such a word w, there is a bound m_w such that $\forall \underset{\sim}{x}(w(\underset{\sim}{x}) = e)$ holds in every normal-valued ℓ-group if it holds in $\{\mathrm{Wr}^n\mathbb{Z} : 1 \leq n \leq m_w\}$ (A gross bound for m_w is obtained by taking 2 plus the sum of the lengths of the w_{ij}'s--if w_{ij} has form $x_1^{-k}x_2x_3^3x_1^k$, then it has length $k + 1 + 3 + k$). Hence the free normal-valued ℓ-group on any number of generators has soluble word problem. Actually, an algorithm is given to decide whether or not a disjunctive sentence holds in a free normal valued ℓ-group.

[81a] This is essentially Bernau's proof [75] but with some considerable shortcuts.

[81b] This is a considerable extension of the ideas in McCleary [80]. For example, it is shown that if $w_1(\underset{\sim}{x})$ and $w_2(\underset{\sim}{x})$ are group words, then $\forall \underset{\sim}{x}(w_1(\underset{\sim}{x}) \vee w_2(\underset{\sim}{x}) \geq e)$ holds in all normal valued ℓ-groups if and only if there is a group word $w(\underset{\sim}{x})$ such that $w_1(\underset{\sim}{x})$ and $w_2^{-1}(\underset{\sim}{x})$ are positive powers of $w(\underset{\sim}{x})$.

[81c] The main result of this paper is that the free ℓ-group on a finite number (> 1) of generators has a doubly transitive representation on $\mathbb{Q}$. It was previously known--Glass [74]--that the free ℓ-group on an infinite number of generators has a doubly transitive representation. Hence Theorem 6.7 holds and a free ℓ-group on more than one generator is not directly decomposable. Also see Kopytov [79].

G. F. McNulty:

[80] This is a refinement of Holland [76b] (i.e., Theorem 11D): viz.: If an ℓ-group $G \notin \mathcal{N}$ and θ is a positive universal sentence that holds in G, then θ is true in every ℓ-group. Equivalently, the only class of ℓ-groups that is closed under ℓ-subgroups and ℓ-homomorphic images and contains a doubly transitive ℓ-group is the class of all ℓ-groups. The last paragraph of the proof of Theorem 11D is shown to yield: any finite set of inequations in any $A(\Lambda)$ can be realised in any doubly transitive G.

R. N. McKenzie and R. J. Thompson:

[73] This is an extremely ingenious construction of a finitely presented group with insoluble word problem. The proof proceeds by coding elements of ${}^{\omega}\omega$ into the symmetric group $\mathcal{H}$ on ${}^{\omega}\omega$ in such a way that if A_f denotes the permutation of ${}^{\omega}\omega$ corresponding to $f \in {}^{\omega}\omega$, then in some finitely presented group $\mathcal{G}_f$, conjugation by the formal symbol A_f naturally corresponds to the function f if f is recursive, and the relations in $\mathcal{G}_f$ all hold in $\mathcal{H}$ under a fixed interpretation. Heavy use is made of those permutations determined by their action on ${}^{n}\omega$ $(n \in \omega)$ and of a definable element R of order 3 which cyclically permutes the top three levels of ${}^{\omega}\omega$. For ℓ-groups, the absence of such an element R is finessed by conjugating a set of elements simultaneously to another set by one element of $A(\mathbb{R})$ (see Chapter 13).

N. Ja. Medvedev:

[77] Let H_1 be $\mathbb{Z}$ wr $\mathbb{Z}$ ordered by: $(\{g_n\},m) > e$ if $m > 0$ or $(m = 0$ & $g_{n_1} > 0)$ and H_2 be $\mathbb{Z}$ wr $\mathbb{Z}$ ordered by: $(\{g_n\},m) > e$ if $m > 0$ or $(m = 0$ & $g_{n_2} > 0)$, where $n_1 = \max\{n \in \mathbb{Z}: g_n \neq 0\}$ and $n_2 = \min\{n \in \mathbb{Z}: g_n \neq 0\}$. Let $\mathcal{W}_i$ be the ℓ-group variety generated by the o-group H_i $(i = 1,2)$. Let N_0 be the free nilpotent class 2 group on x,y ordered by: $x^m y^n [x,y]^k > e$ if $m > 0$ or $(m = 0$ & $n > 0)$ or $(m = n = 0$ & $k > 0)$. Let $\mathcal{N}_0$ be the ℓ-group variety generated by N_0. It is shown that if $\mathcal{V}$ is a representable ℓ-group variety that contains a soluble non-Abelian ℓ-group, then $\mathcal{V}$ contains $\mathcal{N}_0$, $\mathcal{W}_1$ or $\mathcal{W}_2$; i.e., these three ℓ-group varieties cover $\mathcal{A}$ and every representable ℓ-group variety containing a soluble non-Abelian ℓ-group, contains at least one of them. Also see Scrimger [75]. Are there any other representable covers of $\mathcal{A}$?

R. B. Mura and A. H. Rhemtulla:

[77] An elegant and excellent account of various group-theoretic conditions which imply various forms of orderability.

B. H. Neumann:

[60] The analogue of Theorem 10A for o-groups is proved (We have made only minimal changes in the proof to obtain Theorem 10A). Also a finitely generated o-group is constructed o-isomorphic to a proper o-homomorphic image of itself.

B. H. Neumann and H. Neumann:

[59] Several results first obtained for groups by Higman, Neumann and Neumann (*J. London Math. Soc.*, 24(1949), 247-254) using free products with amalgamation are obtained using Wreath products. This method can be adapted to o-groups and ℓ-groups (see B. H. Neumann [60]) whereas the original cannot.

T. Ohkuma:

[55] Theorem 4.2.2 is established. The proof is longer than the one we have given and does not yield the stronger form, Theorem 4.2.7, due to Glass, Gurevich, Holland and Shelah [81]. Dushnik and Miller [40] can be thought of as a precursor of Ohkuma [55]--though Ohkuma's work is considerably more sophisticated and difficult.

K. R. Pierce:

[72] Theorems 10B and 10C are proved as is 10.12. The proofs are essentially those we have given. For an application, see Chapter 12, or Glass and Pierce [80c].

H. L. Putt:

[77] Let (G,Ω) be a permutation group. A partial order $\leq$ on Ω is called G-invariant if $\alpha \leq \beta$ implies $\alpha g \leq \beta g$ for all $g \in G$. Fix $\alpha \in \Omega$. A subsemigroup P of G is said to be an α-cone if $P \cap G_\alpha = \emptyset$ and $PG_\alpha = P = G_\alpha P$--$P$ is allowed to be empty. The set of G-invariant partial orderings on Ω is shown to be in one-to-one correspondence with the set of α-cones. Necessary and sufficient conditions are given for each of the following: (1) Every G-invariant partial order on Ω extends to a G-invariant total order on Ω. (2) Condition (1) plus there are exactly two G-invariant total orders on Ω. (3) There are just three G-invariant partial orders on Ω, the trivial one and two G-invariant total orders. Moreover, $(A(\mathbb{R}),\mathbb{R})$ is shown to satisfy (3), where $A(\mathbb{R})$ is viewed as acting on the *set* $\mathbb{R}$. Applications to right totally ordered groups and special G-invariant partial orderings are also considered.

E. B. Rabinovich:

[75] Theorem 2D is established. The key lemma is the following: Let $g \in A(\Omega)$, where Ω is doubly homogeneous. If $M_g = \{f \in A(\Omega): g^n = f^m \text{ for some } m,n \in \mathbb{Z}\}$, then $C(M_g) = \prod\{A(\Delta_i): i \in I\}$, where Δ_i are pairwise disjoint intervals of Ω maximal with respect to being disjoint from supp(g). It is also shown that if Ω_j are doubly homogeneous and G_j are subgroups of $A(\Omega_j)$ containing $B(\Omega_j)$ $(j = 1,2)$, then any group isomorphism between G_1 and G_2 extends uniquely to a group isomorphism between $A(\Omega_1)$ and $A(\Omega_2)$. Hence $L(\Omega_1)$ is a characteristic subgroup of $A(\Omega_1)$ unless Ω_1 is antiordermorphic to an orbit of $A(\Omega_1)$ in $\bar{\Omega}_1$; in this case, $L(\Omega_1) \cong R(\Omega_1)$ (as groups)--cf. McCleary [78b]. Proposition 18 of McCleary [73a] is shown to be false (also see Holland [75b]). Further results are obtained for good chains--a doubly homogeneous chain Ω is good if whenever Δ_1 and Δ_2 are bounded intervals of Ω having the same initial characters and the same final characters, then Δ_1 and Δ_2 are ordermorphic. The orbits under $A(\Omega)$ of holes in such Ω comprise all holes of the given character. Let $\hat{\Omega}$ be the set of holes in Ω of the same character as the points of Ω. It is proved that if Λ is homogeneous, Ω is good and $\hat{\Omega} \neq \emptyset$, then $A(\Omega) \cong A(\Lambda)$ (as groups) if and only if Λ is ordermorphic or antiordermorphic to Ω or $\hat{\Omega}$.

E. B. Rabinovich and V. E. Feinberg:

[74] Theorem 2.3.2 is claimed without the restriction that Ω has countable coterminality. As we saw in Theorem 2.3.7 this is false. The crucial error occurs in their Proposition 3 (which is false). It is claimed that there is $s \in A(X)$ such that $s(c_\beta) = a_\beta$ and $s(d_\beta) = b_\beta$ $(0 \leq \beta < \alpha)$--see p. 193 line -3 of the English translation. This fails if $\alpha > \omega$, $a_\omega = \sup\{b_n: n \in \omega\}$ and $c_\omega > \sup\{d_n: n \in \omega\}$.

J. A. Read:

[71] The main theorem proved is Theorem 8.1.2. The proof is quite ingenious and is a simplification of that of Lloyd [64] or [67]. The key lemma (also in Lloyd, *op. cit.*) is: If $e < g \in A(\Omega)$ and $\alpha \in \text{supp}(g)$, then either there is $e < h \in A(\Omega)$ with $\text{supp}(h) \subseteq (\alpha,\alpha g)$ or $\{h \in A(\Omega): e \leq h \leq g \;\&\; \text{supp}(h) \subseteq \Delta(\alpha,g)\}$ contains a basic element. The two cases are then examined separately using a criterion of Weinberg's from ℓ-groups (the same is true of Lloyd's proof).

[75] In this paper, non-overlapping ℓ-permutation groups are examined. The study is motivated by the key lemma of Read [71] (see the previous annotation). Theorems 11A (parts (1), (2) and (4)) and 11.2 (parts (i), (v) and (vi)) are established as is Corollary 11B. The proofs given are more ingenious than the ones given here (since they predate the intransitive structure theory of Sections 3.2 and 4.4). The work was done in 1970 and is independent of Wolfenstein [68]. Since the emphasis is on an ordered permutation group concept, the proofs are very different to Wolfenstein's. Warning: typographical errors riddle the paper; in special: Theorem 2.4 (1) should be $hg \leq g^2h^2$; page 718 line -3 should end: let S_γ be; page 719 line 9: change primarily to precisely.

N. R. Reilly:

[69] The major theorem is: Any o-group can be embedded in a simple o-group. The proof mirrors Philip Hall's for groups: Note first that if H is an o-group, then $H \text{ wr } \mathbb{Z}$ is an o-group under $(\{h_n\},m) > e$ if $m > 0$ or $(m = 0 \ \& \ h_{n_0} > e$ in H where n_0 is the least element of $\{n \in \mathbb{Z}: h_n \neq e\})$--cf. Medvedev [77]. Let G be an o-group, $H_0 = G$ and $H_{n+1} = H_n \text{ wr } G$ so ordered. Then the direct limit of $\{H_n: n \in \omega\}$ is the desired group.

[72] Using the analogue of B. H. Neumann's group theory construction "permutational products," it is shown that certain amalgamations are possible in the class of o-groups (ℓ-groups)--cf. Theorems 10.12 and 10C.

[81] An ℓ-group G is representable precisely when it satisfies $(\forall x_0)(\forall u)(x_0^+ \wedge u^{-1}x_0^-u = e)$, where $x_0^+ = x_0 \vee e$ and $x_0^- = x_0^{-1} \vee e$. If F denotes the free group on $\{x_1,x_2,\ldots\}$ and $\mathcal{V}$ is a variety of ℓ-groups, let $F(\mathcal{V}) = \{u \in F: x_0^+ \wedge u^{-1}x_0^-u = e$ is a law in $\mathcal{V}\}$. So if $\mathcal{R}$ is the variety of representable ℓ-groups, $F(\mathcal{R}) = F$. Also $F(\mathcal{L}) = \{e\}$. In general, $F(\mathcal{V})$ is a fully invariant subgroup of F. For N a fully invariant subgroup of F, let $R(N)$ be the variety of ℓ-groups defined by: $x_0^+ \wedge u^{-1}x_0^-u = e$ for all $u \in N$. By making F an o-group and considering $(\mathbb{Z},\mathbb{Z}) \text{ Wr } (F,F)$, it is proved that R is one-to-one, $FR(N) = N$ and $\mathcal{V} \subseteq RF(\mathcal{V})$. Moreover, many nice properties hold; e.g., $\bigcap_{i \in I} F(\mathcal{V}_i) = F(\bigvee_{i \in I} \mathcal{V}_i)$. In particular, since there are towers of fully invariant subgroups of cardinality $2^{\aleph_0}$, there is a tower of varieties of ℓ-groups between $\mathcal{R}$ and $\mathcal{L}$ of size $2^{\aleph_0}$. By a deep result of S. I. Adyan, there is a collection of $2^{\aleph_0}$ pairwise incomparable fully invariant subgroups of F. Hence

there are $2^{\aleph_0}$ pairwise incomparable varieties of ℓ-groups. No other proof of the existence of $2^{\aleph_0}$ varieties of ℓ-groups captures this last fact. (Also see Feil [81] and Kopytov and Medvedev [77].) For a simpler proof for quasi-varieties, see Arora [81].

N. R. Reilly and R. Wroblewski:

[81] The results in this paper are of the same ilk as those in Smith [81]. However, whereas Smith relies on the techniques in Scrimger [75], Reilly and Wroblewski use the more pictorial (ℓ-permutation group) approach of Glass, Holland and McCleary [80], especially "mimicing." Indeed, this is the first application of these techniques to specific varieities of ℓ-groups.

A. Robinson and E. Zakon:

[60] An Abelian o-group G is said to be regularly dense if for each non-empty open interval I of G, $I \cap pG \neq \emptyset$ for each prime p. Any densely ordered Archimedean (Abelian) o-group is regularly dense. It is shown that two regularly dense Abelian o-groups are elementarily equivalent if and only if they are elementarily equivalent as groups; i.e., precisely when they are Abelian and have the same set of Szmielew invariants. Any two discretely ordered Abelian o-groups with all Szmielew invariants 1 are shown to be elementarily equivalent. Y. Gurevich (*Elementary Properties of Ordered Abelian Groups*, Translations Amer. Math. Soc., 46(1965), 165-192) has given a set of invariants which completely determine the elementary equivalence class of an Abelian o-group. He also shows that the theory of Abelian o-groups is decidable.

J. Roitman:

[81] The existence of dense rigidly homogeneous chains of arbitrary outer measure, of infinite inner measure, and of measure 0 are established. The construction is very similar to the one we gave in the proof of Theorem 4.2.7. The main difference is that the construction is achieved by constructing dense rigidly homogeneous chains inside divisible subgroups H of $\mathbb{R}$ with $H \cap \mathbb{Q} = \{0\}$ and $|H| = 2^{\aleph_0}$. Our condition (5) on page 102 is changed to: $(\Omega_\lambda f_\lambda \cup \Omega_\lambda f_\lambda^{-1}) \cap [(D_\mu \backslash \Omega_\lambda) \cup \mathbb{R}\backslash(H \oplus \mathbb{Z})] \neq \emptyset$. The proof is then finessed by dealing with the case that for some $\alpha \in (H \oplus \mathbb{Z})\backslash D_\mu^*$, αf (or αf^{-1}) $\in \Omega_\mu^*$ prior to (ii)--if $\alpha \in (H \oplus \mathbb{Z})\backslash D_\mu^*$ and $\alpha f \in \Omega_\mu^*$, let $h_\mu \in H$ be independent of $D_\mu^* \oplus \mathbb{Q}\alpha$, and let $D_\mu = D_\mu^* \oplus \mathbb{Q}\alpha \oplus \mathbb{Q}h_\mu$ and $\Omega_\mu = \Omega_\mu^* \oplus \mathbb{Q}h_\mu$.

J. G. Rosenstein:

[81] This book makes an extensive study of chains, collecting much of the work scattered in the literature. It will provide an invaluable source and addition to those interested in pursuing ordered permutation groups. Its overlap with our book is small.

T. J. Scott:

[74] For Ω a chain, let $\mathcal{M}(\Omega)$ be the group of monotone permutations of Ω. If H is a subgroup of $\mathcal{M}(\Omega)$, let $G(H) = H \cap A(\Omega)$ and $H' = H \setminus G(H)$. $G(H)$ has index 1 or 2 in H. The first part of the paper is devoted to obtaining results about H when $G(H)$ enjoys certain properties. The second classifies those (H,Ω) for which $H' \neq \emptyset$ and $(G(H),\Omega)$ is a transitive primitive ℓ-permutation group. Note the author uses "regular" for what we call "uniquely transitive."

E. B. Scrimger:

[70] This thesis is the precursor of Sections 3.2 and 4.4 (due to McCleary [76]). First Theorem 10.5 is established. Next, inverse limits of ordered permutation groups are considered. Even though inverse limits of p.o. groups and chains exist, the inverse limit of ordered permutation groups may not. A necessary and sufficient condition is given. The intransitive wreath product is studied next. Let (G,Ω) be an ordered permutation group with congruence $\mathcal{C}$. Choose $\{\alpha_i: i \in I\} \subseteq \Omega$ so that $\{T_i: i \in I\}$ is the set of orbits of $T = \Omega/\mathcal{C}$ under $H = G/L(\mathcal{C})$, where $T_i = (\alpha_i\mathcal{C})G$. Let $(G_i,\Omega_i) = (\hat{G}_{(\alpha_i\mathcal{C})},\alpha_i\mathcal{C})$ and $\Lambda = \bigcup_{i\in I}(\Omega_i \overleftarrow{\times} T_i)$. Form $(W,\Lambda) = \{(G_i,\Omega_i)\}\mathrm{Wr}\,(H,T)$ by: $w = (\{w_\tau\},\bar{w}) \in W$ if $\bar{w} \in H$, $w_\tau \in G_i$ $(\tau \in T_i)$ and $(\alpha,\tau)w = (\alpha w_\tau, \tau\bar{w})$ $(\tau \in T_i;\ \alpha \in \Omega_i)$. (G,Ω) can be o-(ℓ-)embedded in (W,Λ) (cf. orbit Wreath product). Indeed, if (G,Ω) satisfies the hypotheses of Theorem 10.5 it is ℓ-isomorphic to (W,Λ) and is an inverse limit of finite Wreath products of "essentially transitive" such ℓ-permutation groups (except that Ω is ordermorphic to a dense subset of the inverse limit chain).

The last chapter gives necessary and sufficient conditions (on the ℓ-group free product of G and $\mathbb{Z}$) for an ℓ-group G to be ℓ-isomorphic to some $A(\Omega)$. Unfortunately, the free product is not sufficiently manageable to make this useful (see Holland and Scrimger [72]).

[75] Weinberg [65] has shown that the class $\mathcal{A}$ of all Abelian ℓ-groups is the minimal proper variety of ℓ-groups. Let $n \in \mathbb{Z}^+$ and $G_n = \{(\{g_m\},\bar{g}) \in \mathbb{Z}\ \mathrm{Wr}\ \mathbb{Z}: g_m = g_k \text{ if } m \equiv k \pmod n\}$. Each is a two generator ℓ-group. It is proved that if $\mathcal{S}_n$ denotes the variety of ℓ-groups generated by G_n, then $\mathcal{S}_n \neq \mathcal{S}_m$ if

$n \neq m$ and $\{\mathcal{S}_p : p \text{ prime}\}$ is a set of covers of $\mathcal{A}$. Moreover, $\mathcal{S}_n \cap \mathcal{S}_m = \mathcal{A}$ if n and m are coprime.

F. Šik:

[58] Let (G,Ω) be an ℓ-permutation group. Among other things, it is shown that each element of G has one bump if and only if: G is an o-group and if f is a bump of $g \in G$, then $f^n \in G$ for some $n \in \mathbb{Z}$. This paper foreshadows the importance of "bumps" in the more recent work of other authors (e.g., see Section 2.1).

J. E. Smith:

[80a] The main results are (1) $\mathcal{N}$ cannot be written as the join of a finite number of varieties of ℓ-groups strictly contained in $\mathcal{N}$, and (2) If $\{\mathcal{S}_n : n \in X\}$ is any infinite family of Scrimger varieties, then the join is the variety of ℓ-groups generated by $\mathbb{Z}$ Wr $\mathbb{Z}$. As a consequence of (2), the lattice of ℓ-group varieties is not Brouwerian. The proof of (2) can be accomplished more easily by the techniques of Glass, Holland and McCleary [80]; they also show that $\mathbb{Z}$ Wr $\mathbb{Z}$ (or $\mathbb{Z}$ wr $\mathbb{Z}$) generates the variety of ℓ-groups $\mathcal{A}^2$ and that $\bigvee_{n \in \mathbb{Z}^+} \mathcal{A}^n = \mathcal{N}$ (Theorem 11E)--contrast with (1).

[80b] This is a survey of some results on varieties of ℓ-groups with special emphasis on Scrimger [75] and Smith [80a] and [81].

[81] Let $\mathcal{L}_n$ be the variety of ℓ-groups defined by the law: $[x^n,y^n] = e$. The Scrimger variety $\mathcal{S}_n \subseteq \mathcal{L}_n$ $(n \in \mathbb{Z}^+)$. It is shown that $\mathcal{S}_n \neq \mathcal{L}_n$ if $n > 1$ and is composite, and that if $\{\mathcal{L}_n : n \in X\}$ is an infinite nested family of these varieties, $\bigvee_{n \in X} \mathcal{L}_n = \mathcal{N}$. Indeed, ℓ-groups are constructed in $\mathcal{L}_n$ which do not belong to $\mathcal{S}_n \vee \mathcal{A}^k$, where k is the number of (not necessarily distinct) prime factors of n. These generate varieties of ℓ-groups. Joins of these varieties are explored. Also see Reilly and Wroblewski [81]. Note that McCleary (unpublished) has shown that $\mathcal{S}_n \neq \mathcal{L}_n$ if n is prime.

L. B. Treybig:

[63] Let Ω be a dense Dedekind complete chain. Then Ω is shown to be homogeneous if and only if each pair of non-degenerate closed bounded intervals of Ω are homeomorphic. The proof is elementary and relies on: if $\alpha g < \alpha$ and $\beta g > \beta$ for some $\alpha,\beta \in \Omega$, then $\gamma g = \gamma$ for some $\gamma \in \Omega$ between α and β.

As a consequence, a dense Dedekind complete chain is doubly homogeneous if it is homogeneous.

A. A. Vinogradov:

[71] Two groups are exhibited which satisfy the same sentences of group theory, only one of which is lattice-orderable. Hence the class of lattice-orderable groups is not definable by a set of sentences of the language of group theory.

A. A. Vinogradov and A. A. Vinogradov:

[69] An example is given of a group that is not lattice-orderable, but is the union of a tower of finitely generated lattice-orderable groups; i.e., lattice-orderability is not a local property of groups. Holland (unpublished) has recently shown that lattice-orderability is not a residual property of groups; i.e., there is a group G having a family $\{N_i : i \in I\}$ of normal subgroups such that $\bigcap\{N_i : i \in I\} = \{e\}$ and G/N_i is lattice-orderable for each $i \in I$, but G itself is not lattice-orderable.

E. C. Weinberg:

[67] In Holland [63], Corollary 2K was proved assuming G.C.H. (see the proof given in Chapter 10). Lloyd [64] proved Corollary 2K without assuming G.C.H. under certain restrictions. Weinberg's proof avoids the use of G.C.H. by using group rings--see the proof given in Section 2.5.

[80] Whereas the original proofs of Corollary 2K (Holland [63]), Theorem 10B (Pierce [72]) and Corollary 10.4 (McCleary [73a]) depended heavily on G.C.H. (see Chapter 10), Weinberg shows that this dependence can be removed by using Harzheim [64]. Harzheim produced a theory of "minimal" η_α-sets based on order-type rather than cardinality. Let $\aleph_\alpha$ be regular. A chain is said to be order-small if every well-ordered or inversely well-ordered subset has cardinality less than $\aleph_\alpha$. The union of a tower, indexed by ω_α, of order-small chains is said to be almost-order-small. Almost-order-small η_α-sets exist (independent of G.C.H)--they are special models of the theory of dense linear ordering without endpoints--and play the same role that α-sets played in Chapter 10, so removing the dependence on G.C.H. For the details in the case of Theorem 10B, see Weinberg [81a].

[81a] The goal of this paper is to prove Theorem 10B without G.C.H. as promised in Weinberg [80]. Almost-order-small η_α-sets of degree κ are shown to exist (if $\kappa < \aleph_\alpha$) and to be unique (cf. Lemma 10.10). The existence is established in two ways. The first by an inductive construction; the second by a

lexicographic product construction. The analogue of Theorem 10.8 is next obtained from this by essentially the same technique as employed in Chapter 10 and, as a consequence, Theorem 10.12.

[81b] Let G be an ℓ-group. It is shown that (i) G can be ℓ-embedded in $A(\mathbb{Q})$ if and only if G has a countable family of prime subgroups $\{P_n : n \in \omega\}$ such that $\bigcap_{n \in \omega} P_n = \{e\}$ and each chain $R(P_n)$ is countable; (ii) G can be ℓ-embedded in $A(\mathbb{R})$ if and only if G has a countable family of prime subgroups $\{P_n : n \in \omega\}$ such that $\bigcap_{n \in \omega} P_n = \{e\}$ and each chain $R(P_n)$ has a countable subset which is dense in $R(P_n)$. The proofs are by the techniques of Appendix I. Note that the o-group $\mathbb{R}$ cannot be o-embedded in $A(\mathbb{Q})$ by (i); however, it can be embedded in $A(\mathbb{Q})$ as a group (or as a lattice) as is easily seen.

S. Wolfenstein:

[68] By classical ℓ-group methods (as opposed to ordered permutation groups), Theorems 11A (parts (1) and (4)) and 11.2 (parts (i), (ii) and (vi)) are established. In particular, $\mathcal{N}$ is shown to be a variety. Also see Read [75].

[70] Let (G,Ω) be a transitive ℓ-permutation group and H any a-extension of G--see our annotation of Khuon [70a] for the definition. It is shown that there is a chain $T \supseteq \Omega$ such that (H,T) is a transitive ℓ-permutation group. Indeed, if Ω is dense and Dedekind complete, T can be chosen to be Ω. Hence if Ω fits these hypotheses, G has an a-closure. The same conclusion follows if (G,Ω) is doubly transitive--T can be chosen to be a subset of $\bar{\Omega}$. Finally, $A(\mathbb{R})$ is shown to be the unique a-closure of $A(\mathbb{Q})$ and $A(\mathbb{R})$. The proofs are straightforward using only the ideas in Appendix I. That every ℓ-group has an a-closure was proved by Khuon [70b] and McCleary (unpublished).

INDEX

O

P

R

S

T

U

V

W

INDEX OF SYMBOLS

(The characters are listed in order of appearance.)